ERRATA

to accompany

PRESENT AT THE FLOOD
How Structural Molecular Biology Came About
Richard E. Dickerson

The publisher regrets that, because of an error late in production, Greek letters and special symbols in several original figures either failed to print (Figures 3.1, 4.1, 5.2, 5.3, 5.7 and 5.8) or were replaced by the wrong symbol (Figures 4.1 and 6.5). The following corrected figures are printed at the same size as their originals.

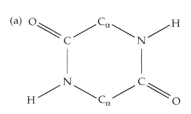

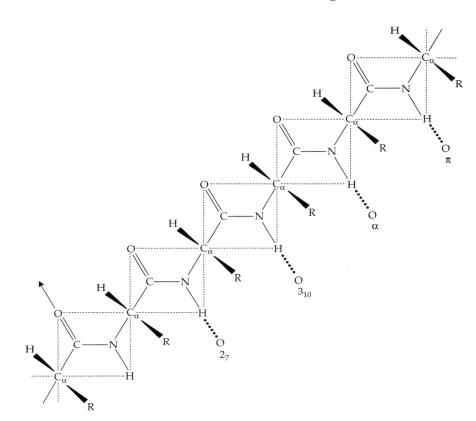

FIGURE 4.1
on page 60

FIGURE 3.1
on page 20

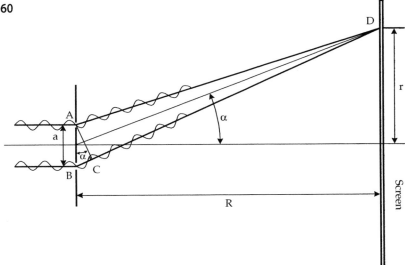

FIGURE 5.2
on page 141

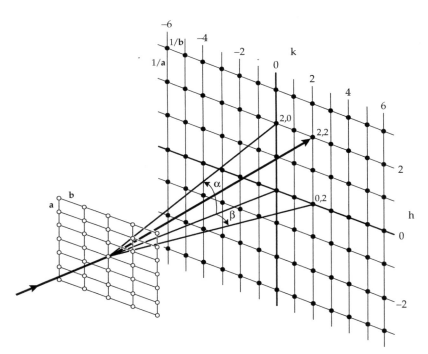

FIGURE 5.3
on page 142

CONVOLUTION THEOREM

(a) $T(A*B) = T(A) \times T(B)$

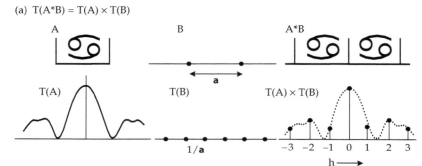

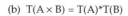

(b) $T(A \times B) = T(A)*T(B)$

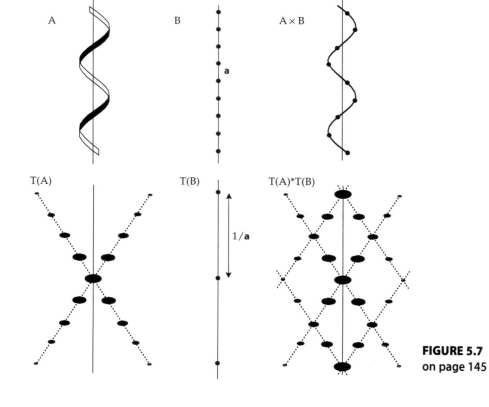

FIGURE 5.7
on page 145

(a) Continous helix

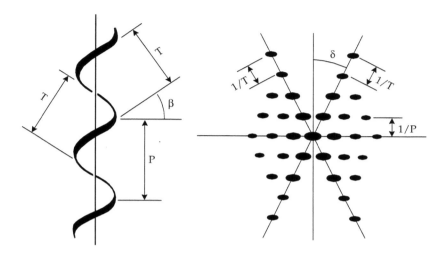

(b) Discontinous helix

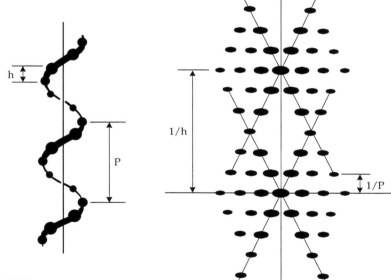

FIGURE 5.8
on page 146

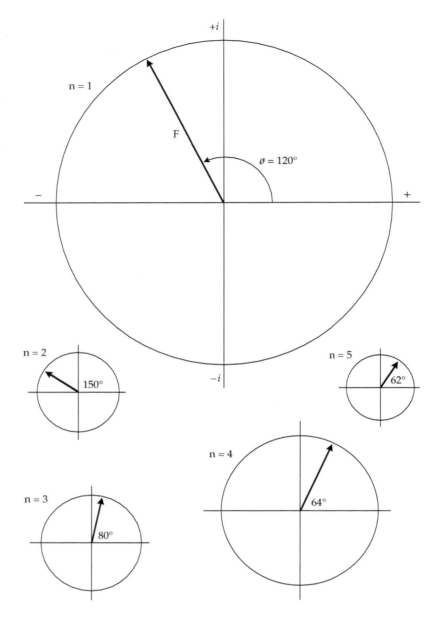

FIGURE 6.5
on page 206

Present at the Flood

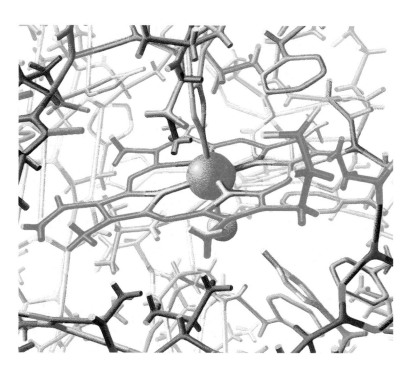

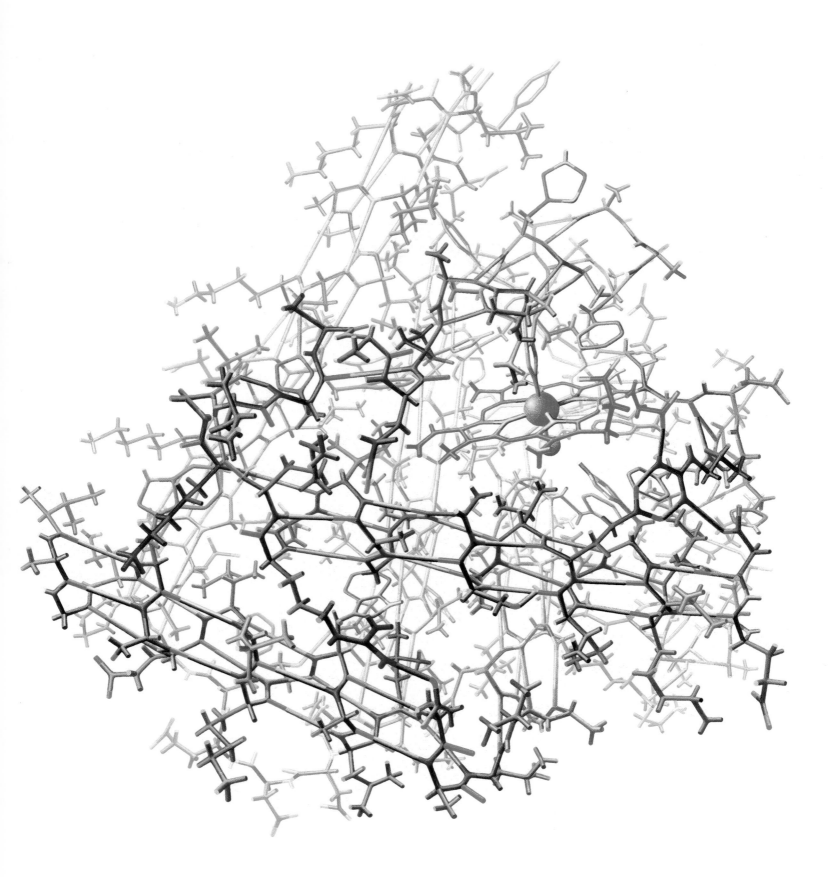

PRESENT AT THE FLOOD
How Structural Molecular Biology Came About

Richard E. Dickerson

Molecular Biology Institute
University of California, Los Angeles

Sinauer Associates, Inc. • Publishers
Sunderland, Massachusetts U.S.A.

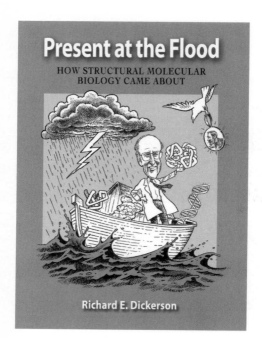

The Cover

In this cartoon by David Granlund, Max Perutz as a modern Noah stands in the Ark, surrounded by macromolecules whose structures were made possible by x-ray methods which he pioneered. "Noah" holds aloft a hemoglobin molecule in triumph, while a dove flies toward the Ark with important news. The title of the book, "Present at the Flood," arises from an essay by the philosopher John Locke on the nature of proof. To paraphrase Locke, eyewitness testimony is superior to circumstantial evidence: "Had I myself seen the Ark and Noah's Flood, as I saw an overflowing of the Thames River last winter, then I should have no more doubts about the reality of the Biblical Flood than about the overflowing of the Thames." As you read the papers in this book, you will become eyewitnesses to the structure revolution.

The Frontispiece

The three-dimensional structure of John Kendrew's sperm whale myoglobin, as painted by Irving Geis for Kendrew's article in the December 1961 issue of *Scientific American*. All atoms are shown, even hydrogens. The red heme group with its orange iron atom sits in a pocket between two alpha helices. In this view, helix E is the one that sits roughly horizontal below the heme, sloping upward slightly to the left. Helix F rises diagonally from left to right at an angle of 45° on the left side of the heme. No structure this complex had ever been determined before. Courtesy of Geis Archives, Howard Hughes Medical Institute.

Library of Congress Cataloging-in-Publication Data

Dickerson, Richard Earl, 1931-

Present at the flood : how structural molecular biology came about / Richard E. Dickerson.-- 1st ed.

    p. cm.

Includes bibliographical references and index.

ISBN 0-87893-168-6 (pbk.)

1. Molecular biology--History--20th century. I. Title.

QH506.D53 2005

572.8'09'04--dc22

2005008575

In memory of Francis H. C. Crick (1916–2004),
who was both a brilliant scientist and a man who epitomized the phrase, "a gentleman and a scholar."
He will be sorely missed; his kind are truly rare.

Contents

Chapter 1 *Introduction 1*

Chapter 2 *Your Cells Are Not Micelles! 3*

REPRINTED PAPER

1934, W. T. Astbury and H. J. Woods, *Phil. Trans. Roy. Soc. London* 10

Chapter 3 *Workers of the World, Cast Off Your Chains! 19*

REPRINTED PAPERS

1936 D. M. Wrinch, *Nature* 26

1937 D. M. Wrinch, *Nature* 28

1939 L. Pauling and C. Niemann, *J. Am. Chem. Soc.* 30

1941 D. M. Wrinch, *J. Am. Chem. Soc.* 38

1984 M. M. Julian, *J. Chem. Edu.* 42

1938 W. T. Astbury, *Trans. Faraday Soc.* 45

Chapter 4 *The Folding and Coiling of Polypeptide Chains 57*

REPRINTED PAPERS

1979 D. Crowfoot Hodgkin, *Ann. NY Acad. Sci.,* 1979 66

1950 W. L. Bragg, J. C. Kendrew and M. F. Perutz, *Proc. Roy. Soc. London* 94

1951 L. Pauling, R. B. Corey and H. R. Branson, *Proc. Natl. Acad. Sci.* 118

1998 M. F. Perutz, In *I Wish I'd Made You Angry Earlier: Essays on Science* 125

1951 M. F. Perutz, *Nature* 128

1951 L. Pauling and R. B. Corey, *Proc. Nat. Acad. Sci.* 130

2003 D. Eisenberg, *Proc. Nat. Acad. Sci.* 135

Chapter 5 *The Race for the DNA Double Helix 139*

REPRINTED PAPERS

1950 E. Chargaff, *Experientia* VI 158

1951 E. Chargaff, *Fed. Proc.* 161

2004 R. D. B. Fraser, *J. Struct. Biol.* 163

1953 L. Pauling and R. B. Corey, *Proc. Nat. Acad. Sci. USA* 169

1953 J. D. Watson and F. H. C. Crick, *Nature* 183

1953 M. H. F. Wilkins, A. R. Stokes and H. R. Wilson, *Nature* 184

1953 R. E. Franklin and R. G. Gosling, *Nature* 187

1974 A. Klug, *Nature* 188

1968 E. Chargaff, *Science* 190

1969 M. F. Perutz, M. H. F. Wilkins, J. D. Watson, *Science* 192

1978 E. Chargaff, in *Heraclitean Fire,* 195

Chapter 6 *How to Solve a Protein Structure 199*

REPRINTED PAPERS

1953 J. Donohue and J. Briekopf, *Max F. Perutz reprint file* 217

* 1947 J. Boyes-Watson, E. Davidson and M. F. Perutz 227

* 1949 M. F. Perutz 228

* 1954 M. F. Perutz 229

* 1954 D. W. Green, V. M. Ingram and M. F. Perutz 230

* 1954 E. R. Howells and M. F. Perutz 231

* 1954 W. L. Bragg and M. F. Perutz 232

* 1958 D. M. Blow 233

* 1961 A. F. Cullis, H. Muirhead, M. F. Perutz, F. R. S. and M. G. Rossmann 234

* 1962 A. F. Cullis, H. Muirhead, M. F. Perutz, F. R. S. and M. G. Rossmann 235

1958 J. C. Kendrew, G. Bodo, H. M. Dintzis, R. G. Parrish, H. Wyckoff and D. C. Phillips, *Nature*, 236

1959 G. Bodo, H. M. Dintzis, J. C. Kendrew and H. W. Wyckoff, *Proc. Roy. Soc.*, 241

* From: *Proc. Roy. Soc. London*

Chapter 7 **High-Resolution Protein Structure Analysis 249**

REPRINTED PAPERS

1960 M. F. Perutz, M. G. Rossmann, A. F. Cullis, H. Muirhead, G. Will and A. C. T. North, *Nature* 257

1960 J. C. Kendrew, R. E. Dickerson, B. E. Strandberg, R. G. Hart, D. R. Davies, D. C. Phillips and V. C. Shore, *Nature* 263

1992 R. E. Dickerson, *Protein Science* 269

Chapter 8 **The Knowledge Explosion 275**

Chapter 9 **Epilogue 279**

Appendix 1 **Pioneers of Structural Molecular Biology, 1933–1963 283**

Appendix 2 **Highly Recommended Reading 284**

Appendix 3 **Irving Geis, the Molecular Vesalius 286**

REPRINTED PAPERS

1997 R. E. Dickerson, *Structure* 287

1997 R. E. Dickerson, *Current Biology* 290

Answers to Questions 293

Credits for the Key Papers 299

Index 301

Preface

This book has been gestating for decades, but owes its birth to a very successful reading/discussion graduate course of the same title held at UCLA in the spring quarter of 2004. I thank David Eisenberg, Mary Kopka, and Todd Yeates for helping keep lively arguments going in class, and later for reading through the manuscript and suggesting improvements.

Special thanks go to Francis Crick for encouraging me to go beyond the published and probable, to the unpublished and never-proven, which turned out to be one of the most interesting aspects of the course. Aaron Klug was especially helpful with background material concerning Rosalind Franklin and DNA, and Jim Watson also shed light on the DNA story in Chapter 5. Gunther Stent helped me work out the logistics of publishing a book of other peoples' reprints, having done the same kind of thing in his annotated Norton edition of Watson's *Double Helix*.

Sandy Geis, representing the Geis Archives and the Howard Hughes Medical Institute, was of immense help in providing beautiful examples of her father's art. Much of the quality of the reprinted papers is because of the cooperation of *Nature* magazine and JSTOR, the internet scholarly journal archive. And particular thanks go to Andy Sinauer, Chelsea Holabird, Christopher Small, and Jefferson Johnson of Sinauer Associates, who put together this complex book of reprints and commentary with enthusiasm and style.

Finally, Bill Gelbart must bear immediate responsibility for inducing the labor that produced this work. In the fall of 2003, as Chair of Chemistry and Biology at UCLA, he commented to me, "Dick, if you have indeed changed your mind and don't intend to retire until spring of 2004, don't you think you should teach a course of some kind this year?" When I asked, "What course?", Bill replied, "Any course you choose. Make up a new course if you like." The rest, as they say, is History.

-1-

INTRODUCTION

Open a current biochemistry or molecular biology textbook, and you would think that everything developed logically from A to B to C, guided by geniuses who were wiser than you and I will ever be. But that's not the way it happened. Back when molecular biology was just "the practice of biochemistry without a license," as Erwin Chargaff waspishly remarked later, people trying to figure out the structures of biological materials were fumbling their way from one idea to the next: some good, some absurd. Who can forget Linus Pauling and Robert Corey's alpha helix, one of the seminal flashes of genius in protein structure? But whatever happened to Dorothy Wrinch's earlier cyclol theory, and why did Pauling and Carl Niemann stumble when they attempted to use thermodynamic arguments to rebut it?

Structural molecular biology was primarily the offspring of two openly competing laboratories: that of Pauling and Corey at Caltech in Pasadena, and Sir Lawrence Bragg's Cavendish Laboratory of Physics at Cambridge. The latter benefited from the excellent scientist Max Perutz, promising graduate students such as John Kendrew and Francis Crick, and a steady stream of smart American (and other) postdoctorals and visitors such as Jim Watson and Jerry Donohue. But why did their great scientific competence itself cause them to miss the alpha helix?

Who can forget the Watson/Crick DNA double helix, after half a century? But who recalls that this was their second attempt, after Bragg took one look at their first model and told them to quit? And who remembers that the *first three* attempts at the structure of DNA, at Cambridge, King's College and Caltech, each involved a *triple* helix?

Too many textbooks hand you canned information, with instructions to open the can and swallow the contents whole. Where the contents came from is frequently not specified on the label. This book is different; you will read the actual words of the people who were working on major problems at the time. You will see when they had flashes of genius, and when they missed the point entirely. You will hear them wax sarcastic or mean-spirited on occasion, against competitors who obviously do not understand what is going on. Hopefully you will come to think of them as real people, who struggled just as hard for their achievements as you will have to do for your own.

The period of time covered by this book is roughly 1933 to 1963 for a good reason. Two great dilemmas of biochemistry were solved during this era, which can be described as:

The Protein dilemma

Are proteins discrete macromolecular compounds, or only colloidal suspensions? Do proteins have fixed molecular weights and definite amino acid compositions? How are their amino acids linked together? If as chains, does a given protein have a specific fixed amino acid sequence, or are sequences random? How can proteins function as molecular machines (e.g. enzymes); are they folded up in some specific manner, and if so, how and why?

The DNA dilemma

How are nucleotides linked together in a DNA chain? Is the order of bases along a chain random, or repetitive, or is the base sequence specified in some manner? And for what purpose? Even if the order of bases was not repetitive, how could anyone think that DNA could be the storehouse of genetic information with its puny four symbols, when proteins have a repertoire of twenty different units? After all, you can't write Shakespeare with an alphabet of four letters!

The great biochemical revolution had been won by 1963, after which it was all a case of "dotting the T's and crossing the eyes." What we have done in the forty years that followed was to work out the implications of that revolution. Everything that we have achieved since 1963 would have been not only impossible, but intellectually incomprehensible, in 1933. It was this thirty-year revolution that made it all happen.

A few words about organization

Each chapter will begin with a list of key papers on the subject at hand. Those which will be discussed and reprinted at the end of the chapter for you to read are listed in bold face and assigned identifying letters A, B, C.... In some of the longer papers, pages irrelevant for our purposes have been deleted. As an example, Astbury and Woods' paper in Chapter 2 runs for a massive 62 pages. But the 8 pages reprinted here contain all of their radically new ideas about protein folding, while the remaining 54 pages are concerned with supporting spectroscopic and mechanical properties of fibers. Unnecessary pages sometimes are included, however, when they give you an insight into the thought processes and the attitudes or biases of the authors. My preference is to print an entire paper as the authors intended it, except when a paper is tediously long. Papers not in bold face and not given a letter code still are important milestones, but cover the same information as in the reprinted papers.

Each of the papers reprinted has been chosen because it is particularly important in the evolution of structural molecular biology. Each will be accompanied by my own background explanations and analysis of its strong points and faults. And just as you are urged to argue in your own mind with what these authors have written, so you should debate the merits of my remarks. Page and figure citations in my text will use the authors' own numbering. Hence "page B.238" means page 238 (in the original journal paging) of paper B. Similarly, "Figure E.5" will indicate Figure 5 of paper E. Illustrations accompanying my own text will be referenced using chapter number followed by figure number (e.g., Figure 2.3). Color plates are grouped in a central color section. Finally, references to other papers not in the "key papers" list will appear at the end of each chapter, and are identified in the body of the text by numbers in the usual way. At the end of each chapter is a short list of study questions, which are designed to help you navigate your way through the reprints. For the faint of heart, answers to the study questions are to be found at the end of the book. Letters in parentheses in these questions point you to the relevant article in the Key Papers list.

Good hunting!

-2-

YOUR CELLS ARE NOT MICELLES!

"And what shoulder, and what art,
could twist the sinews of thy heart?"

WILLIAM BLAKE, *"THE TYGER"*

KEY PAPER

(A) 1934 W. T. Astbury and H. J. Woods, *Phil. Trans. Roy. Soc. London* A232, 333–394. "X-Ray Studies of the Structure of Hair, Wool, and Related Fibres. II. The Molecular Structure and Elastic Properties of Hair Keratin" [8 pages of 62].

One of the live areas of biochemistry in the first decades of the 20th century was that of colloid chemistry. A "colloid" strictly speaking is not a substance, but is a state adopted by substances, intermediate between a liquid and the solid or crystalline state. The name came from the Greek "kolla" or glue, and "eidos" meaning like. Hence colloids were disordered glue-like suspensions of one substance in another. Gelatin, dextrin, proteins, gums, etc. were all regarded as colloidal suspensions of different substances in water, whereas small molecules such as sucrose, glucose, sodium chloride and other salts, were termed "crystalloids" because they contained small discrete molecules or ions which could be crystallized out of solution. The substance which was dispersed in a colloidal suspension was generally not considered to be one discrete, chemically defined molecule, but rather an aggregate which was termed the *discontinuous phase* or *micelles*, whereas the surrounding medium was called the *continuous phase* or *intermicellar liquid*.

The language of colloids was extended to include almost any suspension of one aggregate in another phase. Gold particles dispersed in stained glass, or metal ions dispersed in gemstones to give their characteristic colors, all were regarded as solid-in-solid colloids. The blue haze of forest fires or cigarette smoke would be solid-in-gas colloidal suspensions. Opals were colloidal dispersions of water molecules in a silicate matrix, and pearls a similar dispersion of water molecules in a calcium carbonate medium. Dry out either an opal or a pearl under severe conditions, and it loses its sheen and luster to become only a valueless piece of mineral. The most relevant systems to biochemists were liquid-in-liquid and solid-in-liquid colloidal suspensions; these were thought to describe proteins in solution, and nearly every other fluid component of a living cell.

Virtually the only place where one encounters the term "micelle" today is in the description of soap solutions in water. It is worth looking at these in some detail because they illustrate the way people thought about proteins at the beginning of the 1930's. A soap molecule is the salt (usually sodium) of a long-chain fatty acid, which combines a long hydrocarbon tail with a carboxylic acid head group: —COOH (Figure 2.1). When added to water, soap molecules cluster spontaneously into spheres known as soap micelles. Their

(a) Stearic acid

(b) Sodium stearate (a soap)

or:

(c) Soap micelles in water

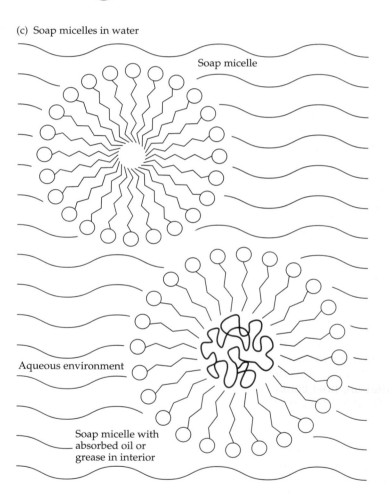

Soap micelle

Aqueous environment

Soap micelle with absorbed oil or grease in interior

FIGURE 2.1   Soap molecules have electrostatically charged head groups and long hydrophobic tails. In aqueous solution they cluster spontaneously into small spherical micelles, with charged head groups on the surface facing the water and hydrophobic tails buried in the interior. Oil or grease in the aqueous environment spontaneously segregates from the water layer by being absorbed into the interiors of micelles. Proteins were once thought to be heterogeneous micellular aggregates of short polypeptides, rather than chemically defined macromolecules.

hydrocarbon tails pack in the interior of the sphere, as far away as possible from water, while their ionized head groups give the outside of the micelle an electrostatic charge. Soaps are effective cleaning agents because any oil or grease present in the aqueous surroundings diffuses spontaneously into the interior of the micelles to get away from water. Furthermore, the charge on individual micelles causes them to repel one another and avoid coagulating into one huge oily mess. They can easily be washed away down the drain.

If one measured the size or diameter of micelles in a soap preparation, one would find not one specific diameter but a range of values, dependent on the length of the hydrocarbon tails and whether a micelle had absorbed any external fats or oils. It would be a legitimate question to ask for the mean and standard deviation of the masses of the micelles. But the query, "What is the *molecular weight* of such a micelle?" would be a foolish question. The concept of molecular weight does not apply to multi-component aggregates like these. One could not even ask the precise molecular structure of micelles, because they could be made from several similar but different fatty acids, and contain varying quantities of absorbed oils or grease.

All this is relevant because it describes precisely how biochemists once thought about proteins. A protein was a colloidal dispersion of organic matter in water. Its dispersed particles were not monomolecular; they were aggregates of amino acids and combinations of amino acids. Even the idea that the amino acids were linked into a linear chain was not universally believed. A widely used textbook of 1929 by Ross Gortner (1) devotes the first 290 of its 757 pages specifically to colloids, and then uses the concept to characterize proteins and nearly every other biochemical substance. Gortner began his textbook with a 1928 quotation by W. J. Mayo, founder of the prestigious Mayo Clinic in Rochester, Minnesota:

The colloid field today presents the most promising realm in medical research.

But by 1972, Marcel Florkin in his history of biochemistry was characterizing this period as "The Dark Age of Biocolloidology" (2). He summed up the situation as follows:

[In Gortner's textbook] the whole presentation is still based on the colloidal theory, and in its subject index the words "colloids," "sols" and "gels" appear in

almost a hundred entries, whereas in any reputable textbook of biochemistry published after World War II, these words are not even mentioned.

The first step in the downfall of "biocolloidology" occurred in 1926, when Sumner crystallized the enzyme urease from jack bean (3). This sent shock waves through the biochemical community. How could a heterogeneous aggregate of anything be crystallized? Northrop followed this in the early 1930's by purifying and crystallizing several digestive enzymes, including pepsin, trypsin and chymotrypsin (4). At this point, J. D. Bernal at the Cavendish Laboratory in Cambridge decided to look at crystalline proteins with x-ray diffraction. He enlisted three promising young colleagues: Dorothy Crowfoot (later Hodgkin), Max Perutz and Isidore Fankuchen. Results were unfavorable at first until it was realized that protein crystals must be sealed in capillaries so they cannot dry out. Suitably protected, wet crystals of many proteins gave wonderful x-ray patterns, and we will return to this matter in Chapter 6.

Two kinds of proteins exist: globular and fibrous. Enzymes, antibodies, oxygen carriers, electron transfer proteins and a great many other proteins are globular, each with its own genetically determined sequence of amino acid residues linked in one or more polypeptide chains that fold into a functional piece of cellular machinery. These are the proteins whose crystallization had such a catastrophic influence on colloid theories. But the first protein structures actually to be established were those of fibrous proteins. These include hair, wool, muscle myosin, fibrinogen, silk and other insect fibers, and collagen. If a protein fiber is stretched, and if a beam of x-rays is passed perpendicularly through it, x-rays are scattered or diffracted so as to produce a pattern on a photographic plate that can tell you in principle the structure of the scattering fibers.

The godfather of fiber diffraction was William T. Astbury at Leeds University in northern England. He and others established two main classes of fibrous proteins: the relatively inextensible collagens (including elastoidin and gelatin) and the more elastic keratin-myosin or "k-m-e-f" group. The latter included a number of chemically and functionally diverse proteins: keratins from hair, nails, horn and quills, myosin from muscle, epidermin, fibrinogen, and silk fibroin and other insect filaments. (For a review, see reference [5].) Keratins produced what Astbury called the "alpha" x-ray pattern, while silk and other insect fibers gave a different "beta" pattern. But if keratin fibers were stretched sufficiently, the pattern changed from the alpha to the beta form, and back again when the tension was removed. It was reasonable to conclude that fibrous proteins were built from extended polypeptide chains made by linking amino acids one after the other.

But was this relevant to the globular proteins which had recently been crystallized? Were they also made from long polypeptide chains? After all, how could one possibly crystallize an endless bundle of fibers? Support for this idea came from the finding that some crystalline proteins could be degraded or denatured by organic solvents, and the denatured chains pulled out into fibers that gave a beta diffraction pattern. In some unknown way, the structures of globular and fibrous proteins must be related. Hence globular proteins probably were also made up of long polypeptide chains, folded in some manner that was destroyed during the denaturation process. Micelles of large numbers of short, heterogenous peptides began to look less and less like a reasonable structure option for globular proteins.

At this point we need to have a closer look at the results of x-ray diffraction experiments. Figure 2.2 shows the x-ray patterns produced by several α proteins, both natural (hair, meromyocin) and synthetic (poly-L-alanine). Figure 2.3 displays β patterns, again from sources both natural (silk) and synthetic (poly-L-alanine). Figures 2.2a and 2.3a compare unstretched α and stretched β states of the same substance, as do Figures 2.2c and 2.3c. The most important aspects of fiber diffraction photos at this point is that their features result from repeats or spacings

FIGURE 2.2 Fiber diffraction patterns of the a family: (a) α-keratin; (b) light meromyocin, and (c, d) synthetic α-poly-L-alanine chains. Fiber axes vertical. The x-ray beam is perpendicular to the fiber axis except for (d), where the fiber has been tilted 31° to record the 1.5 Å reflection at the top of the photo, as shown in Figure 4.2. All photos are at the same scale except for (b), which has been greatly enlarged to show details at the center. Numbers indicate features corresponding to repeat spacings in Ångstrom units in the fiber. From (5).

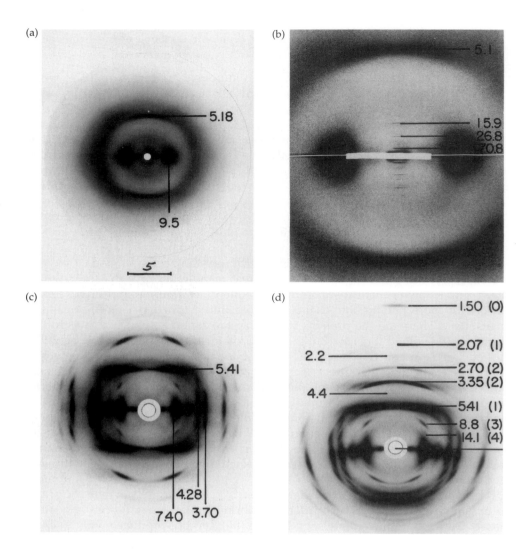

between groups of atoms in the diffracting fiber, and that the x-ray photograph is "reciprocal." That is, features close to the center of the x-ray photograph arise from long distance repeats in the fiber, and features farther out on the periphery of the photo arise from short distances in the diffracting fiber. (We will justify this in Chapter 5.) If one only collects the central portion of the diffraction pattern, one obtains a coarse image of the scattering object ("low resolution"); to obtain a higher resolution picture one must also include data farther out in the x-ray pattern.

In all these photos the fiber axis is vertical. In the α form, a diffraction maximum at ca. 5.2 Å up and down the vertical axis of the photograph tells us that *something* is repeating every 5.2 Å along the fiber. Similarly, for α-keratin a strong maximum at 9.5 Å in the horizontal plane indicates that individual fibers are packing against one another with a separation of 9.5 Å. The more orderly α-poly-L-alanine fibers (Figure 2.2c) show more details of lateral packing, with maxima at 7.4 Å, 4.3 Å and 3.7 Å. Because filaments within a fiber are cylindrically disordered about the fiber axis, these numbers tell you that such lateral packing distances are important, without telling you in which direction around the fiber they occur.

In contrast, the β pattern from silk (Figure 2.3b) indicates repeat distances of ca. 7.0 Å and 3.5 Å along the fiber axis, and lateral spacings between chains of 9.7 Å, 4.7–4.2 Å, and 3.1 Å, again without saying in which direction these spacings occur.

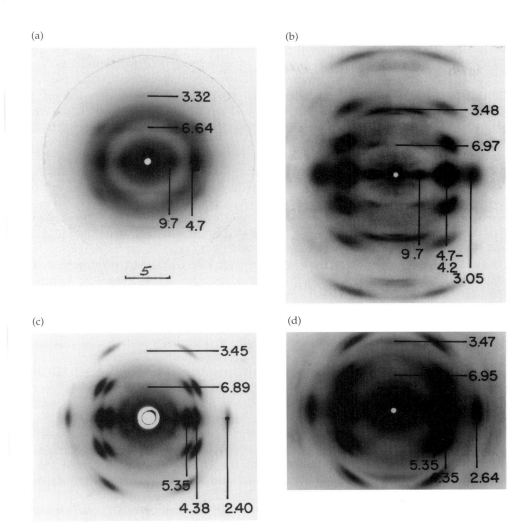

FIGURE 2.3   Fiber diffraction patterns from the β family: (a) α-keratin fibers stretched sufficiently to give the β pattern; (b) β-fibroin from the silkworm *Bombyx mori*; (c) synthetic poly-L-alanine stretched into the β form; (d) tussah silk fibroin. All photos at the same scale. From reference (5).

These are the tools that Astbury worked with at Leeds. The first paper we shall look at, by Astbury and Woods (A), describes what they proposed in 1934 as models for the contracted α and extended β forms of hair keratin. Both the α and the β forms later turned up as important elements of folding in globular proteins, which is why this discussion of fibrous proteins is relevant here. The collagen helix is only marginally related to globular proteins, and will not be considered further.

Paper A is mind-numbingly long, 62 pages, because Astbury and Woods had to prove their interpretations in the face of considerable skepticism, and also included much industrial information about wool fibers that does not concern us. We have only reprinted the eight pages that present the results of their structural findings. The complete paper is available in many libraries, or on-line at the JSTOR internet site*.

Astbury and Woods begin by describing the two reversible states of hair keratin, α and β. From the outset they ruled out the idea that the α/β transition involved only the sliding of filaments past one another, or the rotation of some kinds of micellular bodies. (Note that as early as 1934, when referring to proteins

---

* The JSTOR ("journal storage") internet site is an enormously valuable resource that makes available journal articles in all fields going back more than 75 years. It can be accessed at www.jstor.org. Your college or university needs to be a subscriber for you to use the site, but most educational institutions are. Check with your head librarian.

Astbury sometimes enclosed the word "micelles" in quotation marks, indicating that he didn't really believe in them.) The extension of α into β, and the contraction back to α when tension was released, must represent conversions between two different states of a linear polypeptide chain, which Astbury described as "a kind of molecular centipede."

Astbury presents his models for the two states of the polypeptide chain on page A.343. His extended beta structure on the right is correct. In a zig-zag extended chain with proper geometry about the C and N atoms, the length of one amino acid residue is 3.5 Å, just as suggested by the β x-ray photos in Figure 2.3. But the side-to-side alternation of the chain means that the *true repeat distance* along the extended chain involves two successive amino acids and a distance of 7.0 Å (page A.371). Hence the two layer lines of spots observed as one travels up the axis of the x-ray photograph, listed at 6.89 Å and 3.45 Å in Figure 2.3c. This photo is of β-poly-L-alanine; the β-keratin photo above it has similar features but is not as well defined.

On page A.373 he carries the beta structure one step further, by proposing that adjacent chains run in opposite directions and are connected by NH····OC hydrogen bonds, with spacings of 4.65 Å between adjacent chains. Because adjacent chains run in opposite directions, the true repeat in a direction at right angles to the chain axis is every other chain: first chain with third, second chain with fourth, etc. Hence the true repeat spacing is double the spacing between adjacent chains, or 9.3 Å. This feature of the fiber pattern overlaps with a 9.7 Å spacing that Astbury attributed to the stacking of one sheet atop another. Although his drawing does not show it well, the —R side chains of the individual amino acids were proposed to extend perpendicularly up and down from the plane of the sheet, and hence to be involved in this sheet stacking. Hence silk has covalent bonds in one direction (that of the fiber), hydrogen bonds in a second direction across a sheet, and van der Waals attractions between side chains on neighboring stacked sheets in the third direction. Astbury's insight about the β-sheet was later confirmed by Pauling and Corey, who proposed both an antiparallel and a parallel-chain form of the sheet (Chapter 4). Both are found in globular proteins.

But what about the more contracted α-keratin? Here Astbury suggested a very plausible folded chain which eventually turned out to be wrong. He proposed that three successive amino acids folded into loops with a 5.1 Å repeat along the fiber direction (page A.343), thus explaining the prominent 5.1 Å (or 5.4 Å) diffraction smear along the vertical axis of the photos in Figure 2. As we shall see in Chapter 4, Pauling and Corey ultimately solved the problem correctly with a helix containing 3.6 amino acid residues per turn, and a helix pitch or repeat distance of 5.4 Å. This disagreement between a 5.1 Å repeat in natural hair keratin and the 5.4 Å repeat in synthetic polymers such as α-poly-L-alanine would haunt investigators until it ultimately was explained by Francis Crick in 1952.

But as seminal as Astbury and Wood's paper was, it also suggested a wrong direction that would plague protein structure analysis for nearly a decade: the so-called "cyclol theory." In his figure on page A.343, the rings in the alpha form are closed by hydrogen bonds between N—H and O—C. Astbury called them "pseudo-diketopiperazine rings," but *did not propose that the dashed bonds were covalent.* Indeed, had they been strong covalent bonds, stretching the fiber to the beta form would have been impossible. But now imagine that the nitrogen gives up its H atom and forms a true covalent bond with the carbon atom. That atom replaces its previous double bond to O with a single bond to an OH group, which retrieves the abandoned H atom. That is:

$$\begin{array}{cc} \diagdown \quad \diagup & \diagdown \quad \diagup \\ \text{N—H····O=C} \quad \text{becomes:} & \text{N—C—OH} \\ \diagup \quad \diagdown & \diagup \quad \diagdown \end{array}$$

The result is a cyclol ring, and the argument that protein chains consisted of cyclol rings rather than extended polypeptides raged for years. This curious episode is the subject of the next chapter.

References

1. R. A. Gortner. 1929. *Outlines of Biochemistry*, Wiley, New York.
2. M. Florkin. 1972. In *Comprehensive Biochemistry, Vol. 30, A History of Biochemistry*, Elsevier, Amsterdam.
3. J. B. Sumner. 1926. *J. Biol. Chem.* 69, 435–441. "The Isolation and Crystallization of the Enzyme Urease."
4. J. H. Northrop. 1935. *Biol. Revs.* 10, 263–282. "The Chemistry of Pepsin and Trypsin."
5. R. E. Dickerson. 1964. In *The Proteins* (Hans Neurath, ed.), Academic Press, New York, 2nd edition, Vol. 2, pp. 603–778. "X-Ray Analysis and Protein Structure."

Study Questions

1. How did Astbury's x-ray photographs suggest that fibrous proteins, at least, were made from *extended polypeptide chains* rather than colloidal micelles of short peptides?
2. What made Astbury and others so confident that globular proteins, like fibrous proteins, were also built from long polypeptide chains of amino acids?
3. What experimental fact proved that the dashed bonds at the left on page 343 of Astbury and Woods (A) could not be ordinary covalent bonds, as the cyclol theory would have us believe? Why?

[333   ]

X. *X-Ray Studies of the Structure of Hair, Wool, and Related Fibres.*
II.—The Molecular Structure and Elastic Properties of Hair Keratin.

By W. T. Astbury *and* H. J. Woods, *Textile Physics Research Laboratory,*
The University, Leeds.

(*Communicated by Sir* William Bragg, *O.M., F.R.S.*)

(Received June 30, 1933.)

[Plates 8 and 9.]

Introductory Summary.

In a previous communication* an account was given of a preliminary exploration, chiefly by X-ray methods, of the problem of the molecular structure of animal hairs. The present paper is a natural continuation of the record, in which earlier tentative suggestions are either confirmed or rejected, and an attempt is made to lay bare the general structural principles underlying the properties of the protein, *keratin*. It will be unnecessary here to outline once more the historical development of the subject ; we shall proceed at once to the main point of this introductory section, which is to give what appears to be the solution of the problem before setting out in detail the experimental facts and arguments leading up to it. Such a procedure is advisable because of the complex nature of the properties under discussion ; such a long series of experiments have been involved in their elucidation, that without some sort of preliminary statement of the chief conclusions, the issue is apt to grow confused.

Briefly, the whole argument rests on the discovery† that the X-ray " fibre photograph " which appears to be common to all mammalian hairs, human hair, wool, whalebone, nails, horn, porcupine quills, etc., and which is undoubtedly the diffraction pattern of crystalline, or pseudo-crystalline, keratin, the common fibre substance of all these biological growths, is changed into a quite different fibre photograph when the hair is stretched. The change is a reversible one, recalling that previously discovered by Katz‡ in rubber, because when the hair is returned to its initial unstretched length, the normal keratin photograph reappears. It is clear that the X-ray effects give a diffraction record of a reversible transformation involving not merely an internal slipping of the fibre substance or a rotation of " micelles " into stricter alignment,

---

* Astbury and Street, ' Phil. Trans.,' A, vol. 230, p. 75 (1931) ; referred to later as I. Cf. also Astbury and Woods, ' Nature,' vol. 126, p. 913 (1930).

† Astbury, ' J. Soc. Chem. Ind.,' vol. 49, p. 441 (1930) ; ' J. Text. Sci.,' vol. 4, p. 1 (1931).

‡ ' Chem. Z.,' vol. 49, p. 353 (1925) ; ' Naturwiss.,' vol. 13, p. 411 (1925).

but a definite elastic elongation and contraction of the keratin complex itself. It has been proposed, I, to call the two forms of keratin thus revealed by X-ray analysis α-keratin and β-keratin, the former being the shorter, normal form.

Putting aside for the moment the question of the analysis of the normal fibre photograph (α), it is to be noticed at once that the photograph of stretched hair (β) is closely analogous to that always given by the protein of natural silk*, *fibroin*, whether unstretched or stretched, and there is every reason to believe†, both from X-ray and general physical and chemical evidence, that the fibre substance of silk is for the most part built of fully-extended polypeptide chains of the simple kind postulated by FISCHER. It follows, therefore, that β-keratin is most probably also based on fully-extended polypeptide chains, while α-keratin must be constructed out of the same chains in some shorter, folded form. Natural silk is thus virtually non-elastic, while mammalian hairs, on account of the inherent configurational instability of the extended keratin complex, show long-range elasticity of a most valuable and instructive character.

We may picture a polypeptide chain as a long series of α-amino-acid residues, each of the general formula

$$\begin{array}{ccc} & CO & NH \\ CH & & \\ R & & \end{array}$$

— a kind of molecular centipede whose legs represent the various univalent " side-chains " denoted in the general formula by the letter R ; and in a fibre such as hair, built, as X-rays show, from a system of polypeptide chains all lying more or less parallel to the fibre axis, we can see that the equilibrium form of the protein complex must be decided chiefly by the interactions of the side-chains, both of the same main-chain and neighbouring main-chains. Both the pattern formed by the crumpling or folding of the main-chains, which may or may not be seriously distorted by the interactions or actual chemical linkages, electro-valent and co-valent, between the side-chains, and also the lateral extension of the side-chains, may under favourable conditions be examined by X-ray methods ; and when the results of such an examination are correlated with general physical and chemical properties, we may reasonably expect to be able to draw conclusions of a much more fundamental kind than is possible along more restricted lines of investigation. This has occurred in the analysis of the molecular structure of hair ; correlation of all the available data, both X-ray and physico-chemical, has shown that we must think of it as based on parallel polypeptide chains which are linked laterally by both electro-valent and co-valent bridges, and which are normally in equilibrium in a contracted or folded form. These chains may be pulled out into the straight form

* HERZOG and JANCKE, ' Festschrift der Kaiser Wilhelm Gesellschaft ' (1921) ; BRILL, ' Liebig's Ann.,' vol. 434, p. 204 (1923) ; KRATKY, ' Z. phys. Chem.,' B, vol. 5, p. 297 (1929) ; KRATKY and KURIYAMA, *ibid.*, vol. 11, p. 363 (1931).

† MEYER and MARK, ' Ber. deuts. chem. Ges.,' vol. 61, p. 1932 (1928) ; MEYER and MARK, " Der Aufbau der hochpolymeren organischen Naturstoffe," Leipzig (1930).

by the application of tension, and they may even be contracted still further when certain lateral linkages are broken down, while they can also be " set " in the extended form by building up new lateral linkages.   The elastic properties of hair are almost bewildering in their variety, but they all appear to be based on a molecular mechanism which, in its essentials, is relatively simple.

We shall first, to avoid all confusion regarding the many aspects of the properties under discussion, give a general account of the elastic phenomena and X-ray results, together with a general explanatory scheme.   Thereafter it will be convenient to go into the various details as thoroughly as our present knowledge permits.

General Elastic Properties.

Since the pioneer work of HARRISON,* SHORTER,† and KARGER and SCHMID,‡ the most thorough investigation of the load/extension curve of wool, under varying conditions of humidity, temperature, and time, is due to SPEAKMAN.§   Tension/extension curves adapted from SPEAKMAN's, for Cotswold wool at 25° C. and at humidities ranging from 0% to 100%, have already been given, I.   Broadly speaking, they show that as the moisture content of wool is increased, the fibre stretches more easily and farther. In every case the curves show an initial " HOOKE's law region," where dE/dT is small, up to extensions of about 2%, and then a region of rapid extension for the next 25% or so, which passes into a region where the rate of extension gradually decreases again (see, for example, the right-hand curve of fig. 3).   Fig. 1, also after SPEAKMAN, shows how the form of this load/extension curve changes with rising temperature when the fibre is stretched in water.   It will be seen that the limiting stress of the HOOKE's law region continuously decreases as the extensibility increases and the " shoulder " of the curve becomes less obvious—in other words, raising the temperature accentuates the effect of humidity by making the fibre still more easy to stretch and enabling it to be stretched to still greater elongations.

The limiting extensibility of hair,‖ when the fibres are carefully chosen for uniformity, appears to be of the order of 100% ; the most they can be stretched without rupture, even in steam, is to perhaps a little more than twice their original length.   Fig. 2 shows a set of time/extension curves for human hair and Cotswold wool stretched under constant load in steam and in a 1% aqueous solution of caustic soda.   Each fibre was

* HARRISON, ' Proc. Roy. Soc.,' A, vol. 94, p. 460 (1918).

† SHORTER, ' J. Text. Inst.,' vol. 15, p. T207 (1924) ; ' Trans. Faraday Soc.,' vol. 20, p. 228 (1924) ; ' J. Soc. Dy. Col., Bradford,' vol. 41, p. 212 (1925).

‡ KARGER and SCHMID, ' Z. techn. Phys.,' vol. 6, p. 124 (1925).

§ See, *inter alia*, ' J. Text. Inst.,' vol. 17, p. T457 (1926) ; vol. 18, p. T431 (1927) ; ' Proc. Roy. Soc.,' B, vol. 103, p. 377 (1928).

‖ We shall use the word "hair" for mammalian hairs in general, whatever the source of origin. When hair of a definite type is referred to, it will be named, *e.g.*, human hair, wool, etc.

STRUCTURE OF HAIR, WOOL, AND RELATED FIBRES.   343

knowledge of the elastic properties, we must now conclude that the whole process of extension and recovery in the hair fibre is based on a protein chain-system which, under the proper conditions, is capable of being stretched to twice or contracted to half its normal length. These (approximate) limits rest on exhaustive experimental tests of numerous actual fibres, and also find a complete quantitative interpretation in all the available X-ray data, not only of keratin itself, but of other protein fibres also.* The true starting-point of the line of argument is the observation that *the X-ray photograph of β-keratin (stretched hair) is in all essentials analogous to that of the fibroin of natural silk, which is the same whether stretched or unstretched.* From every point of view we must assume that fibroin is built from *fully-extended* polypeptide chains lying closely side by side to form long, thin crystalline " bundles " or micelles, and that the effective length of each amino-acid residue in such a system is 3·5 A.† It follows therefore that if the postulated analogy between β-keratin and fibroin is sound, the characteristic meridian spacing of β-keratin, I, 3·4 A (approximately), corresponds to the average length of an amino-acid residue in the *fully-extended* keratin chains, so that to explain the occurrence of the normal α-form of hair, we have to decide on a method of folding these chains *which will satisfy both the quantitative requirements of the α-photograph and the* 100% *extension revealed by the generalized load/extension curve.* In addition, the molecular model must give, at least, a qualitative interpretation of the main physico-chemical differences between α- and β-keratin, and also promise a basis for a quantitative treatment of the super-contraction phenomenon. The type of intramolecular transformation which best satisfies *all* these various requirements is shown diagrammatically as follows :—

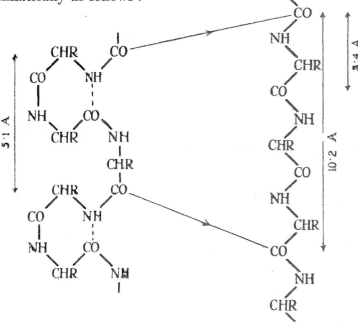

---

* ASTBURY, ' Trans. Faraday Soc.,' vol. 29, p. 193 (1933).

† Footnotes * and †, p. 334.

The β-form is thus represented by fully-extended peptide chains in which each amino-acid residue takes up, on the average, a length along the fibre-axis of 3·4 A, while the α-form is represented by a series of pseudo-diketopiperazine rings which follow each other according to a pattern of length 5·1 A. The unfolding of the rings is clearly accompanied by an elongation of 100%, and the suggested pattern offers an explanation of both the characteristic meridian reflection of the α-form (5·1 A) and of the decrease of resistance of the β-form, as compared with the α-form, to the action of reagents such as steam, etc.

Only a part of the elastic properties of hair are to be interpreted by the application of this principle of intramolecular unfolding ; many of its most striking characteristics are to be referred to the nature and distribution of the side-chains denoted above by the general symbol R. Though there does not appear to be any *sharp* discontinuity in the physical and chemical properties of the keratin complex as a whole, we have to recognize that both the form and the limits of the load/extension curve may be varied over a wide range simply as a result of the changes which take place in the configuration of the side-chains only. As the most convenient example of such side-chain disturbances it will serve for the present to quote the preferential attack of steam which is to be noticed in β-photographs for extensions of 50% and upwards, and which is undoubtedly the cause of the increased capacity for extension. Other changes which are less clear from the X-ray point of view, but which nevertheless are very obvious when examined by more familiar physico-chemical methods, are the freeing of certain side-chain restrictions so as to give rise to the phenomenon of super-contraction, and the " permanent set " of the β-form on prolonged steaming of the fibre in the stretched state. This latter transformation evidently involves the building-up of new side-chain linkages which fix the β-form in the stretched state and preclude once and for all the possibility of ever regaining the normal α-photograph.

Detailed Discussion of the Elastic Properties.

The influence of humidity at ordinary temperatures.—From an examination of load/ extension or tension/extension curves of wool already given,* I, it appears that the main effect of varying the water-content at ordinary temperatures is to alter the scale of the load (or tension) co-ordinates ; the fibre becomes progressively easier to stretch. At the same time it can be stretched farther, so that whereas it seems impossible to stretch a perfectly dry hair by more than about 30% at the very most, a wet hair may readily be stretched in one operation to something of the order of 55%. (In SPEAK-MAN's curves* for wool of 40μ diameter the rate of loading was 1·8 gm. per minute.) Curves obtained under these conditions of fairly rapid rate of loading at 100% R.H. are characterized by a " shoulder " commencing in the region of 20% extension, while

* Footnote §, p. 335.

STRUCTURE OF HAIR, WOOL, AND RELATED FIBRES. **371**

The Structure of the Keratin Complex.

As explained in the Introduction, the fundamental structural basis underlying the whole of these investigations is the existence of an intramolecular transformation considered to take place between a folded polypeptide chain-system (α-keratin) and one in a fully-extended state (β-keratin) analogous to that found normally in the fibroin crystallites of natural silk. Figs. 24, 25, Plate 8, are X-ray photographs of small bundles of unstretched silk and stretched human hair, respectively, taken under similar conditions with Cu Kα rays in a cylindrical camera of radius 3·98 cm. (The hair was stretched in steam to an extension of about 100%.) It will be seen that there is a strong resemblance between the two photographs, in particular with regard to the repetition of pattern parallel to the fibre axis. For silk the period in this direction is about 7 A.,* while for β-keratin it is rather less, something between 6·7 and 6·8 A.† (for the detailed description of the two keratin photographs, see I). It is not at all probable—for keratin at least (see above)—that lengths such as these represent the true period along a polypeptide chain-system, but are rather an expression of the fact that in the simplest formulation of the fully-extended general chain :—

the side-chains project alternately on opposite sides of the main-chain. (This point will be clearer from the models discussed below.) From an examination of all the available evidence, we have to conclude that the submicroscopic crystallites of biological structures are simply bundles, of varying degrees of neatness, of long molecular chains which for fibres such as silk and hair all lie approximately parallel to the direction of the fibre axis‡. Unlike ordinary " laboratory " crystals, therefore, the molecule or molecular complex in a fibre crystallite is in general considerably longer than the length of the crystallographic cell in the direction of the chain axis ; the chains run straight through the geometrical cells, so to speak, in such a manner that the primitive

* Footnotes * and †, p 334.

† Owing to an inherent lack of definition and paucity of reflections, the translations and spacings in X-ray photographs of biological subjects can rarely be measured with any great accuracy. It is not unlikely that the period along the fibre axis of β-keratin is, like that of α-keratin (see above), slightly variable according to previous treatment and the state of tension. The mean value of the spacing of the (020) arc appears to be 3·38 A.

‡ ASTBURY, " The Structure of Fibres " (' Annual Reports of the Chem. Soc.,' for 1931, vol. 28—issued 1932).

372   W. T. ASTBURY AND H. J. WOODS ON STUDIES OF THE

translation along the fibre axis is given by an *intra*-molecular period, while the primitive translations transverse to this are given by the side-to-side separation of the chains.   In silk fibroin the apparent cell given by the X-ray photographs ($a = 9 \cdot 68$ A., $b = 7 \cdot 0$ A., $c = 8 \cdot 80$ A., $\beta = 75° \ 50'$) is associated with a weight equivalent to four glycine residues and four alanine residues,* from which the simplest conclusion seems to be that the chains are for the most part built out of alternate glycine and alanine residues, thus† :—

and that four parallel chains constitute a *crystallographic* group.   From an exhaustive consideration of all the X-ray data, KRATKY and KURIYAMA‡ have shown that the lateral separation of these chains is not less than $4 \cdot 5$ A. and not greater than $6 \cdot 1$ A., a conclusion which agrees well with what we might predict for the polypeptide given above.

The X-ray photograph of β-keratin, I, is most conveniently referred to an orthogonal cell of dimensions, $a = 9 \cdot 3$ A., $b = 6 \cdot 7 – 6 \cdot 8$ A., and $c = 9 \cdot 8$ A. (see footnote†, p. 371), of which b is the most prominent period along the molecular chains, while a and c are " side-spacings."   With regard to the latter two points emerge, (i) that the equatorial " spot " nearer the centre which gives the c-spacing is preserved more or less unchanged when the α-photograph is transformed to the β-photograph, and (ii) that the transformation calls into existence on the equator a very strong spot of spacing $a/2$, *i.e.*, $4 \cdot 65$ A.   From a study of existing X-ray data on proteins§ the interpretation of these results seems clear, that, in fact, the spacing $9 \cdot 8$ A. common to both α- and β-photographs arises from the lateral extension of the side-chains (the R-groups of the general formula given above), while the spacing $4 \cdot 65$ A.‖ represents the distance of approach of the main-chains on those sides free from side-chains.   The controlling factor in this closest approach of neighbouring " backbones " is most probably attraction

* Footnotes * and †, p. 334.

† Much the greatest proportion of the amino-acid mixture obtained by the hydrolysis of silk fibroin is glycine and alanine, but it is not certain that the X-ray photograph is incompatible with arrangements other than strict alternation of the residues of these two acids.

‡ ' Z. phys. Chem.,' B, vol. 11, p. 363 (1931).

§ Footnote †, p. 337.

‖ In I attention was drawn to the fact that this spacing is practically equal to the chief spacing in the X-ray photograph of cystine, the most abundant amino-acid in hair ; but in the light of subsequent evidence, we wish now to withdraw the suggestion that the two spacings have anything more than a numerical relationship.

STRUCTURE OF HAIR, WOOL, AND RELATED FIBRES. 373

between ($=$NH) and ($=$CO) groups, * † ‡, whereby the chains are grouped in pairs, thus :—

Such an arrangement accounts readily for the fact that the a-dimension of the simplest orthogonal cell given above is not 4·65 A., but 9·3 A., represented on the equator by an intense second order (200).

The strongest evidence that the equatorial spacing, 9·8 A. (the reflection (001)), must be associated with the lateral extension of the side-chains comes from an X-ray study of water adsorption and the action of steam. "Quadrant photographs" (see above) of porcupine quill, both α and β, brought first to 0% R.H. by prolonged drying over phosphorous pentoxide and then to 100% R.H., show that though the bulk of the water adsorbed by animal hairs leaves the X-ray photograph unchanged, I, some of it does actually penetrate the crystallites in such a way as to increase the spacing, 9·8 A., by a few per cent. The action of steam, however, as already mentioned, is even more striking. Fig. 23, Plate 8, an X-ray photograph of human hair stretched in steam to twice its original length, shows a marked "spreading" of certain spots along the hyperbolæ ("smear lines"). *The only reflections in the photograph of β-keratin which are unaffected by the action of steam belong all to the zone [001], from which it follows that the spacing disturbance is confined to the zone-axis of this zone, i.e., to the direction of the spacing, 9·8 A., which we have associated with the lateral extension of the side-chains.* This observation must be considered as lending valuable support, of a purely geometrical kind, to the views on protein hydration recently put forward independently by JORDAN LLOYD and PHILLIPS.§

We are thus led to the concept of the average dimensions of an amino-acid residue

---

* Footnote *, p. 348.
† Footnote †, p. 339.
‡ LLOYD, ' J. Soc. Chem. Ind.,' vol. 51, p. T141 (1932).
§ LLOYD and PHILLIPS, ' Trans. Faraday Soc.,' vol. 29, p. 132 (1933).

-3-

WORKERS OF THE WORLD, CAST OFF YOUR CHAINS!

"Amino acids in chains
Are the cause, so the x-ray explains,
Of the stretching of wool
And its strength when you pull,
And show why it shrinks when it rains."

A. L. PATTERSON (QUOTED BY ASTBURY, REFERENCE F)

KEY PAPERS

	1936	D. M. Wrinch and J. Lloyd, *Nature* 138, 758–759. "The Hydrogen Bond and the Structure of Proteins."
(A)	1936	**D. M. Wrinch, *Nature* 137, 411–412. "The Pattern of Proteins."**
	1937	D. M. Wrinch, *Proc. Royal Soc. London* A160, 59–86. "On the Pattern of Proteins."
(B)	1937	**D. M. Wrinch, *Nature* 139, 972–973. "The Cyclol Theory and the 'Globular' Proteins."**
	1937	D. M. Wrinch, *Proc. Royal Soc. London* A161, 505–524. "The Cyclol Hypothesis and the 'Globular' Proteins."
	1939	D. M. Wrinch, *Nature* 143, 482–483. "The Structure of the Globular Proteins."
(C)	1939	**L. Pauling and C. Niemann, *J. Am. Chem. Soc.* 61, 1860–1867. "The Structure of Proteins."**
(D)	1941	**D. M. Wrinch, *J. Am. Chem. Soc.* 63, 330–333. "The Geometrical Attack on Protein Structure."**
	1940	D. M. Wrinch, *Nature* 145, 669–670. "The Cyclol Hypothesis."
	1943	M. L. Huggins, *Chem. Reviews* 32, 195–218. "The Structure of Fibrous Proteins."
(E)	1984	**M. M. Julian, *J. Chem. Edu.* 61, 890–892. "Dorothy Wrinch and a Search for the Structure of Proteins."**
(F)	1938	**W. T. Astbury, *Trans. Faraday Soc.* 34, 378–388. "X-Ray Adventures Among the Proteins."**

The "villain" in the melodrama to be presented here is the so-called cyclol ring. A linear dimer of two amino acids:

$$H_2N—CHR—CO—NH—CHR—COOH$$

is formed by making one —CO—NH— peptide bond between them. If the remaining initial and final groups themselves are joined in a second peptide bond, the result is a diketopiperazine ring (Figure 3.1a). Astbury, in his model for the folded α-keratin chain, proposed that the polypeptide chain loops back upon itself as in Figure 3.1b, with loops held together by bonds *weak enough to be disrupted by stretching the fiber*; what today we would call hydrogen bonds. This weakly closed loop actually is seen today in many globular proteins, where a strand of a β–sheet region doubles back upon itself to run in the opposite direction, and is called a "beta bend." Unfortunately, Astbury referred to it in passing as a "pseudo-diketopiperazine ring" even though it was not closed by a covalent bond.

Dorothy Wrinch, first at Oxford and later at Smith College, championed the idea that loops such as Astbury proposed were actually closed by covalent bonds as in Figure 3.1c, and that these "cyclol rings" were the key to protein structure, not linear polypeptide chains. She found that one could build flat, closed rings in a multi-hexagonal pattern, containing 2, 6, 18, 42, 66 . . . . (18+24n) amino acids, and proposed this as the basis for fibrous proteins in paper A and the longer 1936 *Proceedings* paper that follows it in the list.

Regrettably, paper A has serious problems. Her statement #1 on the first page that proteins have few or no amine groups not arising from side chains, applies to long folded polypeptide chains (1 free amine) just as well as cyclols (0 free amines). Statement #2 about trigonal or threefold symmetry only reflects the small quantity of information then available about protein structures; later work has shown that it is not true. Wrinch overstates the case by saying that the closed covalent cyclol ring "has already proved useful in the structure of α-keratin" [reference to Astbury and Woods, 1934], and on the second page by implying that the ring-closing bonds in Astbury's α structure, one for every three amino acids along the chain, are in fact covalent. (As was mentioned earlier, had these bonds been covalent, the alpha-to-beta stretching transition in hair and wool would have been impossible without destroying the fiber.) Finally, the five concluding points of her *Nature* paper reflect the limited extent of our knowledge about proteins in 1936, rather than reality.

In 1937 Wrinch extended her model to globular proteins in paper B and in a longer *Proceedings* paper that followed it. She previously had suggested that three-dimensional globular proteins could be built by stacking cyclol sheets atop one another. In this paper she proposes instead that the sheets be folded to build complex three-dimensional polyhedral cages which, although intrinsically beautiful, have no basis in reality. But since at that time no one had the faintest idea how the amino acids in a globular protein might be arranged, one should not be too quick to condemn her models. Had Kendrew's Nobel-Prize-winning myoglobin structure been proposed in 1937, without the extensive experimental analysis that led to it, it would have been laughed off the scientific stage.

Wrinch's proposals stimulated widespread controversy, and generated strong supporters and vigorous opponents. Chief among the latter were Linus Pauling and Carl Niemann at Caltech. In 1939 they published a paper, "The Structure of Proteins" (C) which they fully expected would lay the cyclol theory to rest forever. They based their argument chiefly on thermodynamic calculations of bond energies. After a visiting fellowship with Arnold Sommerfeld at Munich in 1926-7, with side trips to see Born, Heisenberg and Oppenheimer in Göttingen, and Bohr in Copenhagen, Pauling returned to the U.S. with a mission: to put the theory of chemical bonding on a rational, quantum-mechanical basis. An important series of papers began with "The Nature of the Chemical Bond" (1) in 1931, and culminated

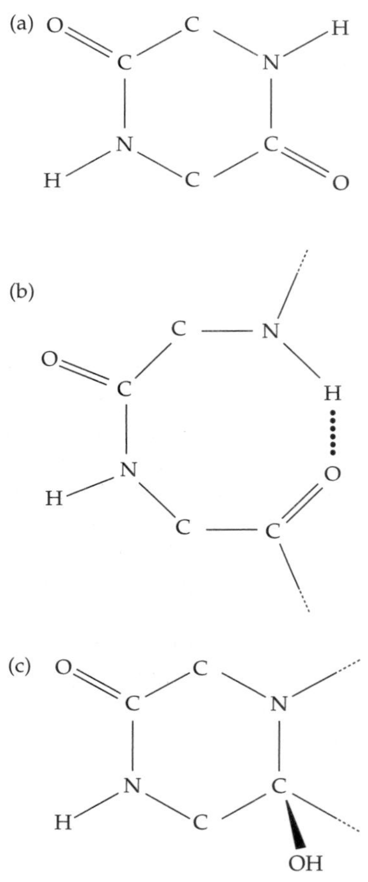

FIGURE 3.1 Three possible rings involving amino acids: (a) Diketopiperazine ring formed by linking CO and NH groups of two amino acids; (b) Hydrogen-bonded "beta bend," seen frequently in globular proteins when a polypeptide chain turns back upon itself to run in an antiparallel direction; (c) Cyclol ring, formed by replacing the hydrogen bond of (b) with a covalent bond between C and N. For simplicity the H atom and amino acid side chain attached to each alpha carbon atom, C_α, are not shown. Dotted line is a hydrogen bond, and dashed lines are continuations of chains.

in a book with the same title in 1939, that would define him as the leader in chemical bonding theories.

One of Pauling's most important concepts was that of the extra stability conferred on a molecule because of what he termed "resonance," which also can be thought of as delocalization of some of the bonding electrons. Figure 3.2a shows the conventional diagram of an amide or peptide bond connecting one amino acid to another. The N atom makes three single bonds and has one unused lone electron pair. The carbonyl C makes a double bond to its O, which has two more unused electron pairs. Pauling proposed that in an alternative structure, Figure 3.2b, the nitrogen lone pair could shift to form a double bond between N and C. This electron shift then would repel the electrons of the C=O double bond, turning it into a single bond and pushing a third lone pair onto the O atom. One consequence would be a full positive charge on the N atom (which would be shared with its hydrogen atom), and a full negative charge on the O atom. Another consequence of the central C=N double bond would be a planar amide, with all six atoms in Figure 3.2b confined to a common plane.

Pauling proposed that reality was somewhere intermediate between these two extremes, with partial double bonds between N and C, and between C and O, and therefore partial positive and negative charges on NH and O. This delocalization or spreading out of electrons would make the resonance hybrid, Figure 3.2c, more stable than the conventional starting structure 2a by 21 kcal/mole. Although he called this behavior "resonance," no flickering between states is implied; the resonance hybrid is simply an intermediate state. Two factors of enormous biological significance follow: All six atoms in the "amide plane" shown in Figure 3.2c are forced by the partial C/N double bond to lie in one plane, drastically limiting the possible conformations of the protein chain. And the partial negative and positive charges on O and NH encourage formation of hydrogen bonds between chains.

This idea of an intermediate bond character between C and N in a polypeptide chain is beautifully borne out by x-ray crystal structure analyses of smaller molecules which contain such bonds. A C—N single bond wherever it occurs typically has a bond length of 1.49 Å; a C=N double bond is of length 1.27 Å, and the intermediate C⋯N bond in a small dipeptide is 1.33 Å long. The link between C and N in a peptide bond does indeed show partial double bond character.

But to return to the cyclol issue, Pauling and Niemann based their case against cyclols on a comparison of relative stabilities of an open chain and one with a single cyclol link, using the concept of bond energies. Pauling had discovered that the energy (actually the enthalpy, ΔH) of a molecule, relative to that of its isolated atoms in the gaseous state, could be calculated accurately by assuming a consistent set of bond energies for all the bonds in the molecule. That is, an O—H bond tends to contribute the same stability to a water molecule, H—O—H, as to a glucose molecule with its six —O—H groups, or an even larger molecule. With a table of bond energies established by examining a great number of organic molecules, Pauling and Niemann calculated the *difference* in enthalpy between an open chain with no cyclol bridges and one with a single bridge (Figure 3.3). Changes in going from the open chain to one with a cyclol link involve (a) the loss of one N—H, one C=O bond and resonance delocalization in one amide group, and (b) the gain of one N—C, one C—O and one O—H bond. As the numbers in Figure 3.3 show, the chain with the cyclol link is *less stable (higher energy) than the unlinked chain by 27.5 kcal/mol.* It may seem strange at first that adding one more covalent bond should make the molecule *less* stable rather than more. But a high price is paid in eliminating the very stable C=O bond, and in losing the energy of resonance delocalization.

Enthalpies of denaturation of a globular protein are, and were known to be back then, small in magnitude but positive. That is, the denatured chain usually is marginally of higher enthalpy than the undenatured protein. The difference is small, we now know, because much of what holds a protein together is internal hydrogen

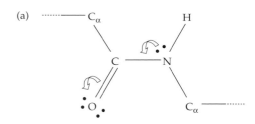

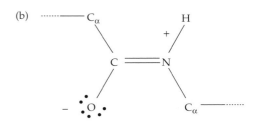

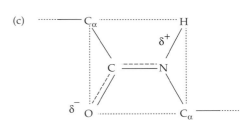

FIGURE 3.2 Pauling's resonance stabilization model of a peptide bond. (a) The classical bond structure; (b) Alternative structure with C=N double bond and O—C single bond. Arrows in (a) indicate shifts of electron pairs that produce structure (b). The true bond structure (c) is assumed to be a resonance hybrid of (a) and (b). The partial double bond character of the central C—N bond keeps all six atoms of the amide planar, and the partial charges on O and NH encourage the formation of hydrogen bonds between chains or between loops or turns of one chain.

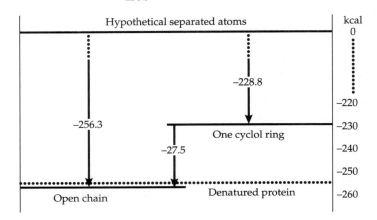

FIGURE 3.3 The Pauling/Niemann comparison of stabilities of an uncyclized polypeptide chain (left) and a chain with a single cyclol bond (right). Differences in bond energies between the two structures are tabulated below, and the resulting relative stabilities are compared on an energy plot at bottom. A denatured protein is roughly one kcal per mole per residue less stable than the folded chain.

Differences in bond energies (kcal/mol)

N—H	–83.3		N—C	–48.6
C=O	–152.0		C—O	–70.0
C≡N and			O—H	–110.2
C≡O	–21.0			
	–256.3			–228.8

bonds, and these are counterbalanced by similar bonds to water molecules in the denatured protein. Most of the remaining stability of the folded protein arises from van der Waals interactions between hydrophobic side chains in the interior. If Pauling and Niemann's calculations were correct, then each cyclol bond would *destabilize* the folded protein relative to the extended denatured chain by roughly 27 kcal. The folded protein would denature explosively. Obviously this doesn't happen, so obviously, they concluded, the cyclol model is wrong.

Wrinch published a vigorous rebuttal (D) which took issue with almost every aspect of Pauling's calculations. How do we know that a denatured protein is an extended polypeptide chain? (But Astbury had in fact answered this by his x-ray photos of fibers from denatured globular proteins.) How do we know that denaturation necessarily involves opening each and every cyclol bond? If only one bond in 27 were broken, then the discrepancy would be reduced to a kcal or two. And why were Pauling and Niemann so sure that their energy calculation as shown in Figure 3.3 was correct?

Nothing was settled. Those who believed in cyclols continued to do so, while those who did not, maintained their position. But the tone of the argument assumed some of the rancor of the 2004 U. S. Presidential campaign. Pauling and Niemann's paper is entertaining reading because it is full of remarks such as:

> **We....have reached the conclusions that there exists no evidence whatever in support of this hypothesis.**

> **We wish to point out that the evidence adduced by Wrinch and Langmuir has very little value.**

> **There can be found in the papers by Wrinch many additional statements which might be construed as arguments in support of the cyclol structure. None of these seems to us to have enough significance to justify discussion.**

> **....the fact that many other parameters were also assigned arbitrary values removes all significance from their argument.**

> **....the arguments are so lacking in rigor, and the conclusions are so indefinite....**

Scientists rarely talk like that in print today. The first person plural pronoun is much in evidence: the expressions: "We have carefully examined....," "We accordingly conclude...," "We agree...," "We conclude that...," "We draw the rigorous conclusion..." and the like occur no fewer than eight times in as many pages.

Wrinch, it must be said, responded in kind in "The Geometrical Attack on Protein Structure" (D), making statements such as:

> **It is unnecessary at the present time to state the case for the cyclol hypothesis**

> **....Pauling and Niemann repeat a number of statements purporting to disprove the theory already made by other writers**

> **....these criticisms have already been discussed, at least so far as their scientific importance appeared to warrant**

> **...the suggested deduction [of Pauling and Niemann] therefore falls to the ground**

> **It must be concluded that no case has been made out for deducing that the cyclol theory is false....**

But by ca. 1940, the cyclol theory was essentially dead. Too much new evidence supported the idea that proteins were built from one or more folded linear polypeptide chains. We know today that the only common covalent crosslinks between protein chains are disulfide bonds from cysteines.

The paper with Niemann was only Pauling's second paper in structural molecular biology, following one with Mirsky three years earlier on native vs. denatured proteins (2). Pauling's subsequent output in structural molecular biology in the 1950's would be an astounding 42 papers, in which he truly reshaped the field. But then his interests shifted, and he subsequently published only three brief reviews on the subject thirty years later (3–5). Interestingly, the first of these three was an appreciation of the work of Dorothy Wrinch, the second was a denigration of Schrödinger's seminal 1944 book, *What is Life?* and the third was an appreciation of one of his own discoveries.

Three players in our drama can fairly be regarded in hindsight as tragic figures: Dorothy Wrinch, Erwin Chargaff and Rosalind Franklin. Wrinch staked her scientific reputation on a theory which, although in advance of other structure theories of the time, ultimately proved to be wrong. Chargaff was a brilliant and meticulous DNA chemist who established that DNA always contains equal quantities of adenine and thymine, and equal quantities of guanine and cytosine. But he never understood the implications of this in terms of DNA structure, and bitterly resented Watson and Crick for their success. And Franklin was an outstandingly talented scientist who produced the finest DNA fiber photographs available, but who hesitated to take the model-building leap that Watson and Crick did, and who lost any chance of a share in the 1962 Nobel Prize by her premature death at age 38. Maureen Julian's paper (E) is an absorbing chronicle of Dorothy Wrinch's scientific and personal career. She struggled against the bias that women in science suffered in that era, losing fellowships early in her career, for example, purely because of her gender. These difficulties probably contributed to her combative style in defending what she felt was her primary scientific legacy.

But when all the dust had settled, the extended polypeptide chain, folded in some as yet unknown manner, was recognized as the foundation of protein structure. In 1938 W. T. Astbury published a Faraday Society lecture, "X-Ray Adventures Among the Proteins" (F) that admirably summed up the state of the art just before we began to understand protein folding. Opening his lecture by quoting Lindo Patterson's limerick "Amino-acids in chains..." is entirely in keeping with Astbury's essentially cheerful, enthusiastic personality, according to people such as Max Perutz who knew him well. The first paragraph of his paper is quite in character: "....scientific thrills are among the best and highest kind of thrills," and the problem of protein structure "....is, I think, the greatest scientific adventure of our times."

Astbury leads us quickly through the entire background of x-ray diffraction by fibers, giving full credit to his predecessors. His summary of silk fibers (α) and hair keratin (α and β) is excellent. His explanation of muscle contraction on page 382 is wrong, but as he would be the first to admit, no one is perfect. He then proceeds to outline the problems to be solved in understanding the structures of globular proteins. At the foot of page 384 he summarizes the critical finding that denatured proteins are extended polypeptide chains:

> On denaturation and coagulation the globular proteins give rise to a common X-ray diagram like that of disoriented α-keratin. In the molecular sense then, at least, they tend to become fibrous; the polypeptide chains are liberated from their specific folded configurations and aggregate to a greater or less extent into extended bundles similar in structure to the sub-microscopic crystallites of α-keratin.

These two sentences essentially confirm Pauling and Niemann's arguments against the cyclol theory.

In his proposal of models for folded polypeptide chains on page 385, Astbury now does appear to invoke covalent bonds to close cyclol-like loops, even though the covalent bond would make impossible his alpha-to-beta stretching conversion in keratins. But the nature and importance of hydrogen bonds at that era were understood by few people other than Linus Pauling. Astbury's review was published a year before Pauling and Niemann's paper appeared, and Astbury did not have access to the bond energy data of Pauling.

At the end of his review, on pp. 387–8, Astbury falls into a numerology fallacy regarding protein size and composition that circulated widely at the time: that the numbers of each kind of residue in a protein could be expressed in terms of powers of 2 and powers of 3. He suggests that "the smallest possible number of residues per [protein] molecule is $2^6 \times 3^2$ or 2×288 (molecular weight ca. 65,000), corresponding to hemoglobin and fibrin." This was rapidly shown to have no basis in reality; many proteins have as few as 100 amino acids and some are even smaller. There is no mathematical formula governing the length of chains in proteins.

But Astbury ends on a prophetic note:

> Exact analyses of the proteins, though always laborious, need no longer be the thankless tasks they have been. Every possible reliable observation now is urgently needed and must sooner or later be fitted into the puzzle. Above all, complete analyses of single proteins are necessary.

In contrast, the paper by Pauling and Niemann a year later (C) contains one prediction that fortunately turned out to be totally wrong:

> It has not yet been possible to make a complete determination with X-rays of the positions of the atoms in any protein crystal; and the great complexity of proteins makes it unlikely that a complete structure determination for a protein will ever be made by X-ray methods alone.

References

1. L. Pauling. 1931. *J. Am. Chem. Soc.* 53, 1367–1400. "The Nature of the Chemical Bond."
2. A. E. Mirsky and L. Pauling. 1936. *Proc. Natl. Acad. Sci. USA* 22, 439–477. "On the Structure of Native, Denatured, and Coagulated Proteins."
3. L. Pauling. 1987. *J. Chem. Ed.* 64, 286. "Dorothy Wrinch and the Structure of Proteins."
4. L. Pauling. 1987. In *Schrödinger: Centenary Celebration of a Polymath* (C. W. Kilmister, ed.), Cambridge University Press, pp. 225–233. "Schrödinger's Contribution to Chemistry and Biology."
5. L. Pauling. 1996. *The Chemical Intelligencer* 2 (1), 32–38 (January). "The Discovery of the Alpha Helix."

Study Questions

1. How did Wrinch misrepresent Astbury's results in her paper A?
2. How did Wrinch obtain three-dimensional globular protein structures, starting with flat cyclol sheets?
3. In the simple "particle in a box" quantum mechanical calculation, an electron or other particle can have only certain discrete energy states, called its quantum levels. Increasing the mass of the particle or the size of the box lowers all the these energy levels and brings them closer together. How is this relevant to Pauling's theory of resonance stabilization of a peptide bond?
4. How is this same principle relevant to the fact that we ordinarily do not think of the speed of a thrown baseball as being limited just to a set of discrete values, even though in principle it is?
5. Why is a closed cyclol loop *less* stable than an open chain, even though it contains a new N—C chemical bond?
6. Why, in Pauling's view, is a peptide or amide link between two amino acids required to be planar?
7. How does Haurowitz' experimental measurements of the number of free —OH groups in proteins (C) argue against the cyclol model?
8. How does the energy of denaturation of proteins (C) argue against the cyclol model?

MARCH 7, 1936   NATURE   411

The Pattern of Proteins

By Dr. D. M. Wrinch, Mathematical Institute, Oxford

ANY theory as to the structure of the molecule of simple native protein must take account of a number of facts, including the following :

(1) The molecules are largely, if not entirely, made up of amino acid residues. They contain —NH—CO linkages, but in general few —NH₂ groups not belonging to side chains, and in some cases possibly none.

(2) There is a general uniformity among proteins of widely different chemical constitution which suggests a simple general plan in the arrangement of the amino acid residues, characteristic of proteins in general. Protein crystals possess high, general trigonal, symmetry.

(3) Many native proteins are 'globular' in form.

(4) A number of proteins[1] of widely different chemical constitution, though isodisperse in solution for a certain range of values of pH, split up into molecules of submultiple molecular weights in a sufficiently alkaline medium.

The facts cited suggest that native protein may contain closed, as opposed to open, polypeptides, that the polypeptides, open or closed, are in a folded

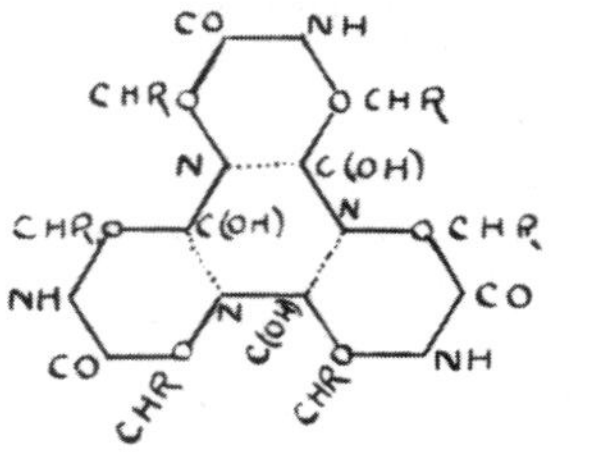

FIG. 1. The 'cyclol 6' molecule.

state, and that the type of folding must be such as to imply the possibility of regular and orderly arrangements of hundreds of residues.

An examination of the geometrical nature of polypeptide chains shows that, *since all amino acids known to occur in proteins are α-derivatives*, they may be folded in hexagonal arrays. Closed polypeptide chains consisting of 2, 6, 18, 42, 66, 90, 114, 138, 162 . . . (18 + 24n) . . . residues form a series with threefold central symmetry. A companion series consisting of 10, 26, 42, 58, 74, 90, 106, 122 . . . (10 + 16n) . . . residues have twofold central symmetry. There is also a series with sixfold central symmetry : others with no central symmetry. Open polypeptides can also be hexagonally folded. The number of free —NH₂ groups, in so far as these indicate an open polypeptide, can be made as small as we please, even zero if we so desire. The hexagonal folding of polypeptide chains, open or closed, evidently allows the construction of molecules containing even hundreds of amino acid residues in orderly array, and provides a characteristic pattern, which in its simplicity and uniformity agrees with many facts of protein chemistry.

The stability of these folded polypeptide chains cannot be attributed to electrostatic attractions between the various CO,NH groups, for the appropriate distance between carbon and nitrogen atoms in these circumstances[2] lies between 2·8 A. and 4·2 A.,

whereas the distance in our case is at most 1·54 A. By using the transformation* suggested by Frank in 1933 at a lecture given by W. T. Astbury to the Oxford Junior Scientific Society,

$$\diagdown C = O \qquad H-N \diagup \quad \text{to} \quad \diagdown C(OH)-N \diagup ,$$

which has already proved useful in the structure of α-keratin[3], the situation is at once cleared up and we obtain (Fig. 1) the molecule 'cyclol 6' (the closed

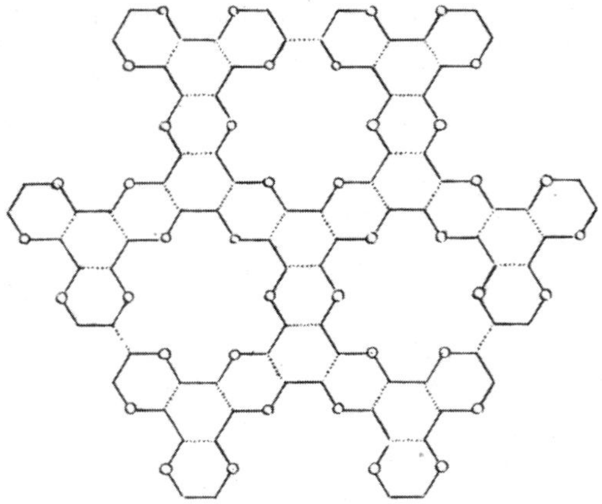

FIG. 2. A 'cyclol 42' molecule.

polypeptide with six residues), 'cyclol 18', 'cyclol 42' (Fig. 2) and so on, and similarly open 'cyclised' polypeptides (Fig. 3).

Hexagonal packing of polypeptides suggests a new *three dimensional* unit, —CHR—$\overset{|}{C}$(OH)—N$\diagup$, which may be used to build three-dimensional molecules of a variety of types. These are now being investigated in detail. At the moment we direct attention only to

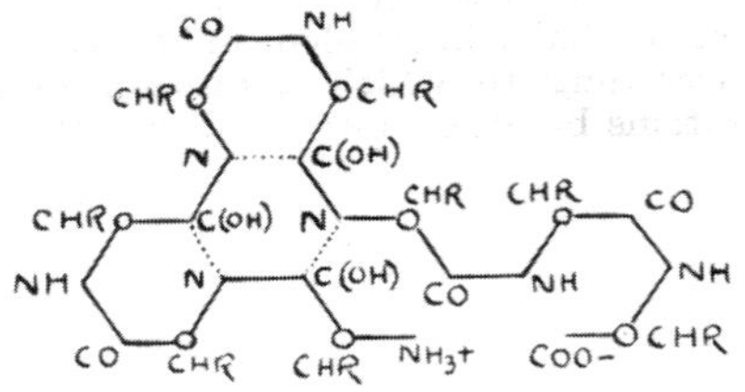

FIG. 3.

single cyclised polypeptides forming hexagons lying approximately in one plane. The cyclol layer molecule is a fabric the thickness of which is one amino acid residue. *Since all naturally occurring amino acids are of lævo type*[4] this fabric is dorsiventral, having a front surface from which the side chains emerge, and a back surface free from side chains. Both front and back carry trios of hydroxyls normal to the surface

---

* The application of this transformation to these molecules was suggested to me by J. D. Bernal.

in alternating hexagonal arrays. Such a layer molecule and its polymers, formed also by the same transformation, can cover an area of any shape and extent. It offers suggestions as to the structure of the solid protein film when it is one amino acid residue thick. In its most compact form, the cyclol layer molecule gives an area per residue of about $9 \cdot 9$ A.^{2}. Less dense layers can be built, for example, from polymers of cyclol 18 and of cyclol 66 respectively, where the corresponding areas per residue are $13 \cdot 2$ A.^{2} and $16 \cdot 2$ A.^{2} respectively. The figures for unimolecular films of gliadin, glutenin, egg albumin, zein, serum albumin, serum globulin range from $1 \cdot 724 \times 10^7$-gm./cm.^{2} for serum globulin to $1 \cdot 111 \times 10^{-7}$ gm./cm.^{2} for serum albumin[5,6,7]. With an average residue weight of 120, these densities give an area per residue ranging from $11 \cdot 48$ A.^{2} to $18 \cdot 82$ A.^{2}. On the basis of the proposed hexagonal packing of polypeptides I therefore suggest that the upper limit of density of which a protein film is capable without buckling, provided that it is only one amino acid residue thick, is one residue per $9 \cdot 9$ A.^{2}; further, that a higher density implies that the film, though it may still be unimolecular, is more than one amino acid residue thick.

Cyclol layers may also be used to build molecules and molecular aggregates with extension in three dimensions, since they may be linked front to front by means of the side chains, in particular, by cystine bridges, and back to back by means of hydroxyls[8]. The single-layer cyclol is a fabric capable of covering a two-dimensional area of any shape and extent; a three-dimensional array can then be built, layer upon layer, the linkage being alternately by means of side chains and hydroxyls. The idea that native proteins consist largely, if not entirely, of cyclised polypeptides therefore implies that some native proteins, including those of 'globular' type, may have a layer structure.

Linkage by means of hydroxyls recalls the structures of graphitic oxide and montmorillonite, etc. Such a structure suggests a considerable capacity for hydration, an outstanding characteristic of many proteins. Further, since alternate layers are held together by means of hydroxyls, and contiguous molecules may also be held together in the same way, a protein molecular aggregate will, on this theory, necessarily be sensitive to changes in the acidity of the medium; in particular, a sufficiently high pH will cause such an aggregate to dissociate into single-layer units or into two-layer units joined by cystine bridges or side chains in covalent linkages. Svedberg's results, according to which a number of different native proteins break up into smaller molecules with sub-multiple molecular weights[1], here find a simple interpretation. The particular sub-multiples which occur may be regarded as affording evidence as to the type of symmetry possessed by the layers out of which the molecular aggregates are built.

The hypothesis that native proteins consist essentially of cyclised polypeptides thus takes account of the facts mentioned in (1), (2), (3), (4) above. Further, it derives support from the case of α-keratin, for with Astbury's 'pseudo-diketopiperazine' structure[3] the polypeptides may be regarded as partially cyclised since they are cyclised at regular intervals, one out of every three (CO,NH) groups being involved. It is also suggestive in relation to a variety of other facts belonging to organic chemistry, X-ray analysis, enzyme chemistry and cytology. I cite the following:

(1) The rhythm of 18 in the distribution of amino acids in gelatin found by Bergmann[9], and the suggestion of Astbury[10] that in gelatin "the effective length of an amino acid residue is only about $2 \cdot 8$ A.".

(2) The low molecular weight not exceeding 1,000 found by Svedberg[11] for the bulk of the material from which lactalbumin is formed.

(3) Secretin[12], a protein with molecular weight of about 5,000, containing no open polypeptide chains.

(4) The nuclear membrane, which, consisting of proteins and lipoids, plays an important part in mitosis on account of its variable permeability.

(5) Bergmann's findings[13] with respect to dipeptidase; these suggest that the dipeptide substrate, upon which this enzyme acts, has a hexagonal configuration.

Finally, the deduction from the hypothesis of cyclised polypeptides, that native proteins may consist of dorsiventral layers, with the side-chains issuing from one side only, suggests that immunological reactions are concerned only with surfaces carrying side-chains. Hence, such reactions depend both on the particular nature and on the arrangement of the amino acids.

Full details of the work, which is to be regarded as offering for consideration, a simple *working hypothesis*, for which no finality is claimed, will be published in due course.

[1] Svedberg, *Science*, **79**, 327 (1934).
[2] International Tables for the Determination of Crystal Structure.
[3] Astbury and Woods, *Phil. Trans. Roy. Soc.*, **232**, 333 (1933).
[4] Jordan Lloyd, *Biol. Rev.*, **7**, 256 (1932).
[5] Gorter, *J. Gen. Phys.*, **18**, 421 (1935); *Amer. J. Diseases of Children*, **47**, 945 (1934).
[6] Gorter and van Ormondt, *Biochem. J.*, **29**, 48 (1935).
[7] Schulman and Rideal, *Biochem. J.*, **27**, 1581 (1933).
[8] Bernal and Megaw, *Proc. Roy. Soc.*, A, **151**, 384 (1935).
[9] Bergmann, *J. Biol. Chem.*, **110**, 471 (1935).
[10] Astbury, Cold Spring Harbor Symposia on Quantitative Biology, **2**, 15 (1934).
[11] Sjogren and Svedberg, *J. Amer. Chem. Soc.*, **52**, 3650 (1930).
[12] Hammersten *et al.*, *Biochem. Z.*, **264**, 272 and 275 (1933).
[13] Bergmann *et al.*, *J. Biol. Chem.*, **109**, 325 (1935).

Association of Technical Institutions

ANNUAL MEETING IN LONDON

THE annual meeting of the Association of Technical Institutions was held in the Goldsmiths' Hall, London, on February 28–29. The Right Hon. Lord Plender was installed as president for 1936. In his presidential address he surveyed such changes in the structure of the business world as the development of large stores, the grouping of railways and the general trend towards amalgamation of small units of business. He referred especially to costing and its scientific application which have been so useful to industry, and he emphasised the need for skilled persons in such branches as accountancy, foreign exchange, advertising, etc.

In a paper on "Technology and the Community" Councillor Wright Robinson touched a question already adumbrated in the recent "Report on Policy

972   NATURE   JUNE 5, 1937

The Cyclol Theory and the 'Globular' Proteins*

By Dr. D. M. Wrinch

A NUMBER of facts relating to proteins[1] suggest that the polypeptides in native proteins are in a folded state[2,3]. The type of folding must be such as to imply the possibility of the regular and orderly arrangement of hundreds of amino acid residues, which to some extent at least is independent of the particular residues in question.

At present two types of folding have been suggested, the cyclol type[3,4] and the hydrogen bond type[5]. The search for other types of folding is being continued. So far, it has not proved possible to discard either theory on the grounds that the type of link postulated is out of the question. It is, therefore, very desirable to test these theories by checking their implications against known facts. Accordingly it is now considered whether the cyclol theory can stand the test of the body of facts relating to the 'globular' proteins, established by Svedberg and his collaborators[6].

THE CYCLOL FABRIC

In previous communications the cyclol postulate has been applied directly to polypeptides, and a number of molecules, cyclol 6, cyclol 18 . . . and the general cyclol fabric (Fig. 1) have been suggested for consideration. These are all of one polyhexagonal type, in which the individual hexagons are alternately '2-way' diazine hexagons sharing opposite sides with triazine hexagons and '3-way' triazine hexagons sharing alternate sides with diazine hexagons. In default of information as to the sides and angles of these hexagons, a mean value is at present adopted for the C—C and C—N distances indifferently, and the tetrahedral angle δ as the valency angle for C. The valency angle for N is also taken as δ, since it has this value in hexamethylene tetramine[7], in which, as in the cyclol fabric, each N atom is joined to three C atoms.

In accordance with these assumptions, the individual hexagons in the cyclol fabric are at present taken to be 'crumpled' as in cyclohexane. The midpoints of the sides of a 'crumpled' hexagon are the vertices of a plane hexagon, and the geometrical problems can be simplified by taking these plane hexagons as the fundamental units in a 'median' network which represents the cyclol in a new way, each 'crumpled' hexagon being replaced by a 'median' hexagon.

In previous communications the cyclols have been considered only in the case when all the median hexagons lie in one common plane. With this limitation there has, of course, been no question of building a closed (that is, a space-enclosing) cyclol. To do so, it is necessary to investigate the conditions under which a cyclol fabric can bend about a line. Evidently it is permissible for two abutting median hexagons to lie on different planes, if the angle between the planes is the tetrahedral angle δ. Thus a cyclol fabric need not have a single median plane, but may turn about certain lines provided that the angle between abutting median planes is δ. The problem of the possible existence of space-enclosing cyclols

* This article is based upon lectures given at the Harvard, Yale and George Washington University Medical Schools and at the Medical Centre, Columbia University, in October and November 1936. Communicated to the Editor of NATURE on March 10, 1937.

can then be stated precisely as follows. Is it possible for a cyclol network to bend across one line after another so that it joins up and thus surrounds a portion of space ? In other words, can a cyclol network be drawn, not on a plane but on the surface

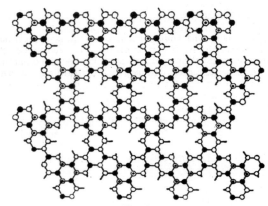

Fig. 1.

THE CYCLOL PATTERN. THE MEDIAN PLANE OF THE LAMINA IS THE PLANE OF THE PAPER. THE LAMINA HAS ITS 'FRONT' SURFACE ABOVE AND ITS 'BACK' SURFACE BELOW THE PAPER.

● = N.

○ = C(OH), PEPTIDE HYDROXYL UPWARDS.

◉ = C(OH), PEPTIDE HYDROXYL DOWNWARDS.

○— = CHR, DIRECTION OF SIDE CHAIN INITIALLY OUTWARDS.

○- = CHR, DIRECTION OF SIDE CHAIN INITIALLY UPWARDS.

of some polyhedron such that the faces at any edge crossed by the network abut at the tetrahedral angle ?

THE CLOSED CYCLOLS

To solve this problem, all the polyhedra in which some at least of the dihedral angles are equal to the tetrahedral angle will be considered in turn. As a first step it is remarked that among the regular and semi-regular polyhedra[8], only four satisfy the conditions. These are the truncated tetrahedron, the octahedron, the truncated octahedron and the skew triangular prism. On this occasion, as an example of this method of building megamolecules, attention is directed to the truncated tetrahedron, on which it has proved possible to draw closed cyclol networks. These networks form a linear series $C_1, C_2, \ldots C_n,$ . . . which comprise 72, 288, . . . $72n^2$, . . . amino acid residues. Figs. 2 and 3 show models of C_1 and C_2 in which the cyclol fabric is represented by the median hexagons. These models have 4 hexagonal faces, 4 triangular faces and 6 slits. The actual distribution of the (C—C—N) groups of atoms in the amino acid residues in the molecule can be inferred from the median hexagons, which are to be regarded simply as a shorthand notation.

The possibility of building closed cyclols on the other polyhedra mentioned above and on polyhedra which are not regular or semi-regular is being investigated. If they can be constructed they will also exist in linear series, each comprising numbers of residues represented by a quadratic function of the natural numbers 1, 2, . . . n, . . .

The cyclol hypothesis therefore predicts the existence of one or more series of 'space-enclosing' protein molecules, each series comprising numbers of residues given by a quadratic function of the natural numbers 1, 2, 3, . . . n, . . . and in particular of a series C_1, C_2, . . . C_n, . . . having the shape of truncated tetrahedra and comprising 72, 288, . . . $72n^2$, . . . residues.

These predictions are, it is claimed, confirmed by the results obtained by Svedberg and his collaborators[6].

(1) It is found that certain proteins are 'globular' molecules which, in appropriate circumstances, are monodisperse. The cyclol theory implies the existence of space-enclosing molecules containing certain specific numbers of amino acid residues : their polyhedral character is in accord with and offers an interpretation of the nature of this 'globularity'.

(2) It is found that certain molecules exist in different degrees of association in solutions of different pH, having a maximum molecular weight in a certain pH range and dissociating reversibly into molecules with submultiple molecular weights on one or both sides of this range. This is interpreted to mean that, in certain types of protein, the molecule will form multiple molecules at appropriate pH values by linkages between peptide hydroxyls or by salt linkages between side chains, the process of dissociation on changing the pH being reversible.

(3) It is found that the molecular weights of proteins are not distributed at random, but fall into a sequence of widely separated classes, the molecular weights in one class varying by as much as 15 per cent from a mean value. This is interpreted to mean that the proteins falling into one of these classes have a common structure as regards the arrangement of the constituent amino acids, and it is further suggested that each class connotes one closed cyclol network or an association of a certain number of such units. The variation in molecular weight within a class is then accounted for by the different selections of residues in the various proteins which can yield

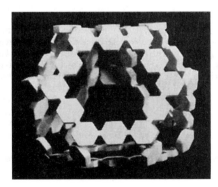

Fig. 2.

an average residue weight varying (say) from 100 to 135, which may also entail the presence of different numbers of water molecules in the molecule[9]. A further modification is also introduced if imino acid residues are present.

These tests are only qualitative. It would be of greater interest to apply more stringent quantitative tests. This will only be possible when data are available which give, for some of the 'globular' proteins, the shape of the molecule, the average value of the weights of the contained amino acid residues, the number of imino acid residues and the numbers of water molecules which form an integral part of the molecular structure.

It is, however, suggested for consideration in the future, that the group of proteins with molecular

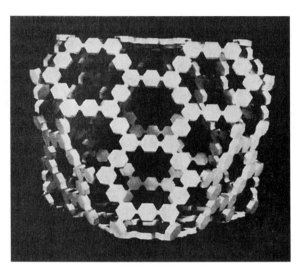

Fig. 3.

weights ranging from 33,600 to 40,500[10] are closed cyclol molecules of type C_2. This molecular weight class is of particular importance, since Svedberg has suggested that very many, possibly most, other proteins have molecular weights which are multiples of (say) 36,000.

Full details of this work will appear shortly.

Postscript added May 28, 1937 : My attention has been directed to recent work by Bergmann and his collaborators, which strongly confirms the conclusion reached above. In particular, it may here be put on record that one of the predictions has now been verified : Bergmann and Niemann (*J. Biol. Chem.*, **118**, 301 ; 1937) deduce from the chemical analysis of egg albumin that this molecule (which belongs to the group of proteins with molecular weights in the neighbourhood of 36,000) consists of exactly 288 residues as predicted by the cyclol hypothesis. Full details of my work are in course of publication in the *Proceedings of the Royal Society*.

[1] Pryde, "Recent Advances in Biochemistry", chap. 1 (1931) : Langmuir, Schaefer and Wrinch, *Science*, **85**, 76 (1937).

[2] Astbury, NATURE, **132**, 593 (1933). *Koll. Z.*, **69**, 340 (1934). Astbury and Street, NATURE, **126**, 913 (1930). *Phil. Trans. Roy. Soc.*, A, **230**, 75 (1931).

[3] Wrinch, NATURE, **137**, 411: **138**, 241: **138**, 651 (1936). *Proc. Roy. Soc.*, A, in the Press.

[4] Astbury, *J. Text. Inst.*, **17**, P. 281 (1936). NATURE, **137**, 803 (1936). *Chem. Weekblad*, **33**, 778 (1936).

[5] Jordan Lloyd, *Biol. Rev.*, **7**, 254 (1932). Mirsky and Pauling, *Proc. Nat. Acad. Sci.*, **22**, 439 (1936). Wrinch and Jordan Lloyd, NATURE, **138**, 758 (1936).

[6] Svedberg *et al.*, *Koll. Z.*, **51**, 10 (1930). *Trans. Far. Soc.*, **26**, 72 and 737 (1930). *Science*, **79**, 327 (1934). *Biol. Bull.*, **66**, 191 (1934). *Chem. Rev.*, **14**, 1 (1935), and a series of papers in *J. Amer. Chem. Soc.*, from 1929.

[7] Dickinson and Raymond, *J. Amer. Chem. Soc.*, **45**, 22 (1923). Wyckoff and Corey, *Z. Krist.*, **89**, 426 (1934).

[8] Andreini, *Soc. Ital. d. Scienze*, **14**, 75 (1907).

[9] Jordan Lloyd, *Biochem. J.*, **14**, 147 (1920); **21**, 1352 (1927); **25**, 1580 (1931). *Biol. Rev.*, **8**, 463 (1933).

[10] Svedberg and Eriksson-Quensel, "Tabulæ Biologicæ Periodicæ", **5**, 351 (1935-36).

1860   Linus Pauling and Carl Niemann   Vol. 61

[Contribution from the Gates and Crellin Laboratories of Chemistry, California Institute of Technology No. 708]

The Structure of Proteins

By Linus Pauling and Carl Niemann

1. Introduction

It is our opinion that the polypeptide chain structure of proteins,[1] with hydrogen bonds and other interatomic forces (weaker than those corresponding to covalent bond formation) acting between polypeptide chains, parts of chains, and side-chains, is compatible not only with the chemical and physical properties of proteins but also with the detailed information about molecular structure in general which has been provided by the experimental and theoretical researches of the last decade. Some of the evidence substantiating this opinion is mentioned in Section 6 of this paper.

Some time ago the alternative suggestion was made by Frank[2] that hexagonal rings occur in proteins, resulting from the transfer of hydrogen atoms from secondary amino to carbonyl groups with the formation of carbon–nitrogen single bonds. This *cyclol hypothesis* has been developed extensively by Wrinch,[3] who has considered the geometry of cyclol molecules and has given discussions of the qualitative correlations of the hypothesis and the known properties of proteins.

It has been recognized by workers in the field of modern structural chemistry that the lack of conformity of the cyclol structures with the rules found to hold for simple molecules makes it very improbable that any protein molecules contain structural elements of the cyclol type. Until recently no evidence worthy of consideration had been adduced in favor of the cyclol hypothesis. Now, however, there has been published[4] an interpretation of Crowfoot's valuable X-ray data on crystalline insulin[5] which is considered by the authors to provide proof[6] that the insulin molecule actually has the structure of the space-enclosing cyclol C_2. Because of the great and widespread interest in the question of the structure of proteins, it is important that this claim that insulin has been proved to have the cyclol structure be investigated thoroughly. We have carefully examined the X-ray arguments and other arguments which have been advanced in support of the cyclol hypothesis, and have reached the conclusions that there exists no evidence whatever in support of this hypothesis and that instead strong evidence can be advanced in support of the contention that bonds of the cyclol type do not occur at all in any protein. A detailed discussion of the more important pro-cyclol and anti-cyclol arguments is given in the following paragraphs.

2. X-Ray Evidence Regarding Protein Structure

It has not yet been possible to make a complete determination with X-rays of the positions of the atoms in any protein crystal; and the great complexity of proteins makes it unlikely that a complete structure determination for a protein will ever be made by X-ray methods alone.[7] Nevertheless the X-ray studies of silk fibroin by Herzog and Jancke,[8] Brill,[9] and Meyer and Mark[10] and of β-keratin and certain other proteins by Astbury and his collaborators[11] have provided strong (but

(1) E. Fischer, "Untersuchungen über Aminosäuren, Polypeptide und Protein," J. Springer, Berlin, 1906 and 1923.

(2) F. C. Frank, *Nature*, **138**, 242 (1936); this idea was first proposed by Frank in 1933: see W. T. Astbury, *J. Textile Inst.*, **27**, 282 (1936).

(3) D. M. Wrinch, (a) *Nature*, **137**, 411 (1936); (b) **138**, 241 (1936); (c) **139**, 651, 972 (1937); (d) *Proc. Roy. Soc.* (London), **A160**, 59 (1937); (e) **A161**, 505 (1937); (f) *Trans. Faraday Soc.*, **33**, 1368 (1937); (g) *Phil. Mag.*, **26**, 313 (1938); (h) *Nature*, **143**, 482 (1939); etc.

(4) (a) D. M. Wrinch, *Science*, **88**, 148 (1938); (b) This Journal, **60**, 2005 (1938); (c) D. M. Wrinch and I. Langmuir, *ibid.*, **60**, 2247 (1938); (d) I. Langmuir and D. M. Wrinch, *Nature*, **142**, 581 (1938).

(5) D. Crowfoot, *Proc. Roy. Soc.* (London), **A164**, 580 (1938).

(6) In ref. 4d, for example, the authors write "The superposability of these two sets of points represented the first stage in the proof of the correctness of the C_2 structure proposed for insulin. ... These investigations, showing that it is possible to deduce that the insulin molecule is a polyhedral cage structure of the shape and size predicted, give some indication of the powerful weapon which the geometrical method puts at our disposal."

(7) A protein molecule, containing hundreds of amino acid residues, is immensely more complicated than a molecule of an amino acid or of diketopiperazine. Yet despite attacks by numerous investigators no complete structure determination for any amino acid had been made until within the last year, when Albrecht and Corey succeeded, by use of the Patterson method, in accurately locating the atoms in crystalline glycine [G. A. Albrecht and R. B. Corey, This Journal, **61**, 1087 (1939)]. The only other crystal with a close structural relation to proteins for which a complete structure determination has been made is diketopiperazine [R. B. Corey, *ibid.*, **60**, 1598 (1938)]. The investigation of the structure of crystals of relatively simple substances related to proteins is being continued in these Laboratories.

(8) R. O. Herzog and W. Jancke, *Ber.*, **53**, 2162 (1920).

(9) R. Brill, *Ann.*, **434**, 204 (1923).

(10) K. H. Meyer and H. Mark, *Ber.*, **61**, 1932 (1928).

(11) W. T. Astbury, *J Soc. Chem. Ind.*, **49**, 441 (1930); W. T. Astbury and A. Street, *Phil. Trans. Roy. Soc.*, **A230**, 75 (1931); W. T. Astbury and H. J. Woods, *ibid.*, **A232**, 333 (1933); etc.

not rigorous) evidence that these fibrous proteins contain polypeptide chains in the extended configuration. This evidence has been strengthened by the fact that the observed identity distances correspond closely to those calculated with the covalent bond lengths, bond angles, and N-H $\cdots$ O hydrogen bond lengths found by Corey in diketopiperazine.

The X-ray work of Astbury also provides evidence that α-keratin and certain other fibrous proteins contain polypeptide chains with a folded rather than an extended configuration. The X-ray data have not led to the determination of the atomic arrangement, however, and there exists no reliable evidence regarding the detailed nature of the folding.

X-Ray studies of crystalline globular proteins have provided values of the dimensions of the units of structure, from which some qualitative conclusions might be drawn regarding the shapes of the protein molecules. An interesting attempt to go farther was made by Crowfoot,[5] who used her X-ray data for crystalline insulin to calculate Patterson and Patterson–Harker diagrams.[12] Crowfoot discussed these diagrams in a sensible way, and pointed out that since the X-ray data correspond to effective interplanar distances not less than 7 Å. they do not permit the determination of the positions of individual atoms;[13] the diagrams instead give some information about large-scale fluctuations in scattering power within the crystal. Crowfoot also stated that the diagrams provide no reliable evidence regarding either a polypeptide chain or a cyclol structure for insulin.

Wrinch and Langmuir[4] have, however, contended that Crowfoot's X-ray data correspond in great detail to the structure predicted for the insulin molecule on the basis of the cyclol theory, and thus provide the experimental proof of the theory. We wish to point out that the evidence adduced by Wrinch and Langmuir has very little value, because their comparison of the X-ray data and the cyclol structure involves so many arbitrary assumptions as to remove all significance from the agreement obtained. In order to attempt to account for the maxima and minima appearing on Crowfoot's diagrams, Wrinch and Langmuir made the assumption that certain re-

gions of the crystal (center of molecule, center of lacunae) have an electron density less than the average, and others (slits, zinc atoms) have an electron density greater than the average. The positions of these regions are predicted by the cyclol theory, but the magnitudes of the electron density are not predicted quantitatively by the theory. Accordingly the authors had at their disposal seven parameters, to which arbitrary values could be assigned in order to give agreement with the data. Despite the numbers of these parameters, however, it was necessary to introduce additional arbitrary parameters, bearing no predicted relation whatever to the cyclol structure, before rough agreement with the Crowfoot diagrams could be obtained. Thus the peak B″, which is the most pronounced peak in the $P(xy0)$ section (Fig. 2 of Wrinch and Langmuir's paper) and is one of the four well-defined isolated maxima reported, is accounted for by use of a region (V) of very large negative deviation located at a completely arbitrary position in the crystal; and this region is not used by the authors in interpreting any other features of the diagrams. This introduction of four arbitrary parameters (the three coördinates and the intensity of the region V) to account for one feature of the experimental diagrams would in itself make the argument advanced by Wrinch and Langmuir unconvincing; the fact that many other parameters were also assigned arbitrary values removes all significance from their argument.

It has been pointed out by Bernal,[14] moreover, that the authors did not make the comparison of their suggested structure and the experimental diagrams correctly. They compared only a fraction of the vectors defined by their regions with the Crowfoot diagrams, and neglected the rest of the vectors. Bernal reports that he has made the complete calculation on the basis of their structure, and has found that the resultant diagrams show no relation whatever to the experimental diagrams. He states also that with seven density values at closest-packed positions as arbitrary parameters he has found that a large number of structures which give rough agreement with the experimental diagrams can be formulated.

We accordingly conclude that there exists no satisfactory X-ray evidence for the cyclol structure for insulin.

(12) A. L. Patterson, *Z. Krist.*, **90**, 517 (1935); D. Harker, *J. Chem. Phys.*, **4**, 825 (1936).

(13) It has also been pointed out by J. M. Robertson, *Nature*, **143**, 75 (1939), that the intensities of 60 planes could not provide sufficient information to locate the several thousand atoms in the insulin molecule.

(14) J. D. Bernal, *Nature*, **143**, 74 (1939); see also D. P. Riley and I. Fankuchen, *ibid.*, **143**, 648 (1939).

3. Thermochemical Evidence Regarding Protein Structure

It is, moreover, possible to advance a strong argument in support of the contention that the cyclol structure does not occur to any extent in any protein.

X-Ray photographs of denatured globular proteins are similar to those of β-keratin, and thus indicate strongly that these denatured proteins contain extended polypeptide chains.[15] Astbury[16] has also obtained evidence that in protein films on surfaces the protein molecules have the extended-chain configuration, and this view is shared by Langmuir, who has obtained independent evidence in support of it.[17] Now the heat of denaturation of a protein is small—less than one hundred kilogram calories per mole of protein molecules for denaturation in solution,[18] that is, only a fraction of a kilogram calorie per mole of amino acid residues. Consequently the structure of native proteins must be such that only a very small energy change is involved in conversion to the polypeptide chain configuration.

It is unfortunate that there exist no substances known to have the cyclol structure; otherwise their heats of formation could be found experimentally for comparison with those of substances such as diketopiperazine which are known to contain polypeptide chains or rings. It is possible, however, to make this comparison indirectly in various ways. A system of values of bond energies and resonance energies has been formulated[19] which permits the total energy of a molecule of known structure to be predicted with an average uncertainty of only about 1 kcal./mole for a molecule the size of the average amino acid residue. The polypeptide chain (amide form) and cyclol can be represented by the following diagrams

Polypeptide chain

Cyclol

The change in bonds from polypeptide chain to cyclol is N–H + C=O $\longrightarrow$ N–C + C–O + O–H. With N–H = 83.3, C=O = 152.0, N–C = 48.6, C–O = 70.0, and O–H = 110.2 kcal./mole, the bonds of an amino acid residue are found to be 6.5 kcal./mole less stable for the cyclol configuration than for the chain configuration. This must further be corrected for resonance of the double bond

which amounts for an amide to about 21 kcal./mole[19]; there is no corresponding resonance for the cyclol, which involves only single bonds. We conclude that the cyclol structure is less stable than the polypeptide chain structure by 27.5 kcal./mole per amino acid residue.

This value relates to gaseous molecules, containing no hydrogen bonds, and with the ordinary van der Waals forces also neglected. It is probable that the ordinary van der Waals forces would have nearly the same value for a cyclol as for a polypeptide chain; and the available evidence[19,20] indicates that the polypeptide hydrogen bonds would be slightly stronger than the hydrogen bonds for the cyclol structure. Moreover, the observed small values (about 2 kcal./mole) for the heat of solution of amides and alcohols show that the stability relations in solution are little different from those of the crystalline substances. We accordingly conclude that the polypeptide chain structure for a protein is more stable than the cyclol structure by about 28 kcal./mole per amino acid residue, either for a solid protein or a protein in solution (with the active groups hydrated[21]).

The comparison of the polypeptide chain and cyclol can also be made without the use of bond energy values. The heat of combustion of crystalline diketopiperazine, which contains two glycine residues forming a polypeptide chain,[22] is known;[23] from its value, 474.6 kcal./mole, the heat of formation of crystalline diketopiperazine (from elements in their standard states) is calculated to be 128.4 kcal./mole, or 64.2 kcal./mole per glycine residue. A similar calculation cannot be made directly for the cyclol structure, because no sub-

(15) W. T. Astbury, S. Dickinson and K. Bailey, *Biochem. J.*, **29**, 2351 (1935).

(16) W. T. Astbury, *Nature*, **143**, 280 (1939).

(17) I. Langmuir, *ibid.*, **143**, 280 (1939).

(18) M. L. Anson and A. E. Mirsky, *J. Gen. Physiol.*, **17**, 393, 399 (1934).

(19) (a) L. Pauling and J. Sherman, *J. Chem. Phys.*, **1**, 606 (1933); (b) L. Pauling, "The Nature of the Chemical Bond," Cornell University Press, Ithaca, N. Y., 1939. The values quoted above are from the latter source; they involve no significant change from the earlier set.

(20) M. L. Huggins, *J. Org. Chem.*, **1**, 407 (1936).

(21) The suggestion has been made [F. C. Frank, *Nature*, **133**, 242 (1936)] that the energy of hydration of hydroxyl groups might be very much greater than that of the carbonyl and secondary amino groups of a polypeptide chain; there exists, however, no evidence indicating that this is so.

(22) R. B. Corey, ref. 7.

(23) M. S. Kharasch, *Bur. Standards J. Research*, **2**, 359 (1929).

stance is known to have the cyclol structure; but an indirect calculation can be made in many ways, such as the following. One hexamethylenetetramine molecule and one pentaerythritol molecule contain the same bonds as four cyclized glycine residues and three methane molecules; hence the heat of formation of a glycine cyclol per residue is predicted to have the value 32.2 kcal./mole found experimentally[24] for $\frac{1}{4}C_6H_{12}N_4(c) + \frac{1}{4}C-(CH_2OH)_4(c) - \frac{3}{4}CH_4(c)$. Similarly the value for $N(C_2H_5)_3(c) + C_2H_5OH(c) - 3C_2H_6(c)$ is 40.2 kcal./mole. The average of several calculations of this type, 36 kcal./mole, differs from the experimental value of the heat of formation of diketopiperazine per residue, 64 kcal./mole, by 28 kcal./mole. This agrees closely with the value 27.5 kcal./mole found by the use of bond energies, and we can be sure that the suggested cyclol structure for proteins is less stable than the polypeptide chain structure by about this amount per amino acid residue. Since denatured proteins are known to consist of polypeptide chains, and native proteins differ in energy from denatured proteins by only a very small amount (less than 1 kcal./mole per residue), we draw the rigorous conclusion that *the cyclol structure cannot be of primary importance for proteins; if it occurs at all (which is unlikely because of its great energetic disadvantage relative to polypeptide chains) not more than about three per cent. of the amino acid residues could possess this configuration.*

The above conclusion is not changed if the assumption be made that polypeptide chains are in the imide rather than the amide form,[3b] since this would occur only if the imide form were the more stable. In this case the experimental values of heats of formation (such as that of diketopiperazine) would still be used as the basis for comparison with the predicted value for the cyclol structure, and the same energy difference would result from the calculation.

It has been recognized[25-27] that energy relations present some difficulty for the cyclol theory (although the seriousness of the difficulty seems not to have been appreciated), and various suggestions have been made in the attempt to avoid the difficulty. In her latest communication[3h] Wrinch writes, "The stability of the globular pro-

teins, under special conditions, in solution and in the crystal, we attribute to definite stabilizing factors;[26,27] namely, (1) hydrogen bonds between the oxygens of certain of the triazine rings, (2) the multiple paths of linkage between atoms in the fabric, (3) the closing of the fabric into a polyhedral surface which eliminates boundaries of the fabric and greatly increases the symmetry, and (4) the coalescence of the hydrophobic groups in the interior of the cage." These factors are, however, far from sufficient to stabilize the cyclol structure relative to the polypeptide chain structure. (1) The hydrogen bonds between hydroxyl groups in the cyclol structure would have nearly the same energy (about 5 kcal./mole) as those involving the secondary amino and carbonyl groups of the polypeptide chain. The suggestion[26] that resonance of the protons between oxygen atoms would provide further stabilization is not acceptable, since the frequencies of nuclear motion are so small compared with electronic frequencies that no appreciable resonance energy can be obtained by resonance involving the motion of nuclei. (2) We are unable to find any aspects of the bond distribution in cyclols which are not taken into consideration in our energy calculation given above. (3) There is no type of interatomic interaction known to us which would lead to additional stability of a cage cyclol as the result of eliminating boundaries and increasing the symmetry. (4) The stabilizing effect of the coalescence of the hydrophobic groups has been estimated[26] to be about 2 kcal./mole per CH_2 group, and to amount to a total for the insulin molecule of about 600 kcal./mole. It seems improbable to us that the van der Waals interactions of these groups are much less than this for polypeptides. The maximum of 600 kcal./mole from this source is still negligibly small compared with the total energy difference to be overcome, amounting to about 8000 kcal./mole for a protein containing about 288 residues.[28]

We accordingly conclude that the cyclol structure is so unstable relative to the polypeptide structure that it cannot be of significance for proteins.

It may be pointed out that a number of experiments[29-31] have added the weight of their evi-

(24) The values of heats of combustion used are $C_6H_{12}N_4(c)$, 1006.7; $C(CH_2OH)_4(c)$, 661.2; $CH_4(c)$, 210.6; $N(C_2H_5)_3(c)$, 1035.5; $C_2H_5OH(c)$, 325.7; $C_2H_6(c)$, 370.0 kcal./mole.

(25) F. C. Frank, *Nature*, **138**, 242 (1936).

(26) I. Langmuir and D. Wrinch, *ibid.*, **143**, 49 (1939).

(27) D. Wrinch, *Symposia on Quant. Biol.*, **6**, 122 (1938).

(28) Other suggestions regarding the source of stabilizing energy which have been made hardly merit discussion. "Foreign molecules" (Wrinch, ref. 27), for example, cannot be discussed until we have some information as to their nature.

(29) G. I. Jenkins and T. W. J. Taylor, *J. Chem. Soc.*, 495 (1937).

(30) L. Kellner, *Nature*, **140**, 193 (1937).

(31) H. Meyer and W. Hohenemser, *ibid.*, **141**, 1138 (1938).

1864   LINUS PAULING AND CARL NIEMANN   Vol. 61

dence to the general conclusion reached in this communication that the cyclol bond and the cyclol fabric are energetically impossible.

4. Further Arguments Indicating the Non-existence of the Cyclol Structure

There are many additional arguments which indicate more or less strongly that the cyclol structure does not exist. Of these we shall mention only a few.

It has been found experimentally that two atoms in adjacent molecules or in the same molecule but not bonded directly to one another reach equilibrium at a distance which can be represented approximately as the sum of certain van der Waals radii for the atoms.[19,32] Two carbon atoms of methyl or methylene groups not bonded to the same atom never approach one another more closely than about 4.0 Å., and two hydrogen atoms not bonded to the same atom are always at least 2.0 Å. apart. It has been pointed out by Huggins[33] that the cyclol structure places the carbon atoms of side chains only 2.45 Å. apart, and that in the C_2 structure for insulin there are hydrogen atoms only 0.67 Å. apart. We agree with Huggins that this difficulty alone makes the cyclol hypothesis unacceptable.

A closely related argument, dealing with the small area available for the side chains of a cyclol fabric, has been advanced by Neurath and Bull.[34] The area provided per side chain by the cyclol fabric, about 10 sq. Å., is far smaller than that required; and, as Neurath and Bull point out, the suggestion[3e,26] that some of the side chains pass through the lacunae of the fabric to the other side cannot be accepted, because this would require non-bonded interatomic distances much less than the minimum values found in crystals.

One of the most striking features of the cyclol fabric is the presence of great numbers of hydroxyl groups: in the case of cyclol C_2 there are 288 hydroxyl groups exclusive of those present in the side chains. Recently Haurowitz[35,36] has subjected the cyclol hypothesis to experimental tests on the basis of the existence or non-existence of cage hydroxyl groups. In the first communication[35] Haurowitz concludes on the basis of his and pre-

vious experiments[37-39] on the acylation and alkylation of proteins that the experimental evidence is in decided opposition to the conception that proteins possess great numbers of hydroxyl groups and therefore to the cyclol hypothesis. It seems to us that the objection raised by Haurowitz[36] is worthy of consideration and it certainly cannot be disposed of on the grounds that the original structure has been destroyed unless some concrete evidence can be submitted to indicate that this is the case. In a second communication Haurowitz and Astrup[36] write that "According to the classical theory of protein structure the carboxyl and amino groups found after hydrolytic splitting of a protein come from —CO—NH— bonds. According to the cyclol hypothesis, however, the free carboxyl and amino groups must be formed, during the splitting, from bonds of the structure $=C(OH)—N=$. The classical theory would predict on hydrolysis no great change in the absorption spectrum below 2400 Å. because the CO groups of the amino acids and of the peptide bonds both are strongly absorbing in this region.[40] On the other hand, the cyclol hypothesis would predict a greatly increased absorption because of the formation of new CO groups. . . . The absorption for genuine and for hydrolyzed protein is about equal. This seems to be in greater accordance with the classical theory of the structure of proteins than with the cyclol theory."

Mention may also be made of the facts that no simple substances with the cyclol structure have ever been synthesized[29] and that in general chemical reactions involving the breaking of covalent bonds are slow, whereas rapid interconversion of polypeptide and cyclol structure must be assumed to occur in, for example, surface denaturation. These chemical arguments indicate strongly that the cyclol theory is not acceptable.[41]

5. A Discussion of Arguments Advanced in Support of the Cyclol Theory

Although a great number of papers dealing with the cyclol theory have been published, we have

(37) J. Herzig and K. Landsteiner, *Biochem. Z.*, **61**, 458 (1914).

(38) B. M. Hendrix and F. Paquin, Jr., *J. Biol. Chem.*, **124**, 135 (1938).

(39) K. G. Stern and A. White, *ibid.*, **122**, 371 (1938).

(40) M. A. Magill, R. E. Steiger and A. J. Allen, *Biochem. J.*, **31**, 188 (1937).

(41) Another argument against cyclols of the C_2 type can be based on the results reported by J. L. Oncley, J. D. Ferry and J. Shack, *Symposia on Quant. Biol.*, **6**, 21 (1938), H. Neurath, *ibid.*, **6**, 196 (1938), and J. W. Williams and C. C. Watson, *ibid.*, **6**, 208 (1938), who have shown that dielectric constant measurements and diffusion measurements indicate that the molecules of many proteins are far from spherical in shape.

(32) N. V. Sidgwick, "The Covalent Link in Chemistry," Cornell University Press, Ithaca, N. Y., 1933; E. Mack, Jr., THIS JOURNAL, **54**, 2141 (1932); S. B. Hendricks, *Chem. Rev.*, **7**, 431 (1930); M. L. Huggins, *ibid.*, **10**, 427 (1932).

(33) M. L. Huggins, THIS JOURNAL, **61**, 755 (1939).

(34) H. Neurath and H. D. Bull, *Chem. Rev.*, **23**, 427 (1938).

(35) T. Haurowitz, *Z. physiol. Chem.*, **256**, 28 (1938).

(36) T. Haurowitz and T. Astrup, *Nature*, **143**, 118 (1939).

had difficulty in finding in them many points of comparison with experiment (aside from the X-ray work mentioned above) which were put forth as definite arguments in support of the structure.

One argument which has been advanced is that the cyclol theory "readily interprets the total number of amino acid residues per molecule, without the introduction of any *ad hoc* hypothesis"[3e] and that "The group of proteins with molecular weights ranging rom 33,600 to 40,500 are closed cyclols of the type C₂ containing 288 amino acid residues."[3e] Now the presence of imino acids (proline, oxyproline) in a protein prevents its formation of a complete cyclol such as C_2, and many proteins in this molecular weight range are known to contain significant amounts of proline: for insulin 10% is reported,[3f] for egg albumin 4%,[42] for zein 9%,[43] for Bence–Jones protein 3%,[44] and for pepsin 5%.[45] Wrinch has stated that "a future modification" (in regard to the number of residues) "is also introduced if imino acids are present"[3c]; "these numbers perhaps being modified if imino acids are present",[3e] and "if certain numbers of imino acid residues are present, these numbers" (of residues) "may be correspondingly modified."[3g] This uncertainty regarding the effect of the presence of imino acids in cyclols on the expected number of residues leaves the argument little force. In fact, even the qualitative claim that the cyclol hypothesis implies the existence of polyhedral structures containing certain numbers of amino acid residues and so predicts that globular proteins have molecular weights which fall into a sequence of separated classes can be doubted for the same reason.

It has been claimed[36] that the cyclol hypothesis explains the facts that proteins contain certain numbers of various particular amino acid residues and that these numbers are frequently powers of 2 and 3,[46] and it is proper that we inquire into the nature of the argument. Wrinch states[27] "An individual R group" (side chain) "is presumably attached, not to just any α-carbon atom, but only to those whose environment makes them appropriate in view of its specific nature. As an example of different environments, we may refer to the

cyclol cages; here the pairs of residues at a slit have 'different environments' and the residues not at a slit fall into sets which again have 'different environments.' We therefore expect characteristic proportions to be associated with aromatic, basic, acidic, and hydrocarbon R groups, respectively, even perhaps with individual R groups. In any case a non-random distribution of the proportions of each residue in proteins in general is to be expected on any fabric hypothesis. On the cyclol hypothesis, for example, α-carbons having equivalent environments occur in powers of 2 and 3. . . . It is difficult to avoid interpreting the many cases which have recently been summarized in which the proportions of many types of residue are powers of 2 and 3 as further direct evidence in favor of the cyclol fabric. This fabric consists of an alternation of diazine and triazine hexagons, with symmetries respectively 2 and 3." Also it has been said by Langmuir[47] that "The occurrence of these factors, 2 and 3, furnishes a powerful argument for a geometrical interpretation such as that given by the cyclol theory. In fact, the hexagonal arrangement of atoms in the cyclol fabric gives directly and automatically a reason for the existence of the factors 2 and 3 and the non-occurrence of such factors as 5 and 7."

On examining the cyclol C_2, however, we find that these statements are not justified. The only factors of 288 are of the form $2^n\,3^m$; moreover, the framework of the cyclol C_2 has the tetrahedral symmetry T, so that if the distribution of side chains conforms to the symmetry of the framework the amino acid residues would occur in equivalent groups of twelve. But in view of the rapid decrease in magnitude of interatomic forces with distance there would seem to be little reason for the distribution of side chains over a large protein molecule to conform to the symmetry T; it is accordingly evident that any residue numbers might occur for the cyclol C_2. We conclude that the cyclol hypothesis does not provide an explanation of the occurrence of amino acid residues in numbers equal to products of powers of 2 and 3.

Although there is little reason to expect that the distribution of side chains would correspond to the symmetry of the framework, it is interesting to note that the logical application of the methods of argument used by Wrinch suggests strongly that sixty residues of each of two amino acids should be present in a C_2 cyclol. This cyclol contains twenty lacunae of a particular type—each surrounded by a nearly coplanar border of twelve diazine and triazine

(42) H. O. Calvery, *J. Biol. Chem.*, **94**, 613 (1931).
(43) T. B. Osborne and L. M. Liddle, *Am. J. Physiol.*, **26**, 304 (1910).
(44) C. L. A. Schmidt, "Chemistry of Amino Acids and Proteins," C. C. Thomas, Springfield, Ill., 1938.
(45) Unpublished determination by one of the authors.
(46) M. Bergmann and C. Niemann, *J. Biol. Chem.*, **115**, 77 (1936); **118** 307 (1937).

(47) I. Langmuir, *Symposia on Quant. Biol.*, **6**, 135 (1938).

1866   LINUS PAULING AND CARL NIEMANN   Vol. 61

rings. Each of these has trigonal symmetry so far as this near environment is concerned. Hence it might well be expected that a particular amino acid would be represented by three residues about each of these twenty lacunae, giving a total of sixty residues. But the number 60 cannot be expressed in the form $2^n 3^m$, it is not a factor of 288, and the integer nearest the quotient 288/60, 5, also cannot be expressed in the form $2^n 3^m$.

One of the most straightforward arguments advanced by Wrinch[3c,d] is that a protein surface film must have all its side chains on the same side, which would be the case for a cyclol fabric but not for an extended polypeptide chain. This argument now has lost its significance through the recently obtained strong evidence that proteins in films have the polypeptide structure,[16,17] and not the cyclol structure.

There can be found in the papers by Wrinch many additional statements which might be construed as arguments in support of the cyclol structure. None of these seems to us to have enough significance to justify discussion.

6.   The Present State of the Protein Problem

The amount of experimental information about proteins is very great, but in general the processes of deducing conclusions regarding the structure of proteins from the experimental results are so involved, the arguments are so lacking in rigor, and the conclusions are so indefinite that it would not be possible to present the experimental evidence at the basis of our ideas of protein structure[48] in a brief discussion. In the following paragraphs we outline our present opinions regarding the structure of protein molecules, without attempting to do more than indicate the general nature of the evidence supporting them. These opinions were formed by the consideration not only of the experimental evidence obtained from proteins themselves but also of the information regarding interatomic interactions and molecular structure in general which has been gathered by the study of simpler molecules.

We are interested here only in the role of amino acids in proteins—that is, in the simple proteins (consisting only of α-amino and α-imino acids) and the corresponding parts of conjugated proteins; the structure and linkages of prosthetic groups will be ignored.

The great body of evidence indicating strongly that the amino acids in proteins are linked together by peptide bonds need not be reviewed here.

The question now arises as to whether the polypeptide chains or rings contain many or few amino acid residues. We believe that the chains or rings contain many residues—usually several hundred. The fact that in general proteins in solution retain molecular weights of the order of 17,000 or more until they are subjected to conditions under which peptide hydrolysis occurs gives strong support to this view. It seems to us highly unlikely that any protein consist of peptide rings containing a small number of residues (two to six) held together by hydrogen bonds or similar relatively weak forces, since, contrary to fact, in acid or basic solution a protein molecule of this type would be decomposed at once into its constituent small molecules.

There exists little evidence as to whether a long peptide chain in a protein has free ends or forms one more peptide bond to become a ring. This is, in fact, a relatively unimportant question with respect to the structure, as it involves only one peptide bond in hundreds, but it may be of considerable importance with respect to enzymatic attack and biological behavior in general.

A native protein molecule with specific properties must possess a definite configuration, involving the coiling of the polypeptide chain or chains in a rather well-defined way.[49] The forces holding the molecule in this configuration may arise in part from peptide bonds between side-chain amino and carboxyl groups or from side-chain ester bonds or S–S bonds; in the main, however, they are probably due to hydrogen bonds and similar interatomic interactions. Interactions of this type, while individually weak, can by combining their forces stabilize a particular structure for a molecule as large as that of a protein. In some cases (trypsin, hemoglobin) the structure of the native protein is the most stable of those accessible to the polypeptide chain; the structure can then be reassumed by the molecule after denaturation. In other cases (antibodies) the native configuration is not the most stable of those accessible, but is an unstable configuration impressed on the molecule by its environment (the influence of the antigen) during its synthesis; denaturation is not reversible for such a protein.

Crystal structure investigations have shown

(48) We believe that our views regarding the structure of protein molecules are essentially the same as those of many other investigators interested in this problem.

(49) H. Wu, *Chinese J. Physiol.*, **5**, 321 (1931); A. E. Mirsky and L. Pauling, *Proc. Nat. Acad. Sci.*, **22**, 439 (1936).

that in general the distribution of matter in a molecule is rather uniform. A protein layer in which the peptide backbones are essentially co-planar (as in the β-keratin structure) has a thickness of about 10 Å. If these layers were arranged as surfaces of a polyhedron, forming a cage molecule, there would occur great steric interactions of the side chains at the edges and corners. (This has been used above as one of the arguments against the C_2 cyclol structure.) We accordingly believe that *proteins do not have such cage structures*.[50] A compact structure for a globular protein might involve the superposition of several parallel layers, as suggested by Astbury, or the folding of the polypeptide chain in a more complex way.

One feature of the cyclol hypothesis—the restriction of the molecule to one of a few configurations, such as C_2—seems to us unsatisfactory rather than desirable. The great versatility of antibodies in complementing antigens of the most varied nature must be the reflection of a correspondingly wide choice of configuration by the antibody precursor. We feel that the biological significance of proteins is the result in large part of their versatility, of the ability of the polypeptide chain to accept and retain that configuration which is suited to a special purpose from among the very great number of possible configurations accessible to it.

Proteins are known to contain the residues of some twenty-five amino acids and it is not unlikely that this number will be increased in the future. A great problem in protein chemistry is that of the order of the constituent amino acid residues in the peptide chains. Considerable evidence has been accumulated[46] suggesting strongly that the stoichiometry of the polypeptide framework of protein molecules can be interpreted in terms of a simple basic principle. This principle states that the number of each individual amino acid residue and the total number of all amino acid residues contained in a protein molecule can be expressed as the product of powers of the integers two and three. Although there is no direct and unambiguous experimental evidence confirming the idea that the constituent amino acid residues are arranged in a periodic manner along the peptide chain, there is also no experimental evidence which would deny such a possibility, and it seems probable that steric factors

might well cause every second or third residue in a chain to be a glycine residue, for example.

The evidence regarding frequencies of residues involving powers of two and three leads to the conclusion that there are 288 residues in the molecules of some simple proteins. It is not to be expected that this number will be adhered to rigorously. Some variation in structure at the ends of a peptide chain might be anticipated; moreover, amino acids might enter into the structure of proteins in some other way than the cyclic sequence along the main chain.[51] The structural significance of the number 288 is not clear at present. It seems to us, however, very unlikely that the existence of favored molecular weights (or residue numbers) of proteins is the result of greater thermodynamic stability of these molecules than of similar molecules which are somewhat smaller or larger, since there are no interatomic forces known which could effect this additional stabilization of molecules of certain sizes. It seems probable that the phenomenon is to be given a biological rather than a chemical explanation—we believe that the existence of molecular-weight classes of proteins is due to the retention of this protein property through the long process of the evolution of species.

We wish to express our thanks to Dr. R. B. Corey for his continued assistance and advice in the preparation of this paper, and also to other colleagues who have discussed these questions with us.

Summary

It is concluded from a critical examination of the X-ray evidence and other arguments which have been proposed in support of the cyclol hypothesis of the structure of proteins that these arguments have little force. Bond energy values and heats of combustion of substances are shown to lead to the prediction that a protein with the cyclol structure would be less stable than with the polypeptide chain structure by a very large amount, about 28 kcal./mole of amino acid residues; and the conclusion is drawn that proteins do not have the cyclol structure. Other arguments leading to the same conclusion are also presented. A brief discussion is given summarizing the present state of the protein problem, with especial reference to polypeptide chain structures.

PASADENA, CALIF.   RECEIVED APRIL 22, 1939

(50) Wrinch recently has suggested[3b] that even if proteins are not cyclols the cage structure might be significant.

(51) H. Jensen and E. A. Evans, Jr., *J. Biol. Chem.*, **108**, 1 (1935), have shown that insulin probably contains several phenylalanine groups attached only by side-chain bonds to the main peptide chain.

330   DOROTHY M. WRINCH   Vol. 68

chloroform melted at 214–216°. The pure ester of the terephthalic acid is recorded[11] as melting at 225°. Neutralization equivalent of the mixture of acids was 84.3; calculated for phthalic acid 83.3. No trace of benzoic acid could be detected among the oxidation products. Apparently the mixture is largely isophthalic acid. The yield calculated on the basis of four propyl chloride molecules to one of the butylpropylbenzene mixture varied from 14 to 38%, an estimation based upon the quantity of high boiling residue left after distilling the butylbenzene from each experiment.

About 6 to 12 g. of a mixture of saturated and unsaturated compounds, boiling in the hexane and nonane range, was obtained from each of these preparations, the quantity being larger the higher the temperature of reaction. Careful fractionation in an eighteen-plate column failed to show any single product. All fractions also showed presence of unsaturated material. Upon ozonolysis of a 63 to 66° cut, decomposition of the ozonides with zinc and water, and careful fractionation of the resulting aldehydes and ketones, definite evidence in the form of the dinitrophenylhydrazone derivative was obtained for the presence of acetaldehyde. The melting point observed was 146–150°; that of an authentic sample was 147.5–148°; and that of the mixture was 148–150°. A mixed melting point with the like derivative of propionaldehyde (m. p. 190°) was 141°. Owing to the small amount of material left after attempts at purification, a satisfactory melting point for 2-butanone-2,4-dinitrophenylhydrazone could

not be secured but the crystals obtained were shown to be identical with an authentic sample in having parallel extinction, pleochroism, centered acute bisectrix, and elongated obtuse bisectrix, and were not in whole agreement with observations on known samples of the like derivatives of acetone and butyraldehyde. A trace of propionaldehyde may also be present, for a few crystals prepared from one of the fractions appeared to have similar optical properties.[12]

Summary

Organosodium compounds from *n*-butyl and *n*-propyl chlorides have been prepared, the ease of reaction being progressively less, the yields lower, and the ratio of di- to monocarboxylic acid greater than observed with the amyl homolog.

n-Butyl- and *n*-propylsodium are progressively less reactive with benzene and toluene than is *n*-amylsodium.

n-Butylbenzene can be prepared conveniently by adding propyl chloride to sodium in toluene at 72°.

Metalation of butylbenzene takes place largely in the meta position.

(12) We are greatly indebted to Mr. Gibb for assistance on the optical studies.

CAMBRIDGE, MASS.   RECEIVED OCTOBER 7, 1940

[CONTRIBUTION FROM THE CHEMICAL LABORATORY OF THE JOHNS HOPKINS UNIVERSITY]

The Geometrical Attack on Protein Structure

BY DOROTHY M. WRINCH

It is unnecessary at the present time to state the case for the cyclol hypothesis, since authoritative accounts have already been given by Langmuir of the way in which the theory accounts satisfactorily for many of the well-known properties of the globular proteins.[1,2] In a recent summary,[3] however, Pauling and Niemann repeat a number of statements purporting to disprove the theory already made by other writers. Attention must therefore be directed to a number of publications in which these criticisms have already been discussed, at least so far as their scientific importance appeared to warrant.[2,4–9] We then proceed to

discuss, necessarily in a preliminary manner, two issues (first raised in my publications[10,11]) to which Pauling and Niemann refer, namely (a) certain short interatomic distances in the original cage structures and (b) the energy of formation of these structures. In particular it is shown that the claim by Pauling and Niemann that the cyclol hypothesis can be disposed of by means of Anson and Mirsky's heat of denaturation of trypsin must be rejected.

(a) In my first studies of possible protein structures, very exacting metrical conditions were adopted, mainly in order to demonstrate in a simple manner the possibility of handling problems of protein structure by strictly mathematical methods. These, it appears from crystallographical data, are unnecessarily onerous. Evidently a discussion of certain short interatomic distances in the original cyclol structures can only

(1) Langmuir, *Cold Spring Harbor Symposia Quant. Biol.*, **6**, 135 (1938).

(2) Langmuir, *Proc. Phys. Soc.* (London), **51**, 542 (1939).

(3) Pauling and Niemann, THIS JOURNAL, **61**, 1860 (1939).

(4) Wrinch, *Cold Spring Harbor Symposia Quant. Biol.*, **6**, 122 (1938).

(5) Langmuir, *Nature*, **143**, 280 (1939).

(6) Langmuir and Wrinch, *ibid.*, **143**, 49 (1939).

(7) Wrinch, *ibid.*, **143**, 482 (1939).

(8) Wrinch, *ibid.*, **143**, 763 (1939); **145**, 660, 1018 (1940).

(9) Langmuir and Wrinch, *Proc. Phys. Soc.* (London), **51**, 613 (1939).

(10) Wrinch, *Proc. Roy. Soc.* (London), **A160**, 59 (19); **A161**, 505 (1937).

(11) Wrinch, *Nature*, **138**, 241 (1936).

be profitable if accompanied by a detailed study of the propriety and the feasibility of relaxing the metrical conditions. Further, it must be remembered that cases are known (*e. g.*, in the terpenes, the sterols and phenanthrene derivatives) in which the accumulated evidence of organic chemistry makes it necessary to place certain nonbonded atoms nearer together than the minimum distances allowed by Pauling and Niemann. These distances in any case refer to interatomic distances in different molecules. Thus the statement that methyl groups in the case of hexamethylbenzene are 4.0 to 4.1 Å. apart refers to intermolecular distances. Within a single molecule of hexamethylbenzene, methyl groups approach as near as 2.93 Å. Further, these distances have been derived from a study of simple molecules which bear no resemblance to the very complex highly organized structures which, on any plausible theory of protein structure, must characterize the globular proteins. Nothing is known as to the nearest approach of groups *within a single molecule* in such cases. (b) It is known that, under certain closely defined conditions, the heat of denaturation of trypsin amounts to less than 1 kcal./mole of amino acid residues.[12] Pauling and Niemann use this figure in an attempt to disprove the cyclol hypothesis as follows: (1) they claim that a cyclol cage structure is about 28 kcal./mole of amino acid residues less stable than the corresponding linear polypeptide containing NH·CO groups: (2) they claim that Anson and Mirsky's denatured trypsin consists of linear polypeptides. They then deduce that the trypsin molecule cannot be a cyclol cage structure, since if it were the heat of denaturation obtained by Anson and Mirsky would have been about −28 kcal./mole of amino acid residues. It is my opinion that both claims (1) and (2) were unfounded, and that the suggested deduction therefore falls to the ground.

(2) Little or nothing is known as to the structure of any denatured protein, nothing as to the structure of the denatured trypsin under discussion. It is assumed by Pauling and Niemann that denaturation on the cyclol theory means the opening of all cyclol bonds. This is not the case. (In particular the difference between trypsin and denatured trypsin in this case cannot consist in opening of all cyclol bonds, since, as Mirsky and Pauling state,[13] there is no change in molecular

weight.) Denaturation, on the cyclol therory, corresponds to any breakdown of the intact structure produced by the opening of any subset of cyclol bonds.[14] The theory thus readily offers an interpretation of the well-known fact that very different denatured products can be obtained from one and the same protein (*cf.* the film studies of pepsins which have been shaken for different times, heated for different times at 65°, etc.[1]), regarding each type of denatured product as due to the opening of some definite subset of cyclol bonds and possibly the re-formation of some of them. A denatured product (particularly when, as in the case under discussion, the denaturation is reversible) may have only a small or even negligible decrease in its complement of cyclol bonds. In the case of a C_2 cyclol cage, such as Pauling and Niemann are assuming for trypsin, a net decrease of one in n cyclol bonds, where n is 12 or even 24, is sufficient to allow a gross change in structure and so may be sufficient to account for denaturation. A degradation product of trypsin may therefore possess $^{11}/_{12}$ or even $^{23}/_{24}$ of its full complement of cyclol bonds. Thus if it were the case that a cyclol cage structure is about 28 kcal./ mole of amino acid residues less stable than the corresponding linear polypeptides, it still would not follow that the heat of denaturation should be of this order. With reversible denaturation the figure should in this event be $28/n$, or say 2.3 or 1.17 if n were 12 or 24, respectively.

(1) But there is, in any case, no reason to suppose that the figure 28 kcal./mole of residues represents the situation even approximately.

We consider first the attempt of Pauling and Niemann to estimate the difference between the heats of formation of a glycyl linear residue and a glycyl cyclol residue by the explicit use of bond energies. Repeating without acknowledgment the calculation (yielding the value 27.3) which I published in 1936 to ventilate the question,[15] Pauling and Niemann write for the heats of formation in the gaseous state

linear glycyl residue − cyclol glycyl residue
$= C{=}O + N{-}H + (\text{resonance in CO-NH bond}) - (C{-}O + O{-}H + C{-}N)$
$= 152 + 83.3 + 21 - (110.2 + 70 + 48.6) = 27.5 \quad (1)$

or say 28, since the standard energy of NH is usually taken to be 83.7.[16] It is my opinion that

(12) Anson and Mirsky, *J. Gen. Physiol.*, **17**, 393 (1934).
(13) Mirsky and Pauling, *Proc. Nat. Acad. Sci. U. S.*, **22** 439 (1936).
(14) Wrinch, *Phil. Mag.*, **25**, 706 (1938).
(15) Wrinch, *Nature*, **138**, 241 (1936). This calculation was brought to the notice of Professor Pauling in January, 1938.
(16) Pauling, "Nature of the Chemical Bond," Cornell University Press, Ithaca. N. Y., p. 53.

no weight is to be attached to this formal evaluation, since little or nothing is known as to the values of several of the bond energies in these particular cases. Particularly I question the validity of (1) assuming that the C–CH₂ bonds have the same energy in both structures, (2) assuming the C–O bonds in cyclols have an energy as small as in the primary alcohols. These uncertainties appear to me to make the calculation valueless, except in so far as it calls attention (as was my reason for publishing my calculation in 1936) to the types of data which are needed, namely, bond distances (and derivatively bond energies) in long linear peptides—particularly those containing residues other than glycyls—and in ring compounds in which a carbon bearing an OH group is linked to two nitrogens in the same ring, and to another carbon bearing a variety of substituents.

However, as Pauling and Niemann rest the greater part of their case against the cyclol hypothesis upon this calculation and upon two others, it is necessary (A) to point out that these two other calculations are strictly equivalent to the first and (B) to consider how far present data offer any guide as to the bond energies involved in the first calculations. (A) Pauling and Niemann state that they have obtained without the use of bond energies other values (in fact 32 and 24) agreeing closely with and so further supporting the value 28 found with the use of bond energies. First, we have to point out that, assuming that the heat of combustion of crystalline diketopiperazine is 474.6, it is erroneous to conclude that the heat of formation of crystalline diketopiperazine from elements in their standard states is 128. The correct value is

$$-474.6 + 4 \times 94.45 + 3 \times 68.37 = 108.3$$

These two calculations, as they stand, are certainly remarkable in that they give values which differ from the first by $+4$ and -4, respectively, or using the correct value for the heat of formation of crystalline diketopiperazine, by -6 and -14, respectively. The reason for these discrepancies is easily seen. For if we write for the heats of combustion of crystalline compounds for which experimental data are available a typical equation such as

$$C(CH_2OH)_4 = 4C_2H_5OH - 3CH_4 \qquad (2)$$

say $A = 4B - 3C$, with the values used by Pauling and Niemann to assess the heat of formation

of cyclol residues, we obtain $1302.8 = 1293.0$. Again the equation

$$C_6H_{12}N_4 + 12C_2H_6 = 4N(C_2H_5)_3 + 6CH_4$$

say $D + 12 E = 4F + 6C$, using again the values used for evaluating the glycyl cyclol residue, we obtain the result $5446.7 = 5405.6$. Plainly such equations are untrue in the case of these compounds. As little reliance can be placed upon them for the evaluation of cyclol residues. Evidently the only way of interpreting such an equation, $P = Q + R$, is to write it in the form

Heat of combustion of $P_c + K(P_c) =$ Heat of combustion of $Q_c +$ Heat of combustion of $R_c + K(Q_c) + K(R_c)$

where $K(X_c)$ is the heat of sublimation of X plus the amount by which the heat of formation of X_{gas} exceeds the sum of bond energies, when the standard values (e. g., $N - C = 48.6$) are used. When written in this form, all such equations of course yield the result $0 = 0$. In other words, such an equation as (2) states simply that, if equal energies be given to the corresponding bonds, then

$$4C—C + 8C—H + 4C—O + 4O—H =$$
$$4(C—C + 5C—H + C—O + O—H) - 12C—H$$

After the same manner, the equations for the cyclol residue constructed by Pauling and Niemann, which may be written

$$\text{linear glycyl residue} - \text{cyclol glycyl residue}$$
$$= \tfrac{1}{2}\ \text{diketopiperazine} - (F + B - 3E)$$
$$= \tfrac{1}{2}\ \text{diketopiperazine} - (A + D - 3C)/4$$

give simply my original calculation. In view of the claim that such equations give estimates without the use of bond energies, it must be emphasized that they are simply bond energy equations. They make no use of the heats of combustion cited and add nothing to my original calculation,[15] to which in fact they are strictly equivalent. (B) The fact that a C–C bond length as small as 1.47 ± 0.03 Å. has been found in the only cyclic peptide so far investigated,[17] whereas in glycine[18] the distance is 1.52 ± 0.02 Å. shows that (1) is not the most plausible assumption. Rather it would seem natural to assume that this bond is shorter and therefore stronger in the cyclol structure. Further, it seems difficult to reconcile the assumption of Pauling and Niemann that the tertiary cyclol OH has its C–O bond as low in energy as the standard value for primary alcohols, with Pauling's statement[19] that this bond in secondary

(17) Corey, This Journal, **60**, 1598 (1938).
(18) Albrecht and Corey, ibid., **61**, 1087 (1939).
(19) Pauling, ibid., **54**, 3570 (1932).

and tertiary alcohols "seems to be 0.3 to 0.5 v. e. more stable than corresponds to this C–O value." Rather it would seem more reasonable to follow Pauling's 1932 lead in attributing say 0.4 v. e. or 9.2 kcal. more energy to this bond, as is in fact suggested by the observed heat of combustion of trimethylcarbinol.[20] These corrections can easily add 12 to 18 to the heat of formation of the glycyl cyclol residue, reducing the balance in favor of the linear peptide in the gaseous state to (say) 13. So far everything said has had reference only to glycyl residues and only to the gaseous state. Now proteins in general have at least one CH_2 group per residue in their R groups. If any serious estimate of the relative stability of the cyclol cage and the extended peptide is to be attempted, this fact must be taken into account. In view of the solubilities of the native proteins it may be presumed that the interarrangement in space of this large number of CH_2 groups is such as to contribute considerably more, say 2, to the heat of formation of the former than to that of the latter, and to facilitate a higher degree of hydration in the former than in the latter. This higher degree of hydration, if it amounts to two more hydrogen bonded water molecules per residue in the cyclol than in the peptide, would add a further 10 or more to the balance in favor of cyclol cages. These two terms thus may further reduce the balance in favor of linear peptides to $13 - 2 - 10 = 1$. Finally, the association of oppositely ionized R groups has to be taken into account. The dipole moments of the native proteins which have been measured are all low (180×10^{-18} for egg albumin, a molecule of diameter 30 Å. or more and 500×10^{-18} for hemoglobin, a molecule of about twice the weight).[21] While it must again be emphasized that the data necessary for an estimate are lacking (since nothing is known as to the arrangement of the ionizable R groups on any native protein) it appears to be illegitimate and certainly unplausible to assume that the coulomb energy terms would be unchanged when the cyclol cage structure opens into linear peptides. The fact that

the stability of a native protein is impaired when the pH goes outside some central "stability" range suggests that these terms make some considerably greater contribution to the heat of formation of the cage cyclol than to that of the linear peptide. In egg albumin, for example, where there are say 27 basic and 27 acidic R groups, an arrangement of these groups in isolated pairs at distances apart of 4 Å. would contribute to the heat of formation of the cyclol cage an amount of the order of 2 or more per residue. This changes the balance in favor of the cyclol to 1 or more.

To Sum Up.—The estimate of a balance in favor of the linear peptide of 28 kcal./mole of residues rests upon many assumptions which lack plausibility. With assumptions which are less unplausible, the estimate can be changed to a balance in favor of cyclol cages of at least 1. However the lack of information makes any such estimates, ranging from 28 to −1 or less, of little usefulness. Since in any case, a reversible denaturation probably destroys only a small part of the fine structure, the heat of denaturation should tally not with the balance in favor of cyclols per residue but with some small fraction of this amount.

It must be concluded that no case has been made out for deducing that the cyclol theory is false on the basis of Anson and Mirsky's figure for the heat of denaturation of trypsin; further, that no measurements of the heat of denaturation of proteins can provide a test of the cyclol hypothesis unless the structure of the denatured products in question is known and the heat of formation of such structures and of cyclol cages can be estimated.

Summary

Arguments against the cyclol hypothesis, which have been collected by Pauling and Niemann in a recent article, are examined. It is found that they do not disprove it. In particular their statements purporting to prove that a protein with the cyclol structure would be less stable than the polypeptide chain structure by a very large amount is examined and found to be unproven.

BALTIMORE, MD.   RECEIVED APRIL 30, 1940

(20) Kharasch, *Bur. Standards J. Research*, **2**, 359 (1929).
(21) Cohn, *Chem. Rev.*, **24**, 210 (1939).

profiles in chemistry

edited by
ROGER R. FESTA
Northeast Missouri State University
Kirksville, Mo. 63501

Dorothy Wrinch and a Search for the Structure of Proteins

Maureen M. Julian

Department of Geological Sciences
Virginia Polytechnic Institute and State University
Blacksburg, VA 24061

Dorothy Wrinch was a Renaissance scholar whose depth, vision, leadership, and energy were resented rather than respected. Her work spanned mathematics, philosophy, sociology, chemistry, physics, and biology. She dedicated her life to the solution of one of nature's most important and challenging mysteries, the architecture of protein molecules. While her theories did not mirror nature in quite the way she predicted, they were a critical stimulus to research. Nevertheless, she herself remained an outsider and a restless, disappointed renegade.

Dorothy Wrinch was born of English parents in 1894 at Rosario, Argentina where her father was an engineer for a Brisith firm. She was educated in England and won a scholarship to study mathematics at Girton College, Cambridge University. Her circle of friends included Bertrand Russell with whom she studied philosophy. After she received her Cambridge BA and MA, she was appointed in 1918 to a lectureship in pure mathematics at University College, London. In 1922 she married John Nicholson, an Oxford physicist known for his work on atomic structure; their only child, Pamela, was born in 1928.

At Cambridge, and later at London and Oxford, Dorothy Wrinch was active in philosophical and mathematical circles. In addition to teaching mathematics, she addressed professional gatherings and wrote prolifically. By 1930, at the age of 35, she had published 16 philosophical papers on the scientific method, and 20 papers in pure and applied mathematics. Her additional credentials included both MSc and DSc degrees from the University of London, and an MA from Oxford University. Then, in 1929, she was the first woman to receive a DSc from Oxford.

That year, after having taught mathematics at Oxford for seven years, she applied for a Rhodes Traveling Fellowship but was unsuccessful in receiving the award. Her sponsors frankly informed her that she failed because she was a woman. She filed two more applications: one for a Rockefeller Fellowship to study mathematics at Göttingen, Germany, and another for a fellowship in sociology to travel in America. Although she was welcomed by the leading mathematician at Göttingen, she failed to get the Rockefeller, largely because the foundation officials learned of the sociology proposal and considered it to be evidence that she was not solely devoted to mathematics. This was in fact true. Since the sociology application was also unsuccessful, she stayed in Oxford.

Nevertheless, in 1930, she published a sociological study, "The Retreat from Parenthood," under the pseudonym, Jean Ayling (1). In the book she sketched a broad plan for reorganizing medical services, home design, child care, and labor laws so that child rearing would be compatible with the professional lives of both women and men. At this time, she made a definitive decision to concentrate her energies on biological architecture. At Cambridge, she was a member of Joseph Needham's Theoretical Biology Club together with J. D.

Irving Langmuir and Dorothy Wrinch examining a cyclol model in 1936. Courtesy of Smith College Archives.

Bernal, J. H. Woodger, C. H. Waddington, and Dorothy Needham. They were interested in what is now known as molecular biology. Dorothy Wrinch (2) describes her new direction

Until 1933, my work was research in mathematics and mathematical physics. I had, however, long had a consuming interest in structural problems in physiology and chemistry, and I had always hoped to find specialists in this field with whom I could develop certain ideas. It proved impossible to arrange such collaboration, since the mathematical point of view was difficult to link up with the point of view of the professional chemist. At this time, then, it became clear to me that I had but two choices, either to abandon the attempt to develop these ideas or to undergo apprenticeship in chemistry sufficiently extensive to enable me to formulate the ideas in a form suitable for development by specialist workers. I chose the latter course and spent a year's leave of absence from Oxford on the continent of Europe, beginning an apprenticeship in many different laboratories. Already in 1934 the work had progressed far enough to attract the Rockefeller Foundation, who in 1935 provided support for my project for 5 years
. . . .

By 1935, Wrinch had produced an original and remarkable theory of the gene. At the Theoretical Biology Club, she suggested that the specificity of genes resides in the specificity of the amino acid sequences (3). Thus, she made the connection between the linear sequence of the genes and that of the amino acids in the polypeptide chain. She then went on to construct a molecular model of the chromosome. The cyclic tetranucleotides were assumed to be at right angles to the chromosome axis. This structure was shown not to be correct by experimental birefringence data. However, the sequence hypothesis was important.

Wrinch then turned her attention to the globular protein molecules and within a year or so produced a structural model (4). Beginning with the single hypothesis that peptide chains would be polymerized into sheets by links between CO and NH groups, Wrinch deduced that the sheets would fold into a series of closed octahedra which she called cyclols. The series was described by the general formula $72n^2$, where n was the integral number of amino acid residues. When Bergmann and Niemann (5) in 1937 deduced that egg albumin had 288 = 72 × 2^2 residues in the molecule, Wrinch felt her hypothesis was proven. Here was an example with $n = 2$. This clever deduction startled the emerging world of molecular biology. Controversy raged with Nobel prize winners lining up on either side; Irving Langmuir was her leading advocate (see figure) and Linus Pauling her leading critic. The embroilment reached disastrous proportions with Wrinch being virtually blacklisted by most of the scientific community. She felt personally attacked. For example, in 1938, after a meeting with Linus Pauling she wrote (2), "The fact that they are against cyclols in any fundamental and *a priori* way in itself rather gives the show away. Because it is undeniable that the theory had NOT yet been shown to be false and therefore only fools or men of evil wishes towards me will be against it *a priori* . . ." However more than a decade was to pass before any correct protein structures were discovered. Although these proteins did not contain cyclols, Wrinch's ideas had nonetheless stimulated much thought and work.

Wrinch's personal life was not happy. In 1937 her marriage was dissolved, and, two years later, she moved with twelve-year-old Pamela to the United States. The Rockefeller Foundation supported Wrinch's work until 1940; she spent the academic year 1940–41 at the Chemistry Department of Johns Hopkins University. In 1939 Pauling and Niemann (6) predicted from bond energy values and heats of combustion that a protein with the cyclol structure would be less stable than one with a polypeptide chain by about 28 kcal/mole of amino acid residues. Although the details of their calculations were incorrect, as Wrinch pointed out at the time (7), they were widely accepted. Wrinch's fellowship was up, she could not find a job, and World War II had begun. She was worried about Pamela who was quite sensitive about the controversy in which her mother was involved. Pamela had written a letter to Pauling himself (7).

Your attacks on my mother have been made rather too frequently. If you both think each other is wrong, it is best to prove it instead of writing disagreeable things about each other in papers. I think it would be best to have it out and see which one of you is really right. There are many quarrels in the world alas!! Don't please let yours be one; it is these things that help to make the world a kingdom of misery!!

Wrinch eventually found a position in 1941. With the help of O.C. Glaser of the Amherst College biology department, she was appointed a visiting professor at Amherst, Smith, and Mount Holyoke Colleges. Not surprisingly, there was a fair amount of local opposition to her appointment. Some months later, Wrinch and Glaser were married, and they settled permanently in Massachusetts.

After her year as a jointly appointed visiting professor, during which she gave seminars at all three colleges, Wrinch was given a research position at Smith, nominally in the physics department. Her association with Smith lasted for 30 years, until her retirement in 1971. There she had a few graduate students, conducted seminars for students and faculty, lectured, and continued her research. In the summers, she taught and lectured at the Marine Biological Laboratories at Woods Hole, Massachusetts. Her work flowed at an incredible rate. In the 1940's she concentrated on developing techniques for interpreting X-ray data of complicated crystal structures, and applying these techniques to protein X-ray data that she was able to obtain from experimentalists. Numerous papers and an important monograph "Fourier Transforms and Structure Factors" (9) were published. She studied mineralogy, hoping to deduce significant features of protein structures by drawing analogies between the morphologies of their crystals and certain minerals. Notebook after notebook was filled with her criticisms and ideas on papers covering many branches of science. Her total list of publications eventually reached 192. When Wrinch's husband died in 1952, she moved to a faculty residence on the Smith College campus. After her retirement in 1971, she moved to Woods Hole.

In the period following World War II, Wrinch's life had almost taken a new turn. John von Neumann had invited her to become a consultant in his pioneering work on computers at Princeton because she had persuaded him that one of the major uses of computers would be to aid in the interpretation of the X-ray patterns of protein crystals. Wrinch was enthusiastic and full of ideas, but her hoped-for move to Princeton never came about. Surely if von Neumann really wanted her, he could have used his vast personal influence and power to obtain necessary funds (8).

It appears that no one but Wrinch believed that proteins had the cyclol structure. For her part, she was convinced that the available experimental evidence had been incorrectly interpreted. Then one evening in 1954, a chemistry professor at Smith came across an article written two years earlier by the Swiss chemist, Arthur Stoll (10) which claimed that the cyclol structure had been found in the ergot alkaloids. Ergot is a parasitic fungus that thrives on cereals, especially on the ears of rye (11). Ergot of rye is the starting material for many pharmaceutical preparations. The uterotonic property of ergot inspired Stoll to prepare a number of partially synthetic compounds which have found a variety of unexpected clinical applications. The structure of the peptide portion of the ergot alkaloids gave Wrinch's theories experimental verification. The linkage between amine groups of neighboring peptide chains confirmed cyclol theory. The ergot alkaloids are, in a sense, simple versions of proteins, and Wrinch was absolutely delighted with this example (12). Thirteen years after Pauling and Niemann had "proved" that the cyclol bond was too unstable to exist, it had been actually found! Alas, the scientific community had moved on to other fronts and it paid little attention.

Wrinch once told Marjorie Senechal (2), "First they said my structure couldn't exist. Then when it was found in nature they said it couldn't be synthesized in a laboratory. Then when it was synthesized, they said it wasn't important anyway." However she persisted and was able to get funding from the Ziskind Foundation and, later, the National Science Foundation. She spent the rest of her life working out the details of her theory. Her daughter, Pamela Wrinch Schenkman, died tragically in a fire in November 1975; Wrinch died three months later at the age of 82.

What are the conclusions to be drawn from Dorothy Wrinch's life? She was brilliant, witty, ambitious, hard working, and at times was considered jealous, abrasive, and aggressive. Her considerable talents bridged many fields. She accomplished much but was restless and dissatisfied. What might have happened if she was less embittered, if her crossing of the boundaries between disciplines had been better accepted, if her cyclol structure had appeared more in nature, if she had not felt so strongly that she was fighting the male-female interface, or if her funding and job situation had been better, it is impossible to guess. Perhaps the zeal for attacking impossible and important problems, together with an ineptitude for politics of science, would force conditions of disappointment on anyone. Michael Polanyi wrote to Dorothy Wrinch in 1948 (8), "You and I have much in common in the manner we managed to make our scientific careers less dull than usual." Indeed hers was.

Literature Cited

(1) Ayling, Jean (pseudonym), "The Retreat from Parenthood," Kegan Paul, Trench, Trubner & Co., Ltd., London, **1930.**
(2) Senechal, M., "A Prophet without Honor: Dorothy Wrinch, Scientist, 1894–1976," *Smith Alumnae Quarterly* (April 1977).
(3) Wrinch, D., *Nature*, **134**, 978 (1934); *Nature*, **135**, 788 (1935).
(4) Wrinch, D., *Proc. Roy. Soc.*, **160A**, 59 (1937).
(5) Bergmann, M. and Niemann, C., *J. Biol. Chem.*, **118** 310 (1937).
(6) Pauling, L., and Niemann C., *J. Amer. Chem. Soc.*, **61**, 1860 (1939).

(7) Wrinch, D., *Nature*, **145**, 669 (1940).
(8) Senechal, M. (*Editor*), "Structures of Matter and Patterns in Science Inspired by the Work and Life of Dorothy Wrinch, 1894–1976," Schenkman Publishing Company Inc., Cambridge, MA, **1980.**
(9) Wrinch, D., "Fourier Transforms and Structure Factors," American Society for X-ray and Electron Diffraction, Monograph No. 2, Cambridge, MA, **1946.**
(10) Stoll, A., *Prog. in Chem. of Org. Nat. Products*, **9**, 114 (1952).
(11) Stoll, A., Hoffman, A. and Petrzilka, T., *Helv. Chim. Acta*, **34**, 1544 (1951).
(12) Wrinch D., *Nature*, **179**, 536 (1957).

contemporary history series

edited by:
LEONARD FINE
Columbia University
New York, NY 10027
ERIC S. PROSKAUER

Antimetabolites of Coenzyme Q and Their Potential Clinical Use as Antitumor Agents

A New Biochemical Approach to Cancer Chemotherapy

Karl Folkers and Thomas H. Porter
Institute for Biomedical Research, The University of Texas at Austin, Austin, TX 78712

Cancer includes more than 100 disease entities that present a variety of histopathology and clinical symptoms. It is the third leading cause of death in the United States after accidents and cardiovascular disease. One of every four Americans will develop cancer, and most will eventually die of their disease. It is estimated that in 1975 alone there were 665,000 new cases and 380,000 deaths from cancer in America (*1*).

Roughly half of all cancer deaths are caused by cancer of three organs: the lung, the large intestine (colon), and the breast. Lung cancer, a disease of the 20th century for which there has been little improvement in life expectancy, kills more Americans than any other type of cancer—an estimated 72,000 persons in 1975—and its incidence has more than doubled since 1947 in both men and women (*1*).

Investigators experienced with cancer epidemiology suggests that about 80% of all human cancers are causatively related to environment factors. It is estimated that some 50% of all female cancers and about 33% of all male cancers in the Western world are related to nutritional factors (primarily high fat, low fiber diets) (*2*), while some 20% of cancers in the U.S. are caused by occupation-related chemical hazards (*3*). Many now believe that one of the most promising approaches is preventive medicine—to identify those cancer-causing factors by epidemiological studies and eliminate them or lower their environmental levels.

For treatment of cancer patients, there are now four major modalities: (1) surgery; (2) radiation therapy; (3) immunotherapy; and (4) chemotherapy. For a long time chemotherapy was considered as a treatment of last resort—more palliative than curative, often producing unpleasant and sometimes dangerous side effects (*4*). Recent advances in chemotherapy now offer promise of cancer cures. Appropriate combinations of new drugs and drugs following surgery now offer greater chance of success. For instance, more than 90% of patients with early Hodgkin's disease and about 70% with advanced disease are surviving five years (*5*). It is firmly established that chemotherapy combined with surgery and radiotherapy is curative in 80–90% of patients with Wilm's tumor, a previously fatal childhood cancer of the kidney. Markedly improved survival rates of patients with Ewing's sarcoma, choriocarcinoma, various types of non-Hodgkin's lymphomas, and uterine cancer have been achieved (*5*). In all there are 10 human cancers which are highly responsive to chemotherapy, and 50% of these patients should achieve normal life expectancy (*6*). Unfortunately these successes do not include the major cancer killers such as breast, colon, or lung cancer. For example, the median survival rate of all lung cancer patients from diagnosis to death remains less than six months (*7*).

Rational approaches to the design of new cancer chemotherapeutic agents are needed. We believe one such approach is the development of antimetabolites of coenzyme Q_{10}.

Discovery of Coenzyme Q_{10} in Cancer Tissue and Early Research on Antimetabolites of Coenzyme Q_{10}

Coenzyme Q_{10} (CoQ_{10}) became of interest in cancer research shortly after it was isolated in pure form and structurally elucidated. In 1968, Scichiri et al. (*8*) showed apparent low levels of coenzyme Q_{10} in rat and human neoplastic tissues. The concentration of CoQ_{10} in the mitochondria of ascites hepatoma was one-half that of normal liver and regenerating liver. In mitochondria of human cancer tissues, the concentrations of CoQ_{10} were significantly lower than that of control tissues taken from the same individuals. The activity of the succinate-cytochrome C reductase was very low in rat ascites hepatoma and was related to the concentration of CoQ_{10}.

A lower concentration of CoQ_{10} in human neoplastic tissue in comparison with normal tissue coupled with a lower compliment of mitochondria per cell seems to offer one advantage of selectivity in the potential use of antimetabolites of CoQ_{10} for antitumor activity in patients. These data seem to suggest that analogs of coenzyme Q_{10} which are antagonists or inhibitors of either functional or biosynthetic CoQ_{10}-enzymes in tumor tissues could become clinically useful chemotherapeutic agents for cancer that are novel and different in structure from existing agents and that are based upon a

378   X–RAY ADVENTURES AMONG THE PROTEINS

THE FOURTH
SPIERS MEMORIAL LECTURE.

X-RAY ADVENTURES AMONG THE PROTEINS.

BY WILLIAM THOMAS ASTBURY, M.A., Sc.D., F.INST.P.

> " Amino-acids in chains
> Are the cause, so the X-ray explains,
> Of the stretching of wool
> And its strength when you pull,
> And show why it shrinks when it rains."
>
> A. L. PATTERSON.

All scientific research is an intellectual adventure of course, and scientific thrills are among the best and highest kind of thrills. But nowhere do we feel this sense of adventure so much as when investigating living things and the complex bodies that take part in the life processes. Far and away the most important and complex of these bodies are the proteins, and the problem of their structure and properties is, I think, the greatest scientific adventure of our times.

The X-ray analyst was bound to be attracted sooner or later, but ten or a dozen years ago the prospect of his being able to contribute anything of value did not appear at all bright : in spite of all his previous triumphs he seemed to have met his match. Pessimism was premature though, and much exciting ground has been covered since.

Naturally, I cannot in one hour retrace every step of this ten years' adventuring among the proteins, but when the Faraday Society honoured me by inviting me to deliver this lecture, I though it would be acceptable, in memory of one who helped so much to set on foot that most successful scientific adventure, the Faraday Society itself, if I tried to give you a short connected account of the more exciting episodes and show you a glimpse of the new country that seems to be opening out to us.

Since the days of Hofmeister and Emil Fischer it has been believed that the proteins are in some fundamental fashion chain-like, that in fact they are constructed from *polypeptides* of the general formula :—

$$\ldots \; -CO-\underset{\underset{R'}{|}}{\overset{\overset{R''}{|}}{CH}}-NH-CO-CH-NH-CO-\underset{\underset{R'''}{|}}{CH}-NH-CO-\underset{\overset{R''''}{|}}{CH}-NH- \; \ldots$$

in which —R', —R", etc., stand for various univalent groups—twenty-two different kinds are now known—which act as " side-chains " to a common " main-chain " or " backbone." A polypeptide chain is thus a sequence of groups all of the form $-CO-\underset{\underset{R}{|}}{CH}-NH-$, α-amino-acid *

* The α-amino-acids from which the proteins are formed, and into which they are resolved again on digestion, have the general formula $NH_2-\underset{\underset{R}{|}}{CH}-COOH.$

They must actually exist in the proteins as " residues " ; that is, as molecules minus the elements of water.

"residues," as they are called; and the problem at the outset is to see whether chains of this kind pre-exist at all as such in the proteins, or whether they are artifacts of chemical manipulation.

Broadly speaking, there would seem to be two kinds of proteins, the non-fibrous and the fibrous, which is roughly the same as the crystalline and the non-crystalline—or at least, the not-obviously-crystalline. Of the crystalline proteins, well-known examples are egg albumin, hæmoglobin and insulin, while of the fibrous proteins there are silk, hair, muscular and connective tissue, for instance. The fibres invite attention first, not merely because they are more stable and readily obtainable, but because they are fibres. Natural fibres mean something much more than long thin cylinders. We know now, particularly from X-ray and other studies of the chief vegetable fibre, cellulose, that they are fibrous in a deeper sense: their molecules are "fibrous" too, long thread-like bodies that build up the structure of the visible fibre as a yarn is built from the fibres themselves. The significance of this generalisation for proteins lies in the idea that they are somehow constructed from polypeptide chains; for if such chains have any real existence, then surely they must be observed in the protein fibres at least. The very fact that proteins form so many fibrous structures is a most powerful argument in favour of their molecules being essentially chain-like. What shape or disguise they assume in the crystalline proteins is a different matter, about which we shall say more later. At first sight the crystalline proteins, especially since it has been shown directly that their molecules are massive and often roundish bodies, would appear difficult to explain in terms of chain-molecules.

The principles and technique of X-ray analysis are now sufficiently well known not to necessitate further discussion here—indeed, they form the subject matter of the previous Spiers Memorial Lecture *—and we may assume that it will be appreciated that the X-ray methods, especially in such an involved study as this of protein structure, can conceivably make up in part for what is lacking in present chemical methods. Even when we know the exact proportions of all the amino-acids given up by a protein on hydrolysis, and we are in no such happy state as regards the vast majority of proteins, that is still only a beginning: we still have to find out in what order the amino-acids are arranged, the actual configuration the polypeptide chains take up, and how they themselves are linked one to another—a rather frightening task, when we stop to think of the vast number of permutations and combinations that are theoretically possible to the hundreds or thousands of residues (although there are only twenty-two different kinds) that are known to take part in the make-up of the usual protein molecule. But the speciality of X-ray analysis, so to speak, is the detection and measurement of *regularities* of structure—atomic and molecular patterns; and it is particularly because he is armed with this power that the X-ray physicist has the hardihood to enrol in this great biological adventure.

It was first shown by Meyer and Mark that the X-ray photographs (obtained by Herzog, Brill, Kratky and others) of natural silk, which consists of perhaps one of the simplest proteins, *fibroin*, could be interpreted very satisfactorily on the basis of the polypeptide chain theory. To explain these photographs and the general properties of the silk fibre it does not seem necessary to postulate anything more complicated than

* "Molecule Planning" (Sir W. H. Bragg, 1934).

380   X–RAY ADVENTURES AMONG THE PROTEINS

that the fibre is a sort of molecular yarn made up of polypeptide chains lying approximately parallel to the fibre axis, provided it is assumed also that the chains are fully stretched out ; that is, that they have the configuration expressed conventionally by the formula set out above. The X-ray diagram agrees well with a molecular pattern along the fibre axis of the kind that chemistry suggests, and the polypeptide chains apparently form more or less regular bundles that may be looked upon as submicroscopic, and not very perfect, crystallites of fibroin.

So far so good : the spoils of the first adventure are at least recognisable, and the Snark is not a Boojum after all. Not so with the other natural protein fibres, which give X-ray photographs quite unlike that of silk. They tally with the silk photograph in that they also indicate regular bundles of chain-molecules lying along, or simply related to, the fibre axis ; but there are discrepancies in the repetition of molecular pattern in and between these chain-molecules : it seemed impossible to explain their X-ray diagrams *and* that of fibroin on the basis of polypeptides fully stretched out. Hair in particular, and more important still, muscle, seemed intractable from this point of view.

The correct, or rather the more comprehensive, point of view—it is a natural extension of the theory—would appear to be that the chains are *not* always fully extended ; and that is what X-rays suggest directly from the experiment of stretching hair and taking the photograph again. It is then that we find a diffraction pattern recalling that of silk, with once more a repetition along the fibre axis that would fit in with fully-extended polypeptide chains. This new photograph fades away again however when the hair is allowed to resume its unstretched length, as it does very quickly when released in the presence of water, and the normal hair photograph reappears—but it is possible always to obtain either photograph at will by stretching the hair or releasing it. The inference is obvious, that the diffraction pattern of the normal hair photograph arises from regularly folded polypeptide chains lying in the direction of the fibre axis, and that these chains are pulled out straight when the hair is stretched but fly back again when it is released. In other words, the hair protein *keratin* is capable of existing in two forms— two " mechanical stereo-isomers," if you like, since one is obtained from the other simply by pulling—and the reversible passage from one form to the other is the molecular basis of the remarkable *long-range* elasticity of hair. We may call the two forms of hair keratin, α and β, the former being the normal (folded) form.

Thus the second X-ray adventure among the proteins brings new knowledge of their stereochemistry and teaches that not only can the polypeptide chains of Fischer conceivably exist in folded forms, but that they actually do exist in such forms : and it brings too new understanding of the extraordinary elasticity and changes of shape so characteristic of living things by showing how the giant molecules used in their construction are susceptible of gross changes of configuration conditioned by their physico-chemical environment. The X-ray interpretation of the elastic properties of keratin has now been worked out in some detail, but it all comes to this, that the " backbone " of a polypeptide chain takes up a shape in conformity with the nature, distribution, and state of interaction of its various side-chains. It may sometimes be forced out of this equilibrium shape by mechanical means, as when hair is stretched and the α-form of keratin extends to β-keratin ; or it may change " voluntarily " with changes of interaction of the side-chains with one another or with accessory molecules, as seems to happen when a muscle alters its length. The intramolecular movements, it will be understood, take place by virtue of rotations about valency bonds linking consecutive atoms—really it is because protein molecules contain so many atoms in chains that they are endowed with such potentialities for change of shape.

According to the indications of X-ray analysis, the keratin molecule is a grid-like structure built up by means of a system of cross-linkages between the side-chains of neighbouring main-chains. In equilibrium, the main-chains are folded or " buckled " out of the plane of the grid, but the latter is drawn out flat when it is pulled at the ends. The sub-microscopic crystallites of keratin appear to be simply piles of such grids.

The intramolecular stresses generated when the keratin molecule is pulled, which must be the driving force of contraction on release, are dissipated on exposure to steam or hot water : the stretched fibre tends to remain stretched. This method of " setting " stretched or bent hair has probably been known and used for ages, but it has been one of the more amusing rewards of X-ray adventuring among the proteins first to perceive how it comes about. The essential molecular change underlying the " permanent wave " or the creasing of trousers is revealed in striking fashion in X-ray photographs of wool or hair—they are fundamentally the same—that has been steamed in the stretched state. It consists of a disruption in one direction only, the direction in which the side-chains point, and it is a natural inference that this breakdown relieves the stresses in the stretched grids and destroys or inhibits their power of elastic recovery. As a matter of fact it only inhibits it ; more prolonged steaming of the fibre in the stretched state is required to " set " it effectively. The initial breakdown of cross-linkages is followed by a process of re-union in unstressed configurations, as was first shown by Woods.

If this rebuilding is not allowed to take place, the fibre is found to have acquired an increased range of elastic recovery : it will contract in steam till it is actually no more than about two-thirds of its original unstretched length. We may call this important new phenomenon " supercontraction," to distinguish it from the ordinary recovery of stretched hair just to its initial unstretched length. Its importance is twofold : in the first place it implies the existence of polypeptide chains in other, more folded states than the one we have called α which is characteristic of normal unstretched hair ; and in the second it gives a clue to the elastic mechanism of muscle.

With regard to the first point, supercontracted keratin unfortunately gives no new regularity of molecular pattern along the fibre axis that can be recognised as such by X-rays, and this is not surprising in view of various circumstances attending its production that need not be gone into here ; but its general elastic behaviour and that of the denatured proteins to be discussed below, the changes that are actually observed in the X-ray photographs, and a whole body of evidence that has emerged from studies of this kind, all conspire in suggesting that the supercontraction of hair does indeed arise from a " superfolding " of the keratin chains beyond that which defines the α-state. To what extent the new folds are also regular folds, as the α-folds appear to be, is another matter which we may refer to again later.

The second point introduces what in some ways is the most exciting major adventure among the proteins, that of the elastic mechanism of

382   X–RAY ADVENTURES AMONG THE PROTEINS

muscle. This quest, even after hundreds of X-ray photographs and many secondary experiments,* is still really only at the beginning, but a convenient epoch has been reached and certain main conclusions, relatively simple in principle, may now be stated quite briefly : The chief solid component of muscle, the protein *myosin*, is also a " fibrous" protein like fibroin and keratin and lies in the form of polypeptide chains along the length of the muscle fibres.   That is as it should be and confirms directly what may also be inferred indirectly from a number of physico-chemical investigations of the constituents of muscle—the deep significant contribution of X-rays is the recognition of the remarkably close analogy between myosin and the supercontracting form of keratin. In a relaxed muscle the myosin chains are in the α-form, like the keratin chains in unstretched hair ; that is to say, they are not, as was originally thought, fully stretched out but are already folded and grouped together in " buckled grids " just as keratin is.   From this configuration they may be pulled out into the β-form or they may be contracted further into an apparently super-folded condition—and we might proceed to enumerate a series of affinities, but it will be sufficient here to summarise what will be set out in detail elsewhere by saying that all the chief properties, both X-ray and elastic, of the supercontracting form of the hair protein keratin find their counterparts in those of the muscle protein myosin.   It seems impossible to avoid the conclusion that they belong to a common family with essentially similar elastic properties, and hard indeed not to infer that, in particular, the contraction of muscle corresponds to the supercontraction of hair.   The changes in the shape of the molecule that have been achieved in the laboratory by cruder means in the case of both keratin and isolated myosin are brought about in the muscle system by a delicate and reversible cycle of interactions with accessory molecules. The myosin grids are the movable parts of the engine, the chemical cycle the motive power.

The comparative insensitivity and stability of the keratin structure would appear to be bound up in some way with its high proportion of residues of the sulphur-containing amino-acid, *cystine*.   The resemblance is not between myosin, which does not contain much sulphur, and keratin, which does, but between myosin and the supercontracting modification of keratin, in which certain cross-linkages of the grid have been disrupted. Recent work by Speakman has shown that these disrupted cross-linkages revealed by X-rays are almost certainly, or include at least, the —S—S— bridges between the two halves of the cystine residues which are incorporated in neighbouring main-chains, which suggests that when we bring keratin into the supercontracting condition and thereby invoke properties like those possessed by myosin in the normal way, we are doing the next best thing to reducing the cystine content ; we are destroying its effectiveness.   The impression gradually evolving out of these investigations and others now in progress is that keratin lies at one end of a series of similar proteins of which myosin is at or near the beginning—or in more popular language, keratin has the appearance of being actually a kind of " vulcanised " adaptation of myosin.

An adaptation of the muscle apparatus as a whole that has tempted a side venture with X-rays is the electric organ of certain fishes, which these animals can stimulate at will to deliver a shock—a dangerous

* W. T. Astbury and S. Dickinson, " X-ray Studies of the Molecular Structure of Muscle " (in preparation).

discharge, in fact, in the case of the electric eel.   Except in the electric catfish, the histological derivation of the organ from muscle tissue is clear; and besides, the two have similar time relations and a number of biochemical features in common.   The fish sets off its electrical apparatus as it would its muscles, though from the former it delivers electrical energy instead of performing mechanical work.   The inevitable question is: Where does the myosin come into the story?   And the answer given by X-rays is that it does not; at least, it does not in the regular oriented form in which it occurs in muscle.   Only in the most primitive electric organ can myosin be detected by X-rays, though it is clear enough in immediately neighbouring and closely related muscle cells; and as we pass on up to examine the electrically more powerful organs of other species it is lost sight of altogether.   It is as though the moving parts of the muscle engine had been progressively removed, leaving the chemical cycle nothing to operate.   The energy, or some of the energy, that would have been handed on to the myosin chains and made them contract had they been there is now set free as electrical energy.   In popular language again, the pistons have been removed, so the engine blows off steam.

It is surely a fact of biological importance that X-rays have so far revealed only one other fundamental type of natural protein fibre besides the keratin-myosin group.   The latter includes also the fibrous protein of the epidermis, as has been shown by Rudall and others, and also, in its β-sub-group, the keratin common to feathers, tortoise-shell, and similar reptilian growths.   Feather keratin gives a particularly beautiful X-ray photograph, corresponding to polypeptide chains not quite fully extended, but constricted by about 7 per cent.: from this approximate β-configuration they can, however, be stretched continuously and reversibly up to their theoretical maximum length before the structure breaks.   Another interesting point about the feather is that though in the main part of the quill the polypeptide chains run roughly parallel to its length, there is an outer layer in which they run round.   The feather quill is a two-ply structure of very considerable strength—it is even more than two-ply really, since though the longitudinal chains lie in parallel layers, they show a pronounced angular dispersion *in* the various planes of these layers.

The second great group of natural protein fibres comprises the tendons, connective tissue, and collagenous fibres in general.   All give the same type of X-ray photograph, which is quite different from that given by the keratin-myosin group, whether in the α- or the β-form. The same photograph is given also by gelatin, an artificial modification of the collagen fibres.   In all these fibres the polypeptide chains presumably form similar configurational groups, even though they are not strictly identical, and it seems probable that they are stereochemically fully extended, though not in the sense of the fully extended (or β-) configuration of keratin or myosin.   A possible interpretation that would fit in with the X-ray data is that the chains are partly in a *cis*-form, as opposed to the *trans*-form of the β-fibres (see below).

In hot water above certain fairly sharp temperatures, or in certain reagents, the collagenous fibres contract spontaneously by sometimes as much as three-quarters of their normal length, and while in this state they too show long-range elasticity like that of keratin and myosin. Their condition then seems to correspond rather to the supercontracted state of the latter—in any case, to folded polypeptide chains.

384   X-RAY ADVENTURES AMONG THE PROTEINS

The important concept emerging from all this, apart from the biological implications of the X-ray classification of the fibrous proteins, is that of long-range elasticity based on intramolecular unfolding and refolding of polypeptide chain systems. The tendency of such systems to fold given the correct conditions and appropriate inter-chain linkages —that is to say, their capacity to assume various configurations and set up all sorts of molecular fields—underlies the manifold physical and chemical functions of the protein molecule in living things. X-ray and supporting studies of the keratin-myosin group offer in the protein fibres the first direct demonstration of the property.

But what of that other half-world of proteins, those that are just as obviously crystalline as the structures we have discussed so far are obviously fibrous? Where do they fit into the scheme? For there must be only one great scheme at bottom for all proteins, to judge by the chemical and physico-chemical evidence. Surely the fibres alone are telling us the answer through their peculiar elastic properties. The elastic properties of muscle and hair and the contractility of the collagen group suggest the possibility of even higher degrees of folding: they are pointing directly at the architecture of the crystalline proteins.

The molecules of the crystalline proteins, as shown by Svedberg's ultra-centrifuge and such X-ray studies as have so far been achieved on unaltered crystals of these unstable substances, are massive bodies— " globular " is a convenient term, though not a particularly good one in view of the fact that though many are round, or effectively so, others are decidedly not—and at the moment we have little or no direct knowledge of how they are constructed. We can conceive very readily, though, that they are probably based on a coiling of polypeptide chains that is merely an elaboration of the keratin folds, and can imagine a general plan that would include the fibrous proteins as the special case corresponding to folds progressing in one direction only. To put it another way, it is a very fair inference that the contracted states of the fibrous proteins are but intermediate steps between the fully-extended configuration and the multiple folds of the " globular " proteins.

Further experimental evidence in favour of this interpretation comes from the X-ray study of protein denaturation. The word " denaturation " explains itself : it seeks to describe the degenerate change that proteins undergo almost universally, and often under the slightest provocation. There are only slight changes in chemical composition, but profound changes in constitution leading to irreversible loss of specific properties. Heat is the simplest denaturing agent, and the most familiar case is when an egg is boiled. The soluble egg albumin loses the special properties that distinguish it from other proteins and coagulates into an insoluble mass.

On denaturation and coagulation the globular proteins give rise to a common X-ray diagram like that of *disoriented* β-keratin. In the molecular sense then, at least, they tend to become fibrous : the polypeptide chains are liberated from their specific folded configurations and aggregate to a greater or less extent into extended bundles similar in structure to the sub-microscopic crystallites of β-keratin. It follows that it should be possible to " spin " denatured protein into artificial macroscopic fibres, and this is so. When the crystalline hemp seed globulin, edestin, for instance, is dissolved in strong urea, the huge globular molecules of the protein break down under the influence of the urea molecules and in time form a viscous solution that can be spun into

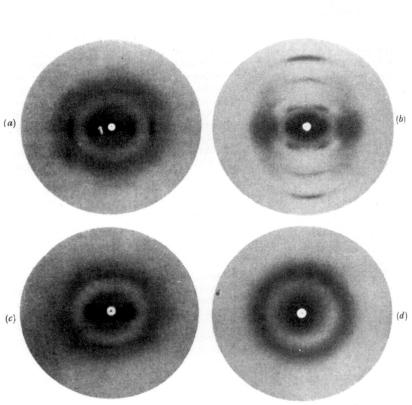

FIG. 1.—The four principal types of X-ray photograph given by fibrous proteins : (*a*) β-type (stretched keratin) ; (*b*) normal collagen type (tendon); (*c*) α-type (unstretched keratin) ; (*d*) " Supercontracted " collagen type (elastoidin contracted in hot water. *Cf.* Champetier and Fauré-Fremiet, *J. Chim. Physique*, 1937, **34,** 197). (Specimen-to-film distance = 3·2 cm.CuK$_\alpha$.)

[*To face page* 385.

a coagulating bath as cellulose solutions are spun in the manufacture of artificial silk. The resulting fibres are highly elastic and give on stretching an X-ray diagram of the type of *oriented* β-keratin (stretched hair)—thus bringing to a successful issue one of the more prophetic of X-ray adventures among the proteins, and one that shows a first glimpse of their domain as a whole.

The view revealed is still sadly lacking in details, of course, but it has the aspect, or the promise, of an underlying simplicity. The proteins appear now as a class of substances all based on the polypeptide chain *and the ways of folding it.* Exactly what such a generalisation means in terms of intra- or inter-molecular linkages we do not yet know, though it is clear that the interactions and combinations of the side-chains with one another and with accessory molecules must be the predominating factor; and chief among these interactions and combinations must be the formation of disulphide bridges through the cystine residues and salt-like or hydrogen bonds between the acidic and basic side-chains. It is not sufficient, as the properties—or rather lack of properties—of the denatured proteins show, to have a given sequence of amino-acid residues in chains : the chains must be folded in some special manner, and certain side-chains must be linked, exposed, or combined with other molecules, as the case may be, before the specific properties of the protein are manifested at full strength. No wonder that such large and complicated systems have such a multiplicity and sensitivity of properties and collapse so readily into a mere jumble of polypeptide chains—for that is all that the denatured proteins are, to judge by their X-ray photographs, their frequent long-range elasticity, and their increased viscosity.

The task of protein studies now is to try to find out the form or forms of protein folds. A possible solution for keratin and myosin that fits in well with the X-ray and elastic data is that they are hexagonal and closed by a lactam-lactim or a keto-enol interchange, thus : —

Lactam-lactim Interchange.

Keto-enol Interchange.

On this basis the four principal states of fibrous proteins may be represented as in Fig. 2, corresponding to the four principal types of X-ray photograph shown in Fig. 1.

Whether the kind of hexagonal fold postulated for the fibrous proteins provides the theme of the globular proteins too, or whether, even, it is really correct for either, it is also not yet possible to decide. It is a good working hypothesis at the moment, and in any case it is most reasonable to assume that, quite apart from the question of the actual form of the folds, the same kind or kinds of folds will be used for all sorts of proteins. This is equivalent to saying that the fibrous proteins are simply the linear prototypes of the globular proteins, and probably the analogue of the globular molecule is the fibre crystallite. The evidence for this interpretation has now been accumulating for some time, for the X-ray photographs of the fibrous proteins, in particular that of feather keratin, show some very large spacings indicative of a " super-structure " over and above the pattern of polypeptide chains. They are indeed rather like the photographs of the recently isolated tobacco-mosaic virus, the units of which appear to be elongated protein bodies of such relatively great size and internal regularity of structure that, like the diamond, they are both molecules and crystals. On this view the virus protein units are not essentially different in type from the protein fibre crystallites.

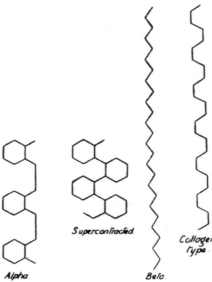

FIG. 2.—(*Cf.* Fig. 1.) Suggested scheme for the four principal states of fibrous proteins.

So, having begun this lecture with a broad classification of proteins into the fibrous and non-fibrous, we finally come round to the feeling that probably after all there is no real distinction between the two, and that the structure of keratin, the first complex protein to be examined in detail by X-rays and to reveal the phenomena of long-range elasticity in relation to regular intramolecular folding, may perhaps serve as a model for them all.

The latest X-ray step towards this hoped-for fusion of ideas has been taken with the aid of analytical chemistry. As is well known, protein analyses present formidable difficulties, and the physicist is handicapped for want of chemical data. Such data as are available are not only insufficient, but are often conflicting and out of tune with stoichiometrical principles. Block and others had succeeded in showing in certain cases that some of the amino-acid residues are present in fairly constant proportions, but it was not until the recent work of Bergmann and Niemann that stoichiometrical hopes began to rise really high. These authors have drawn the conclusion, from a study of egg albumin, cattle hæmoglobin, cattle fibrin, and silk fibroin, that the total number of amino-acid residues in a protein molecule and also the numbers of each of the various kinds of residue are expressible in the form $2^n 3^m$. They find that the minimum molecular weights of the four proteins mentioned correspond to 288, 2 × 288, 2 × 288, and 9 × 288 residues, respectively; in other words, are multiples of a common unit whose weight on the average will be something of the order of 36,000. In other words again, Bergmann and Niemann, by purely chemical analytical reasoning, arrive at a similar conclusion to that already arrived at by Svedberg from ultra-centrifugal studies of the globular proteins: and their findings to date cover not only two globular proteins, egg albumin and hæmoglobin, but two fibrous proteins also, fibrin and fibroin—for both these latter give X-ray fibre photographs of the β-type.

Here then is new evidence of a most stimulating kind that fibrous and globular proteins are constructed to a common plan, or at least that some common factor is involved in their method of synthesis. And X-rays have been able to push the argument further. The method of Bergmann and Niemann has a difficulty in that it requires a knowledge of the average residue weight of all the amino-acids incorporated in the protein under investigation; and, of course, it is not possible to estimate this quantity directly without complete analytical data, which are not forthcoming as yet. It is possible to calculate it from X-ray data and the density, however, without making use of the amino-acid proportions given by ehemistry. In β-keratin, for instance, X-ray reasoning indicates that the average size of an amino-acid residue is $9.7 \times 4.65 \times 3.33$ cubic Ångstrom units, while the density is found to be nearly 1.3 gm. per c.c. The average mass of a residue is therefore

$$9.7 \times 4.65 \times 3.33 \times 1.3 \times 10^{-24} \text{ gm.,}$$

corresponding to an average residue weight of

$$(9.7 \times 4.65 \times 3.33 \times 1.3)/1.65 \approx 118,$$

since the mass of a hydrogen atom is 1.65×10^{-24} gm. It follows from this that there must be $100/118 \approx 0.85$ gram-residues in 100 gm. of keratin, and the number 0.85, if the proposition of Bergmann and Niemann is true, should give the numbers of gram-residues of the various acids simply by dividing by values of $2^n 3^m$. The following table shows the results of some of the calculations for wool keratin.

We may not enlarge here on either the reliability or significance of these and other data of the kind, but it will be seen that the measure of agreement is quite good; and though it is hard yet to prove that only powers of 2

Acid.	Frequency.	Gm.-res. in 100 Gm. Wool.	
		Calc.	Obs.
Glutamic .	8 (2^3)	0·106	0·103
Arginine .	16 (2^4)	0·053	0·059
Aspartic .	16 (2^4)	0·053	0·054
Tyrosine .	32 (2^5)	0·027	0·027
Lysine .	48 ($2^4 . 3$)	0·018	0·019
Tryptophane .	96 ($2^5 . 3$)	0·008$_9$	0·009
Histidine .	192 ($2^6 . 3$)	0·004$_5$	0·004
Amide-N .	9 (3^2)	0·094	0·098

and 3 are involved, there is sufficient evidence to warrant the belief that keratin also will be found to conform to a stoichiometrical scheme, and quite possibly to a single common scheme such as that proposed by Bergmann and Niemann. Already the above figures indicate, for

388   X–RAY ADVENTURES AMONG THE PROTEINS

instance, that one-eighth of all the residues are glutamic acid residues, one-sixteenth arginine residues, and so on, and suggest that the smallest possible number of residues per molecule is $2^6 \times 3^2$, or 2×288 (molecular weight about 68,000), corresponding to hæmoglobin and fibrin. However that may be, one thing seems clear enough, that we are now on the verge of something very fundamental indeed in protein theory, and the moral value alone of such stoichiometrical discoveries as these is immense. Exact analyses of the proteins, though always laborious, need no longer be the thankless tasks they have been. Every possible reliable observation now is urgently needed and must sooner or later be fitted into the puzzle. Above all, *complete* analyses of single proteins are necessary. . . .

It is part of the accepted technique of the adventure story to leave each episode on a note of suspense, and the stage at which we must now leave this great adventure of the proteins is no exception to the rule. To my mind the story is at a pre-eminently exciting stage, and the next few years should prove critical, one way or another, and bring us to the heart of the matter. In a sense, we have just completed a cycle of adventures and are about to set off for more. There was first the adventure of the fibre made from straight polypeptide chains, then that of the fibre made from regularly folded chains ; and after that came the globular proteins with their multiple folds. Then came the unfolding of the globular proteins by denaturation and the recovery once more of straight polypeptide chains. Now we should like to fold these chains up again, or at least to find out how they were folded before we undid them. For the moment, though, we can only unwrap the mysterious protein parcel in a most untidy manner—so untidy that we still cannot see what is inside before it comes undone. But one thing we may be sure of, it is something that holds the secret of our health and happiness—our very life itself.

PRINTED IN GREAT BRITAIN AT THE ABERDEEN UNIVERSITY PRESS, ABERDEEN

-4-

THE FOLDING AND COILING OF POLYPEPTIDE CHAINS

"Up the Crick without a Wrinch — a truly apPauling tale, yet one to Bragg about"

KEY PAPERS

1938   J. D. Bernal, Isidore Fankuchen and Max F. Perutz, *Nature* 141, 523–524. "An X-ray Study of Chymotrypsin and Haemoglobin."

1941   D. Crowfoot Hodgkin, *Chem. Reviews* 28, 215–228. "A Review of some Recent X-Ray Work on Protein Crystals."

(A) 1979   D. Crowfoot Hodgkin, *Ann. NY Acad. Sci.* 325, 121–148. "Crystallographic Measurements and the Structure of Protein Molecules as They Are."

(B) 1950   W. L. Bragg, J. C. Kendrew and M. F. Perutz, *Proc. Roy. Soc. London* A203, 321–357. "Polypeptide Chain Configurations in Crystalline Proteins" [24 pages of 37].

(C) 1951   L. Pauling, R. B. Corey and H. R. Branson, *Proc. Natl. Acad. Sci. USA* 37, 205–211. "The Structure of Proteins: Two Hydrogen-Bonded Helical Configurations of the Polypeptide Chain."

(D) 1998   M. F. Perutz, In *I Wish I'd Made You Angry Earlier: Essays on Science.* Cold Spring Harbor Press, pp. 189–191.

(E) 1951   M. F. Perutz, *Nature* 167, 1053–1054. "New X-ray Evidence on the Configuration of Polypeptide Chains: Polypeptide Chains in Poly-γ-benzyl-L-glutamate, Keratin, and Haemoglobin."

1951   L. Pauling and R. B. Corey, *Proc. Nat. Acad. Sci. USA* 37, 251–256. "The Pleated Sheet, a new Layer Configuration of Polypeptide Chains."

(F) 1951   L. Pauling and R. B. Corey, *Proc. Nat. Acad. Sci. USA* 37, 729–740. "Configurations of Polypeptide Chains with Favored Orientations around Single Bonds: Two New Pleated Sheets" [5 pages of 12].

1952   W. Cochran, F. H. C. Crick and V. Vand, *Acta Cryst.* 5, 581–586. "The Structure of Synthetic Polypeptides. I. The Transform of Atoms on a Helix."

1952   F. H. C. Crick, *Nature* 170, 882–883. "Is α-Keratin a Coiled Coil?"

1953   F. H. C. Crick, *Acta Cryst.* 6, 685–689. "The Fourier Transform of a Coiled-Coil."

1953   F. H. C. Crick, *Acta Cryst.* 6, 689–697. "Packing of α-Helices: Simple Coiled-Coils."

1953   J. Donohue, *Proc. Nat. Acad. Sci. USA* 39, 470–478. "Hydrogen Bonded Helical Configurations of the Polypeptide Chain."

1955   R. B. Corey and L. Pauling, *Istituto Lombardo* 89, 10–37. "The Configuration of Polypeptide Chains in Proteins."

(G) 2003   D. Eisenberg, *Proc. Nat. Acad. Sci. USA* 100, 11207–11211. "The Discovery of the α-helix and β-sheet, the Principal Structural Features of Proteins."

When it became possible to crystallize pure preparations of small globular proteins such as the enzymes trypsin, chymotrypsin and pepsin, John Desmond Bernal at Cambridge decided to set up a research group devoted to the x-ray crystal structure analysis of proteins. Dorothy Crowfoot (later Hodgkin) arrived as a graduate student in 1933 and Max Perutz in 1936. Crowfoot elected initially to work on pepsin and then insulin, while Perutz chose hemoglobin. Progress was slow, and diffraction patterns were poor, until they realized that protein crystals usually contain around 50% aqueous solvent, and must be protected against dehydration. Crystals sealed in Lindemann glass capillaries gave excellent patterns which extended out far enough to suggest that complete atomic structure analysis of a protein might be possible (*pace* Pauling).

In 1937, Bernal moved from Cambridge to Birkbeck College, London, and was replaced by William Lawrence Bragg ("Sir Lawrence"), the son in the father/son duo who earned a Nobel Prize in 1915 for inventing x-ray crystal structure analysis. Bragg persuaded Perutz to remain as a senior scientist, and after World War II, ex-serviceman John Kendrew joined the laboratory as a graduate student. This team of Bragg, Perutz and Kendrew was to revolutionize our knowledge of protein structure. (As an example of how small and inbred the scientific world was in Britain, Bernal had served during World War II as a science advisor to General Mountbatten in the Far Eastern theater of operations. There Bernal met a young officer named Kendrew, perceived both his abilities and his interest in structure, and recommended most strongly to him that he join Bragg's laboratory to earn his doctorate when the war was over.)

Dorothy Crowford moved from Cambridge to Oxford upon completion of her degree, and set up a productive crystallographic group that led to the structures of penicillin, vitamin B_{12}, other biologically important molecules, and the protein insulin. In 1979 she published a retrospective (A) that is so comprehensive, and so clearly written, that it is included here even though it begins earlier and ends later than our time frame. At this point I can only say: Stop now, have a careful look at her review, and then let us resume.

Hodgkin's review contains a wealth of information about how proteins first were studied by x-rays, how diffraction patterns from globular proteins were improved enormously by sealing crystals in capillaries to keep them from drying out, and how the two most important protein structure groups came into being at Cambridge and at Caltech. Single crystal photographs of the proteins insulin, excelsin, lactoglobulin, hemoglobin and chymotrypsin revealed diffraction patterns containing thousands of reflections. The intensities of these reflections depended upon the structure of the protein. But how could one possibly work out the structure by using only these thousands of spots? We will find out in Chapter 6. One extremely important bit of information was that, although crystals of globular proteins were roughly 50% water, the protein molecules themselves possessed a structural integrity. Protein molecules shifted slightly in the crystal when the water content was changed, but the structure of the molecules themselves remained unaltered. This meant that it was reasonable that the structure of a protein determined in a crystal also represented the structure of the same protein when in solution in a biological setting. Protein crystallography had a purpose.

Hodgkin in passing makes an intriguing statement about Max Perutz and his protein research:

> **He was interrupted by internment at the beginning of the war, and later by work on Habbakuk in Canada (Habbakuk was a gigantic aircraft carrier made of ice).**

We will come back to his internment in Chapter 6. She gives brief mention to Dorothy Wrinch and her cyclol theory, and then turns to the stunning proposal by Pauling, Corey and Branson of the α-helix as the element of protein structure. This brings us to the topic of the present chapter.

Astbury's extended chain structure for β-keratin and silk was generally accepted, but his folded chain for the α form was not. Both Cambridge and Caltech realized that a better structure was needed, and both set out to find it. A critical assumption was that the *local* environment of each amino acid in the structure should be the same, and that it was the side chains that "did the work." If any polymer is made up of a string of subunits, each with the same local environment, then the result inevitably is a helix. A circle can be thought of as one limit of a helix in which the rise between one turn and the next is zero. Similarly, a fully extended chain can be regarded as the other limit of a helix. In between these two extremes lie a family of helices, defined by (a) their pitch or rise, P, which is the distance between one turn and the next, (b) the number of subunits per 360° turn of helix, n, and (c) the *rise per residue* along the helix axis: $h = P/n$.

Bragg, Kendrew and Perutz in Cambridge embarked on a thorough analysis (B) intended to examine *all of the possible helices* that an alpha polypeptide chain could adopt, using the most accurately established bond lengths and angles. As a boundary condition, Astbury had established that the rise per turn of helix in the α-keratin structure was 5.1 Å. Reprint B contains the most relevant 24 pages of the 37 in the resulting paper. The authors considered various twofold helices (two equivalent steps per turn of helix) in Figures B.5–9, threefold helices in Figures B.10 and 11, and fourfold helices in Figure B.12. Fivefold and higher-symmetry helices were eliminated, as one could not easily fit so many amino acids into a helix with a rise of only 5.1 Å.

All of the helices described were stabilized by hydrogen bonds between C=O and H—N groups on adjacent turns of the helix, as shown by dashed lines in Figure B.6.

A particular helix was characterized by S_R, where S is the number of amino acids per turn of helix (our n, above), and R is the number of atoms in the ring of atoms closed by a hydrogen bond. In the 2_7 helix of Figure B.6, one such ring contains atoms O, C', N, C, C', N1 and H, the latter being hydrogen-bonded to the initial O. In the upper part of the 3_8 helix of Figure B.10, the ring of backbone atoms closed by one hydrogen bond contains the eight atoms H, N1, C1, C1', N2 (behind C2), C2, C2' and O2, the latter being hydrogen-bonded to the initial H. The 4_{13} chain in Figure B.12 is more difficult to follow, but one H-bonded ring involves the thirteen atoms O, C', N, C, C', N1, C1, C1', N2, C2, C2', N3 and H.

They favored a planar amide bond as Pauling had proposed, but did not absolutely require it. In the 2_{13} helix of Figure B.8 the amide is twisted and nonplanar. It is more nearly planar in 2_{14}, Figure B.9. In 3_8, Figure B.10, bonds around the nitrogen atom clearly are pyramidal rather than planar. But in 4_{12}, Figure B.12, the amide bond once again appears to be planar.

More than one helix could be drawn having the same number of amino acids per turn, but differing in how the turns were hydrogen-bonded. Among twofold helices, their Table B.1 lists two variants of 2_7, and one each of 2_8, 2_{13} and 2_{14}. Threefold helices included 3_7, 3_8 and 3_{10}, and fourfold offered two possibilities: 4_{11} and 4_{13}. Note that, if one keeps the hydrogen bonding pattern the same in a series of helices, increasing S by one amino acid adds 3 to the number of atoms in the ring, R. One such series of similarly-bonded helices is: 2_7, 3_{10}, 4_{13} and 5_{16}. In slightly changed form, these would turn out later to be the biologically relevant helices in globular proteins. Bragg, Kendrew and Perutz favored those helices in which every possible hydrogen bond between turns was formed, as the 2_7, 2_8, 3_8 and 4_{13} helices. But they did not rule out others in which only one-third of the possible H-bonds were formed, such as 2_{13} and 2_{14}.

Unfortunately, they were unable to choose among possible helices. Attempts to fit these to the observed Patterson vector maps (not included in our reprint) were inconclusive. None of these structures was felt to be completely satisfactory. In their abstract they said, "The evidence is still too slender for definite conclusions to be drawn...," and in their Discussion, "The conclusion that the chains are of a folded

FIGURE 4.1   The Linus Pauling "analogue computer" for generating polypeptide helices. Copy this diagram onto a transparency. Then roll it around a vertical axis until the oxygen atom with an arrow at lower left overlaps the oxygen labeled "α" The result is an alpha helix. If the cylinder is tightened until the arrowed oxygen overlaps the oxygen labeled "3_{10}," a 3_{10} helix results. Tightening still further yields a 2_7 helix, and loosening the alpha helix by one amino acid residue creates the more open pi helix. This type of diagram, on scratch paper, is what led Pauling to the concept of the alpha helix.

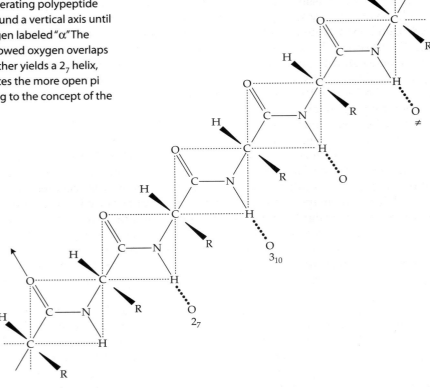

coplanar form resembling the 2_{13} or 2_{14} type would be more convincing if there were any indication that the form has obvious advantages over others." The mammoth paper, intended to define the field, ended on a note of frustration.

In the meantime, Linus Pauling was also considering the same problem, and illustrating his famous ability to "think outside the box" He didn't use x-ray diffraction patterns or any other information; he simply worked with a pencil and a sheet of paper. He spent 1948 as a visiting Professor at Oxford, and while there came down with a typical English cold (1). In bed with nothing to do, he sketched an extended diagonal polypeptide chain with all the N—H bonds pointing down and all the C=O groups pointing up (Figure 4.1). He then began rolling the paper into a cylinder to see if he could bring N—H groups systematically close to C—O groups on the next turn of the helix. As Dorothy Hodgkin described it on page A.147:

> **The alpha helix was just a stroke of genius on the part of Linus Pauling himself. Again I know because I was there. He was having a cold or flu in Oxford the year he was there and was rolling bits of paper around his fingers, which took him away from the crystallographic repeats and allowed the helix to go its own way.**

You can try this yourself by photocopying Figure 4.1 onto a transparency, and then rolling it into a cylinder with axis vertical. Roll it until the oxygen atom with the arrow at lower left just overlaps the oxygen atom marked "α" at the right, with its arrow pointing toward the hydrogen. Adjust the overlap until neighboring N-H and C=O are also aligned, and fasten the cylinder with Scotch tape. The result is an α-helix! Tightening up the cylinder by one residue results in a 3_{10} helix, and further drastic tightening leads to the 2_7 helix. Short segments of both of these occur in some globular proteins. Loosening the helix by one residue creates the more open π-helix or 4.4_{16} helix, which was discovered not by Pauling and Corey but by Barbara Low (2). This has not been found in globular proteins; presumably it is too open and unstructured.

Pauling's only assumptions with the α-helix were that (a) hydrogen bonds should be reasonably straight and bridge from one turn of helix to the next, (b) all hydrogen bonds should point in the same direction along the helix, (c) every N—H and C—O should be involved in such a bond, and (d) the amide bonds between amino acids should be planar. An integral number of residues per turn was not required. The result was a helix with 3.7 amino acids per turn (now generally agreed to be 3.6, or a 100° rotation from one residue to the next), held together by H-bonded rings containing 13 atoms.

The α-helix remained unpublished in Pauling's files for the next three years. Only when the Bragg/Kendrew/Perutz paper appeared, did he write up his model and publish it along with Robert Corey and H. R. Branson, a scientific colleague at Caltech and a visiting professor, respectively. Their paper, (C), is one whose importance is difficult to overstate. The α-helix was accepted, not because it fitted Patterson maps or any other x-ray data, but because it *looked* right: regular, repetitive, with straight H-bonds of the proper length, and planar amides. In the BKP notation, the α-helix would be called a 3.6_{13} helix. The paper presented not only the α-helix but a more open 5.1_{17} or γ-helix. The latter has never been found in nature, and the Caltech team later attributed this to the fact that the γ-helix was hollow, with a central cavity running down its length, whereas the α-helix was close-packed.

Their drawings of the two helices, in Figures C.2 and 3, are curiosities. In both of them the amino acids are in the D-conformation, and the 3.6_{13} helix is left-handed. It was not yet fully appreciated, even by Pauling, that biological amino acids with rare exceptions always exhibit the L-conformation, and that the α-helix formed from them must be right-handed. So Figure C.2 is a mirror-image of reality. But it at least is possible. As you can see from his drawing, and from the unrolled Figure 4.1, each carbonyl oxygen lies on the *other* side of the ascending chain from the amino acid side group, R, on the adjacent Cα atom. Interchanging —R and —H on the alpha carbon would bring the side chain close enough to clash with the carbonyl oxygen. Hence a right-handed α-helix with L-amino acids is normal, a left-handed helix with D-amino acids (as Figure C.2) is possible although not on our planet; but a right-handed helix with D-amino acids and a left-handed helix with L- are impossible because of steric hindrance. None of this was realized in the early papers by Pauling and coworkers.

The Caltech paper undoubtedly was triggered by publication of the Cambridge results. But why had Pauling not published his α-helix in 1948? Probably because the helix does not really fit Astbury's fiber diffraction photos of α-keratin! Using the best available bond lengths and angles (most of which had been established by Corey and Pauling at Caltech from x-ray studies of individual amino acids and short peptides), their α-helix exhibited a rise per turn of helix of 5.44 Å. But the axial reflections in Astbury's photos of α-keratin indicate a repeat distance of only 5.1 Å! Why the discrepancy? What is wrong with the α-helix? This uncertainty probably made Pauling cautious about going on the record with an erroneous model. Only the threat of losing the race with Cambridge made him take the chance.

The Caltech paper reverberated throughout Cambridge. In the words of Max Perutz (D), "I was thunderstruck by Pauling and Corey's paper. In contrast to Kendrew's and my helices, theirs was free of strain; all the amide groups were planar and every carbonyl group formed a perfect hydrogen bond with an amino group four residues further along the chain. The structure looked dead right. How could I have missed it?...On the other hand, how could Pauling and Corey's helix be right, however nice it looked, if it had the wrong repeat?" Perutz went immediately into the x-ray laboratory, obtained some poly-γ-benzyl-L-glutamate and α-keratin fibers, and began collecting fiber diffraction data.

Perutz knew that a fiber photograph taken with the incoming x-ray beam perpendicular to the fiber axis, as in Figure 4.2a, only records the central part of the full

FIGURE 4.2   Because individual polypeptide strands within a protein fiber are rotationally disordered, the diffraction pattern or Fourier transform of the fiber is cylindrically symmetical about the fiber axis. A photographic plate at right, or any other collection device, does not record the entire pattern, or even a plane section through it. Instead, it records that part of the full cylindrical diffraction pattern that intersects a sphere around the crystal of radius $1/\lambda$, where λ is the wavelength of the x-rays. This is termed an Ewald diagram, after the physicist Paul Ewald who first proposed this manner of depicting diffraction around 1914. The diffraction pattern is regarded as being located, not at the fiber itself, but at the point where the direct x-ray beam exits the Ewald sphere. As Figure 2d of Chapter 2 demonstrated, features of the diffraction pattern not visible when the fiber axis is perpendicular to the x-ray beam (top) can be seen if the fiber axis is tilted (bottom). From (3).

(a)

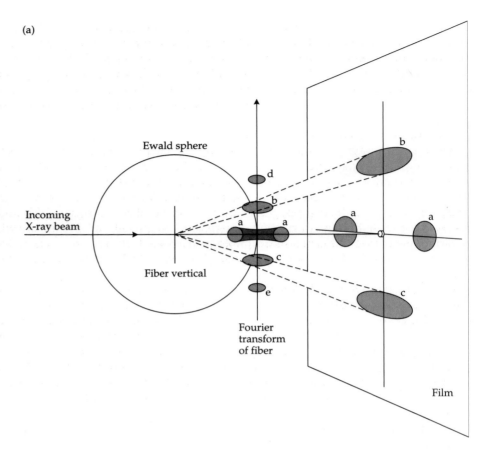

(b)

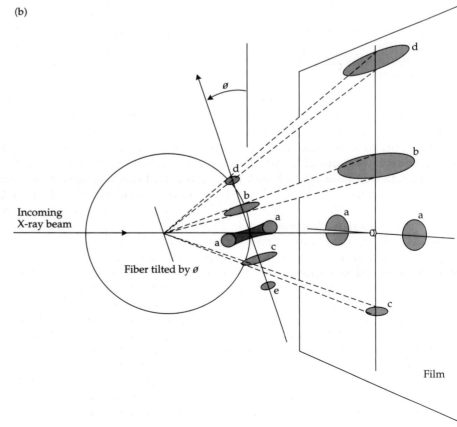

diffraction pattern. The pattern that one sees on the photographic plate is that part of the full fiber transform that intersects a sphere around the crystal with radius dependent on the x-ray wavelength. This is commonly called the "sphere of reflection" or the "Ewald sphere," after the x-ray pioneer who first worked out the geometry of diffraction. But by tilting the fiber axis relative to the x-ray beam as in Figure 4.2b, one could bring in details that were lost in the standard, perpendicular-axis setting.

If the rise per turn of α-helix was 5.44 Å, then there should be strong diffraction maxima up and down the axis of the photograph at distances corresponding to 5.44 Å. This was true for synthetic polypeptides, but for biological keratin fibers the repeat spacing was only 5.1 Å. Leaving aside this discrepancy for the moment, another natural repetition occurred along the α-helix. With 3.6 amino acid residues every turn of 5.44 Å, the rise along the axis *per residue* (not per turn) should be: h = 5.44 Å/3.6 = 1.5 Å. Because diffraction patterns are "reciprocal," a 1.5 Å feature will be farther away from the center of the photo than a 5.4 Å feature; so far, in fact, that it fails to intersect the Ewald sphere (as spot "d" in Figure 4.2a), and is not recorded in the x-ray photo.

But Perutz realized that the 1.5 Å reflection, if it existed, could be made visible by tilting the fiber axis relative to the x-ray beam in the manner shown in Figure 4.2b. He tried it with a 31° tilt, and to his excitement the 1.5 Å reflection actually was there! (See Figure 2.2d of Chapter 2). He telephoned the editor of *Nature*, told him what he had done, and asked how rapidly it might be published. The editor gave him a reply that you will not hear from *Nature* or any other journal today: "We can put it in the next issue if you get it to us right away." The result is reprinted as paper E. Linus Pauling may have proposed the α-helix, but it was Max Perutz who proved it.

As Max continues his account (D):

> **On Monday morning I stormed into Bragg's office to show him my X-ray diffraction picture. When he asked me what made me think of this crucial experiment, I told him that the idea was sparked off by my fury over having missed building that beautiful structure myself. Bragg's prompt reply was, 'I wish I had made you angry earlier!' because discovery of the 1.5 Å reflection would have led us straight to the α-helix.**

In retrospect they almost did find the α-helix. Take the 4_{13} helix of Figure B.12, twist it and tighten it until there are only 3.6 residues per chain, not 4. The hydrogen bonds then become more nearly parallel to the helix axis, all of the bond lengths and angles become canonical, and the amide linkages can become planar. You have just made an α-helix out of a fourfold helix, without making or breaking a single bond! Perutz and colleagues failed to give the 4_{13} helix any special attention because it was somewhat distorted, and there was no special reason to favor it. But the rise of 1.5 Å per residue was there. Had Perutz looked for it in the x-ray pattern, and found it, then the battle would have been won.

There is yet another more subtle reason why Cambridge missed the α-helix. Bragg, Kendrew and Perutz all were trained and experienced x-ray crystallographers, accustomed to working with single crystals, not fibers. In a true single crystal, if a helix runs through a series of identical unit cells, then it must repeat an integral number of times per turn so that each unit cell is like all the others. One can have a crystal with twofold axes, or threefold, or fourfold, but not 3.6-fold. It was their very scientific competence in single crystal structure analysis that induced them to overlook helices with nonintegral numbers of residues per turn.

The contradiction between the 5.1 Å helix pitch in α-keratin, and the 5.44 Å pitch in synthetic polypeptides and in the Pauling α-helix again was ultimately explained in Cambridge, not Pasadena. Francis Crick showed in the three 1952–3 papers listed at the head of this chapter that individual right-handed α-helices in keratin were wound in a left-handed supercoiled bundle of three, to build what he termed a coiled coil. There is a perfectly sound logical reason for this. Take apart an ordinary multi-strand rope. If the final winding is right-handed, then each of the wound strands will themselves be internally left-handed. If component fibers of this strand are themselves twisted, they will be seen to be right-handed. This alternation of helical twist direction from one level of structure to the next prevents the rope from fraying, unwinding and coming apart. We, or rather evolution, thought of this device long before ropemakers ever did. So a wool or hair fiber is not simply a bundle of parallel coiled helices; it is a coiled-coil, and the wrapping of three α-helices around one another shortens the repeat distance from 5.44 Å to 5.1 Å. No such supercoiling is present in synthetic polyptide fibers, so their repeat is the expected 5.44 Å.

Pauling and Corey followed up this triumph in 1951 with no fewer than eleven more papers proposing structures for every class of fibrous protein known. The most important of these were the parallel and antiparallel strand β-pleated sheets (F). The antiparallel sheet, Figure F.6, was an improvement of the sheet proposed by Astbury in 1934. But the use of planar amides and accurate bond lengths and angles led to a sheet that was not flat, but which had ridges like corrugated tin roofing. The side chains of the residues stuck out to either side from the crests of the ridges. Hydrogen bonds between antiparallel chains were essentially parallel. The parallel-chain sheet (Figure F.7) was new. It, too had corrugated ridges with side chains emerging from their crests, but the cross-chain hydrogen bonds were angled in a zig-zag manner.

The antiparallel structure fitted the fiber pattern for Bombyx mori silk beautifully. The distance along an extended chain from one residue to the next was around 3.34 Å, and because of the zig-zag nature of the chain, the true crystallographic repeat along a strand was twice this, or ca. 6.68 Å. (Compare the diffraction maxima up and down Figure 3b of Chapter 2, which are measured as 3.48 Å and 6.97 Å.) Adjacent strands of the helix are 4.7 Å apart, and if neighboring chains run in opposite directions, the true repeat in a direction normal to the chains would be twice this, or 9.4 Å. Both spacings are visible to the left and right of center in Figure 3b of Chapter 2. Finally, antiparallel sheets are stacked atop one another with side chains packed together. The alternating –Gly–Ala–Gly–Ala–Gly–Ala– sequence of silk places all the glycines on one side of a sheet and all the alanines on the other. When sheets are stacked, Gly against Gly and Ala against Ala, the sheet spacings are 3.5 Å at the Gly contacts and 5.7 Å at the Ala contacts, making the true repeat distance across two sheets the sum of these, or 9.2 Å. This produces a diffraction maximum which overlaps with the 9.4 Å reflection from intrasheet repetition.

The Pauling/Corey β-sheet structures fitted β fibrous protein patterns precisely, and were accepted immediately. Their collagen structure proposed was incorrect, and some of their other proposals could be faulted. But the α-helix, and the parallel and antiparallel β-pleated sheets, were triumphs.

As the previous chapter was summarized by a brief biographical sketch of its main protagonist, Dorothy Wrinch, so it is fitting to end this chapter with an evocative essay by Eisenberg about Linus Pauling's triumphs (G). Much of the essay covers the same grounds as this chapter, but the last two sections are of particular interest. He mentions the uncertainty about helix sense and amino acid configuration, and the fact that the important 3_{10} helix was never mentioned, although the unimportant γ-helix was. At the same time that Pauling was publishing his seminal work on protein chain conformations, he was also increasingly active in the public arena, arguing against war and most especially against nuclear war. This was a period when some people were actually urging a preemptive strike, nuclear if necessary, against the Soviet Union before they developed their own Weapons of Mass Destruction. As Eisenberg states,

On the day after Pauling and Corey submitted their seven protein papers for publication, the House Un-American Activities Committee named Pauling one of the foremost Americans involved in a "Campaign to Disarm and Defeat the United States". The press release read: His whole record....indicates that Dr. Linus Pauling is primarily engrossed in placing his scientific attainments at the service of a host of organizations which have in common their complete subservience to the Communist Party of the USA, and the Soviet Union.

It is not at all clear how Pauling's chemical bond theories, planar amides and α-helices would have contributed to Soviet WMD's. But it serves to illustrate the truism, still valid, that stupidity is eternal!

References

1. A. Serafini. 1989. *Linus Pauling: A Man and His Science*. Paragon House, New York, p. 131.
2. B. W. Low and H. J. Grenville-Wells. 1953. *Proc. Natl. Acad. Sci. USA* 39, 785–802. "Generalized Mathematical Relationships for Polypeptide Chain Helices. The Coordinates of the Pi Helix."
3. R. E. Dickerson. 1964. In *The Proteins* (Hans Neurath, ed.). Academic Press, New York, 2nd edition, Vol. 2, pp. 603–778. "X-Ray Analysis and Protein Structure."

Study Questions

1. Why did J. D. Bernal's group find it impossible to get good diffraction patterns from single crystals of globular proteins unless they first sealed the crystals in glass capillaries (A)?
2. The modern alternative to this is to freeze crystals rapidly and then collect diffraction data at low temperatures. How does this accomplish the same goal?
3. In 1950 Bragg, Kendrew and Perutz (B) intended to write the definitive treatise on all possible helical polypeptide conformations. Why was it that their own competence and experience in x-ray crystal structure analysis caused them to miss the alpha helix entirely (G)?
4. Why was the rhinovirus of inestimable value to Linus Pauling in 1948 while he was visiting in Oxford (A)? What two ideas, neither of which were realized by Bragg and coworkers, allowed Pauling to succeed where they had failed (C)?
5. How did Max Perutz in Cambridge prove that Pauling was correct using fibers of α-keratin (D, E)? How rapidly did Perutz publish his proof? [Use detective work and arrive at an upper time limit (C–E).]
6. If the alpha helix was such a brilliant idea, why did Pauling wait three years before publishing it? What was he worried about? What prompted him finally to publish it in 1951?
7. Which of the structures in Bragg, Kendrew and Perutz (B) can be converted into an alpha helix without making or breaking a single bond, and what must one do to convert it?
8. What was the explanation for the discrepancy in the pitch of the alpha helix that had worried Pauling? Who, in Cambridge, finally explained it correctly?
9. Why are we confident that all alpha helices are right-handed, at least in life on this planet? What would it take for life to use left-handed alpha helices on some other planet? How do we know that the alpha helix in Figure 2 of paper (C) must come from an extraterrestrial?
10. Why are the parallel and antiparallel beta sheets called "pleated" sheets (F)?

Crystallographic Measurements and the Structure of Protein Molecules as They Are

DOROTHY CROWFOOT HODGKIN

Chemical Crystallography Laboratory
Oxford, OX1 3PD
England

IN THE HISTORY of the study of the structure of proteins by x-ray crystallographic analysis there are three dates that stand out as of special importance:

(1) 1912, when the first x-ray diffraction patterns of any crystal, copper sulphate in fact, was obtained by Laue, Friedrich, and Knipping;

(2) 1934, when J. D. Bernal recorded the first x-ray diffraction pattern of a protein crystal, pepsin, in its mother liquor;

(3) 1953, when Max Perutz took an x-ray photograph of the mercury benzoyl derivative of horse methemoglobin and observed measurable and interpretable changes in the intensities of the x-ray spectra compared with data from the unmodified protein.

These dates have a common characteristic—they are dates of great promise on our way towards the vision of the actual arrangement of the atoms in protein molecules.

In this paper, I am principally concerned with the period between 1934 and 1953. I propose, however, to begin with a brief account of the history and prehistory of protein crystallography that determined our thinking in the thirties and forties.

PREHISTORY—EARLY OBSERVATIONS ON PROTEIN CRYSTALLOGRAPHY

During the nineteenth and early twentieth century a number of crystals were observed in animal and plant cells under the light microscope and were identified as protein in nature. The earliest of these were probably hemoglobin crystals seen by Baumgaertner in 1830.[1] Hünefeld, in 1840,[2] certainly described hemoglobin crystals growing in a drop of blood from the earthworm between glass plates (Barbara Low and I repeated his experiment in Oxford rather more than a hundred years later). K. E. Reichert in 1849[3] observed tetrahedral crystals growing in the fetal membranes of a guinea pig, six hours after its death. Detailed studies followed, notably by Preyer.[4] By 1909 a great volume of 600

0077-8923/79/0325-0121 $01.75/0 © 1979, NYAS

microphotographs of hemoglobin crystals from 106 species had been collected by E. T. Reichert and A. P. Brown.[5] A number of other protein crystals were also observed. Cohn, "Über Proteinkrystalle in den Kartoffeln," in 1860,[6] described experiments on sharp edged, cubic crystals, which split in two in the presence of water and were readily permeable to liquids and dyes. Ammonia dissolved them from outside in, acetic acid from inside out, leaving a hole in the center. The morphology of these crystals was described by A. F. W. Schimper in a summary paper in the *Zeitschrift für Kristallographie* in 1880[7] together with others, which appear later in protein crystal history, such as excelsin, for example:

1. Art.—Typus: Krystalloide der Paranuss.*

   Optisch einaxige Krystalloide von hexagonal rhomboëdrisch-hemiëdrischer Symmetrie.

   Axenverhältniss: a:c = 1:2.4
   Auftretende Formen: $R(10\bar{1}1), -\frac{1}{2}R(01\bar{1}2), oR(0001)$.

   Doppelbrechung: +, sehr schwach.

In a number of experiments Schimper demonstrated swelling and shrinking of the crystals in water, acids or alkalis. He found that with excelsin swelling does not occur normal to the principal axis, e.g.,

	vor.	nach.	auf 1 bez.
Gr. diag.	8	12	1.5
Kl. diag.	5	5	1.0

He observed that the crystals from *Musa hillii*, when dry, showed almost no birefringence between crossed nicols but brightened when brought near water, passing through the different birefringence colors, violet, orange, yellow, blue-green, green.

That the molecules in these crystals were large and that this conferred peculiar properties on them was realized. Preyer had estimated the molecular weight of hemoglobin as 13,000 in 1871 from the iron analysis (later corrected by Zinoffsky[8]). Schimper comments on variations in metal content observed in crystals of the same form: "Ist es sehr möglich dass in Molekülen, die wie die jenigen der Eiweisskörper aus vielen Hunderten von Atomen bestehen, einige Affinitäten auf sehr verschiedene Weise gesättigt werden können, ohne dass die Krystallform eine wesentlich andere werde." It was also generally the opinion in the late nineteenth and early twentieth century that the packing units in crystals were chemical molecules and the types of arrangement they might have were in process of being worked out in space group theory. But nothing more

---

* *Bertholletia excelsa.*

precise could be found out about protein crystals until x-ray diffraction had been discovered.

1912–34. THE DISCOVERY OF X-RAY DIFFRACTION OF CRYSTALS: FIRST APPLICATIONS TO THE STUDY OF PROTEINS

Any time after the discovery of x-rays in 1895 it would have been possible to discover x-ray diffraction: Röntgen himself tried an experiment of passing a beam of x-rays through a crystal of calcite but the beam was too weak, the crystal too thick for the observation of diffraction effects. The successful experiment in 1912[9] came about almost by accident, through a series of conversations between Paul Ewald, M. v. Laue, W. Friedrich, and P. Knipping in Munich, all physicists, and quite young, working in the laboratories of Sommerfeld and Röntgen; a full account has been given of the circumstance in *Fifty Years of X-Ray Diffraction*.[10] The diffraction pattern given when x-rays were passed through copper sulphate, the first crystal examined, was too complicated to understand and the investigators turned to zinc blende, where at least it was possible to correlate the cubic crystal symmetry with that of the diffraction pattern. Friedrich then carried out a series of experiments, passing x-rays through other crystals and through partially ordered and amorphous materials such as beeswax and paraffin.[11] Within a few months, the experiments had been repeated in England and in Japan and W. L. Bragg in England had solved the first crystal structures by measuring and comparing the diffraction effects from sodium and potassium chloride.[12]

Professor Herman Mark, who took part in the early measurements on fiber structures has written for me the following account of the next stage, the experiments in Japan and afterwards in Germany, which it seems worth quoting in some detail:

Already in 1913—only one year after the Bragg-v. Laue discovery—Nishikawa and Ono[13] in Japan irradiated a randomly selected number of fibers with x-rays; natural silk was amongst them (and I think also hair).† They got a few broad spots and concluded that there must be some kind of molecular order in these materials. Of course, their samples were not properly purified and their x-ray tube was primitive (1913!); it was a *pioneering* effort and was not followed up. Next—quite a few years later —in 1920 Professor R. D. Herzog, Director of the Kaiser Wilhelm Institute for Faserstoffchemie in Berlin-Dahlem, decided that he would initiate a *systematic* study of fibers with x-rays in his Institute. In 1920[14]

† Silk, wood, bamboo, and asa (*Cannabis sativa*).—D.H.

124 ANNALS NEW YORK ACADEMY OF SCIENCES

he published a paper on several fibers—including silk—together with W. Jancke. As a result of better sample preparation and improved x-ray equipment they obtained "fiber diagrams" of cellulose and silk. They concluded that there is a "crystalline" component in these fibers but made no attempt at a quantitative evaluation. In 1921 another coworker of Herzog, Michael Polanyi, evaluated the fiber diagram of cellulose more numerically and determined its crystallographic basis cell.[15] This encouraged Herzog to attempt the same step with natural silk, which, of all proteinic material, gave at that time the clearest diagrams [FIGURE 1]. At his request, Polanyi asked a graduate student—Rudolf Brill—to prepare well-oriented and purified samples and to get as good x-ray photos as possible for a quantitative study. At that time (1922) I was also working, as a postgraduate, at Herzog's Institute. Rudi and I worked together in a makeshift laboratory but had—for that time— rather good x-ray tubes, which would give some 10 mA at about 40 kV for several hours. Rudi worked on silk and hair, I studied metal wires (single crystals) and cellulosics. When we got reasonable looking patterns we would show them to Polanyi and Weissenberg (both so well known and liked!) and they would explain to us what it all meant and how we should evaluate the pictures; we had a wonderful time!

After a year Rudi had finished his work and published it as his Ph.D. thesis.[16] He determined the basis cell (correctly) and concluded that there were 2 possible "structures" for silk fibroin—either a small glycyl-

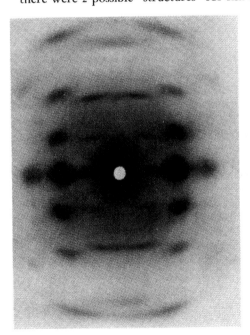

FIGURE 1. X-ray photograph of silk fibroin from *Bombyx mori* taken by Astbury.

D-alanine peptide *or* a long polypeptide chain. This *alternative* was, of course, the *last* the chemists wanted to hear because it was the time of the controversy between small aggregates and macromolecules; but, as in the case of cellulose (Polanyi) the x-ray diagrams of these days were not yet good enough for a more thorough analysis (space group, intensities, etc.).

A year later Dr. Brill took a position in the Ammoniak Laboratory of the IG Farben in Oppau where he studied mainly catalytically active inorganic materials. Again, a year later I joined the Hauptlaboratory of the IG Farben in Ludwigshafen. The Director of the laboratory was Professor K. H. Meyer, a very distinguished scientist, who felt that another, more quantitative study of the structure of fibers should be initiated. At that time Dr. G. von Susich had rather elaborate equipment —x-ray tubes, cameras, etc.—and Dr. H. Hopff helped us in the preparation of highly purified and oriented samples of "bombyx mori." Some diagrams were good enough to determine the space group and to carry out a preliminary intensity evaluation. Using the Bragg atomic radii, K. H. Meyer and I proposed a chain structure for silk fibroin, which was published in 1928[17] [FIGURE 2].

Meyer and Mark actually suggested two alternative unit cells and chain arrangements for silk fibroin consistent with their observations and different probable space groups. They did not attempt to place the atoms

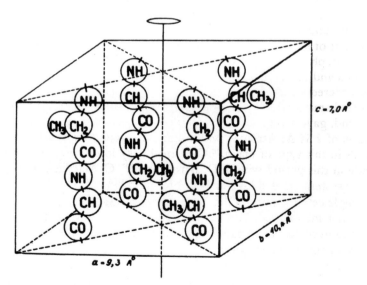

FIGURE 2. Extended peptide chain arrangement (one version) proposed by Meyer and Mark for silk fibroin in 1928. (From Meyer & Mark.[17])

126   ANNALS NEW YORK ACADEMY OF SCIENCES

precisely owing to the limited data given by the photographs. They
assumed that regularity, such as it was, occurred in limited regions within
micelles between which the main chain passed in a less orderly fashion.
(Compare Weissenberg's theories in 1926.[18]) And among the "Bragg
radii" they used were those for oxygen and nitrogen, partly derived from
the first organic crystals analyzed, hexamethylene tetramine by Gonell
and Mark[19] and by Dickinson and Raymond,[20] and urea by S. B. Hendricks.[21]

The observations of the Berlin school were known to W. H. Bragg.
After his Christmas lectures at the Royal Institution on "Old Trades and
New Knowledge" in 1924 he planned a new lecture, "The Imperfect
Crystallisation of Common Things" in 1926. He asked W. T. Astbury to
take some photographs for him of fibers like wool and hair (first recorded
earlier by Herzog and Jancke). Astbury (as Bernal recorded) took to
them from the beginning; they led to his first formal appointment as
lecturer in Textile Physics at Leeds in 1928 and became his life's work.
With H. J. Woods and A. Street, he discovered that wool gave two
types of x-ray photograph, α when unstretched, β when stretched.[22, 23]
He deduced that the α photograph, which gave only two features, an
equatorial "reflection" of spacing 10 Å and a meridional "reflection" of
spacing 5.1 Å, represented some type of folded peptide chain that could
be reversibly stretched to a straight chain, giving the β photograph. The
β photograph, which showed two equatorial reflections of 10 Å and
4.65 Å and one meridional one, 3.4 Å, clearly, to his eyes, was the same
type as that of silk fibroin. As the years passed, he found the same types
of photograph given by many proteins, keratin, myosin, epidermin,
fibrinogen and called them the "kmef" group. Occasionally much more
highly ordered structures giving many reflections, were observed, such
as feather keratin, which still conformed to this type.[24] Collagen, on the
other hand, gave a very different kind of photograph with a meridional
reflection of 2.84 Å; Astbury, with others, speculated on possible different folds in this type of chain.[66] One small very interesting observation
he made in this period with J. D. Bernal and T. C. Marwick in a quite
different system, guided his thinking later. He showed that the cell wall
of the single cell organism valonia ventricosa, the sea grape, gave an x-ray
pattern that indicated cellulose chains crossed one another at an angle
that varied over the cell surface, suggesting the chains were wound over
the surface as over a ball of wool.[25]

The x-ray photographs of the natural fibers were too limited in themselves to give precise information about the atomic arrangement in fibrous
proteins (FIGURE 1). J. D. Bernal began a first paper (1931)[26] on the
crystal structures of the natural amino acids and related compounds with

the words "a knowledge of the crystal structure of the amino acids is essential for the interpretation of the x-ray photographs of animal materials: silk fibroin, keratin, collagen, proteins, etc., which have been studied by this method for the first time in the last few years." He gave preliminary x-ray data and some guesses about the structures of the crystals of 15 substances kindly prepared for him by Dr. A. Leese at the Biochemistry Laboratory, Cambridge. He ended with a request for more crystals of related compounds, of 0.01 mm or over in size, for extended examination. In a further paper he gave some x-ray data on the very interesting crystals of cuprous glutathione prepared by N. W. Pirie in Cambridge.[27]

It was clear that detailed x-ray analysis of the kind needed was likely to be slow even for such simple crystals. But new ideas of how to proceed were being developed with the use of Fourier analysis, as Bernal mentioned. In 1928 B. Warren and W. L. Bragg published the crystal structure of the feldspar diopside,[28] a structure involving the determination of 27 parameters to place 10 atoms. They used a marvelous series of intricate arguments involving the relative scattering contributions of the different atoms to the different observed intensities of the x-ray spectra. With amplitudes derived from their measured intensities and phase constants from their found atomic positions, Bragg calculated a projection of the electron density on the main crystal plane (010). Not till very long afterwards did any one notice that most of the phase constants, 0° and 180°, were here determined by the calcium ion contribution alone. The history of the introduction of the direct approach to crystal solving through the use of Fourier series in electron density calculations, which W. L. Bragg and West discussed in relation to this structure, is a little odd. W. H. Bragg (W. L. Bragg's father) first suggested the application of Fourier methods in 1915.[29] In a theoretical paper by Duane in 1925,[30] followed by Havighurst's practical demonstration on sodium chloride,[31] it was pointed out that in the alkali halide structures, the chlorine ion contributions determined the phase constants necessary for the calculation, here signs, plus or minus; in 1927 Cork[32] demonstrated with the alums that the same information could be derived through the study of isomorphous crystals. Yet, as first, the electron density calculation was used mainly in the representation and refinement of crystal structures as with diopside or the first organic crystal seen in an electron density map, hexachlorbenzene (Lonsdale[33]). W. L. Bragg used to quote the structure of copper sulphate found by Beevers and Lipson in 1934 as one of the first crystal structures actually solved by the "direct" Fourier approach.[34] ‡

‡ In this analysis, Beevers and Lipson also devised "strips" we all used in our first calculations of Patterson and electron density maps for proteins.

As for the protein crystals themselves, every one who worked seriously on proteins in this period knew they existed. Their number had been added to very dramatically in the twenties by the first crystallization of several enzymes by Sumner, Northrop, and Kunitz; pepsin, trypsin, and chymotrypsin and others, and of insulin by J. J. Abel. There were many attempts by crystallographers to obtain x-ray photographs of protein crystals in this period, e.g., by W. H. George[35] and by Clark and Corrigan[36] on insulin (Clark and Corrigan recorded some difficult-to-interpret reflections with long spacings), and A. L. Patterson on hemoglobin. All observed almost nothing but vague blurs for reasons that became obvious in 1934.

1934–53. THE FIRST X-RAY CRYSTALLOGRAPHIC MEASUREMENTS ON SINGLE PROTEIN CRYSTALS AND RELATED STUDIES

As with the discovery of x-ray diffraction itself, there were both accidents and purposes contributing to the taking of the first x-ray photographs of protein crystals by J. D. Bernal in the spring of 1934. John Philpot was working at that time on the purification of pepsin at Uppsala; a preparation he left standing in a fridge while he was away skiing produced very large—2 mm long—and beautiful crystals of the hexagonal bipyramidal form obtained earlier by Northrop and Kunitz. They were seen by Glen Millikan, a passing visitor and friend of Bernal's, who knew of his interest in proteins and his appeal for crystals. John Philpot gave Millikan a tube of crystals in their mother liquor to take back to Cambridge.

Bernal had an old fashioned classical Cambridge education in mineralogy and a very large petrographic microscope§ with rotating nicols, which made it easy for him to see the crystals in the tube in which they grew and that they were moderately birefringent. As Schimper had done for the seed globulins before him, he recorded the axial ratio, $c/a = 2.3 \pm 0.1$, and the birefringence, positive, uniaxial. He first took a crystal out of its mother liquor, noticing as he did that the birefringence fell. He mounted it and took an x-ray photograph, which showed nothing but vague blackening. He then thought of drawing crystals in their mother liquor into thin walled capillary tubes of Lindemann glass—fortunately available in the laboratory since he and Helen Megaw had been growing ice crystals in them to study the expansion of ice. The following photographs, taken on a 3 cm radius cylindrical camera in a series of 5° oscillations, showed hundreds of x-ray reflections, rather large, corresponding

§ Now on the working bench of Max Perutz.

with the size of the crystals, and noticeably even in intensity distribution.[37] The a axis was easily measured as ~67 Å but the reflections defining the long c axis were not completely resolved owing to the large size of the crystals, the experimental conditions used, and some disorder that introduced smearing. The c axis was recorded as $n \times 154$ Å—a value derived from the axial ratio. Much later experiments by Max Perutz (1949)[38] and very recently by Tom Blundell at Birkbeck have provided accurate data for the pepsin crystal lattice constants, $a = 67.9$ Å, $c = 292$ Å, space group $C6_122$, $n = 12$. Tom Blundell's photographs are shown in FIGURE 3. Since no one can now find them, it seems almost certain that the original pepsin photographs were destroyed in the bombing of Birkbeck College during the war.

The large lattice constants observed for the pepsin crystals were in general agreement with the magnitude of the molecular weight of the pepsin molecule, about 40,000, derived from ultracentrifuge measurements. The unit cell dimensions suggested a packing of oblate spheroidal molecules, 25 Å × 35 Å. The communication to *Nature* comments on the intensity distribution, "From the intensity of the spots near the centre we can infer that the protein molecules are relatively dense globular bodies, perhaps joined together by valency bridges but in any event separated by relatively large spaces which contain water. From the intensity of the more distant spots it can be inferred that the arrangement of atoms inside the protein molecule is also of a perfectly definite kind, though without the periodicities characterising the fibrous proteins." The latter remark very much worried Astbury who had obtained diffuse reflections at $4\frac{1}{2}$ Å and 10 Å spacing, β keratin-like, from dried crystalline pepsin powder[39] given him by Northrop. For a moment it suggested very radical changes might occur on drying a protein crystal, polymerization between the molecules to form chains.

The proper exploitation of the x-ray photography of pepsin was too large an undertaking for Bernal's very small research group, heavily otherwise engaged in the summer of 1934. But the observations led immediately to various events important in protein crystallography. Professor Herman Mark visited Bernal's laboratory and saw the photographs. He was so excited he forgot to arrange, as intended, for his research student, Max Perutz, to work with Hopkins and arranged instead for him to work with Bernal. Bernal lectured in Manchester that summer about his results and I. Fankuchen heard him and asked after the lecture if he might work with him. And after I had returned to a research fellowship in Oxford in the autumn, Professor Robert Robinson put a sample of insulin crystals he had been given into my hands.

It is interesting that the observations on pepsin crystals were not

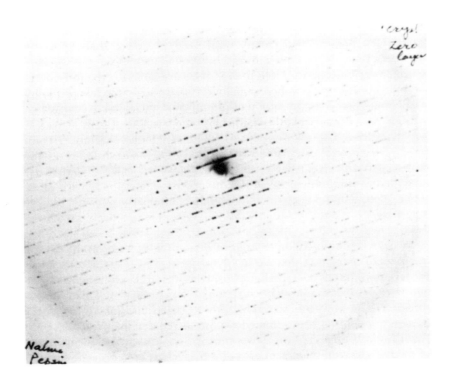

FIGURE 3. (Top) Precession photograph of pepsin taken in 1978. (Bottom) Crystals of pepsin, 1978. (From T. L. Blundell.)

immediately taken up in other countries although at the time there was considerable interest both in proteins and in x-ray diffraction. The research workers most immediately implicated all turned their efforts in other directions. A. L. Patterson had earlier (1930) considered working on protein crystals but decided the methods available for the interpretation of x-ray diffraction phenomena were inadequate. He left the Johnson Foundation where he was working in 1931 and took a year off on his own resources—which extended to two—to investigate at M.I.T. the properties and uses of Fourier series. Out of this came the understanding of Fourier series calculated with the squared amplitudes of the x-ray spectra as coefficients. These series, which we all call by Patterson's name, define distributions in which the density corresponds with interatomic vectors in the crystal.[40] No protein crystal structures could have been solved without their use. R. W. G. Wyckoff had been introduced to x-ray analysis and space group theory by Nishikawa in 1918 and had later worked specifically on crystals of biological interest, on the crystal structures of urea and thiourea. In 1936 with R. B. Corey he took x-ray powder photographs of a number of protein crystals kept in their mother liquor[41] and also of Stanley's preparation of "crystalline" tobacco mosaic proteins.[42] The x-ray photographs showed sharp lines of long spacing but clearly could be of no immediate use in the detailed study of proteins. Wyckoff turned to the development first of small ultracentrifuges and then of electron microscopes—with marvelous effect on the separation of individual proteins and observation of their external shapes. Corey was invited to Pasadena by Linus Pauling, to work on the exact crystal structures of the amino acid constituents of proteins, more immediately amenable to x-ray analytical methods than the protein crystals themselves. His work and that of E. W. Hughes and their students established the structures of diketopiperazine, glycine, alanine, and diglycylglycine, which provided evidence extremely helpful in the eventual understanding of protein structures.

Bernal and Fankuchen were themselves diverted from the main line of protein crystallography by the preparation of the tobacco mosaic virus given to them by Bawden and Pirie (1936).[43] They developed techniques for producing very fine intense monochromatic x-ray beams and studied the diffraction effects obtained from the virus preparations under a variety of conditions, in solutions of varying concentration, salt concentration, and pH, still or flowing. The strong lines in their patterns were similar to those found by Wyckoff and Corey. They showed it was possible to separate a wide angle pattern, which stayed unchanged through a number of concentration variations, from small angle patterns, which varied with concentration. The wide angle pattern appeared very like fiber patterns of the complexity of that obtained from feather keratin; it clearly indicated structure within the virus particles themselves. From the small angle pattern and the relations between the different liquid crystal states it was possible to derive the shape and approximate size of the virus particles; they were clearly rod shaped and 150 Å across; rough estimates of their lengths as about ten times their width agreed well with the value 1500 Å shown in the first electron microscope pictures. From the wide angle pattern the authors concluded that the particle had a structure not unlike that of a small protein crystal with a spiral close packed arrangement of subunits $11 \times 11 \times 11$ Å, rather smaller than the units they thought existed in protein molecules, "in some ways simpler." The paper in which they recorded their many fascinating observations on the organization of the virus particles in different liquid crystalline states, positive and negative tactoids, isotropic and anisotropic gels, was written just before the war and published in the *Journal of General Physiology* in 1941,[44] after I. Fankuchen had returned to the United States. (Bernal passed over the final preparation of the text to me and the drawing of some of the illustrations.)

Before war broke out in 1939 single crystal x-ray photographs were taken of five different protein crystals, insulin (1935, 1938, 1939)[45-47]; excelsin (1936)[48]; lactoglobulin (1938, two forms)[49]; hemoglobin (1938)[50]; and chymotrypsin (1938).[50]

The insulin crystals first photographed were rhombohedral in form, similar to the first variety obtained by J. J. Abel in 1926. They were grown from 10 mg of a microcrystalline sample of Boots insulin, following D. A. Scott's prescription (1934) from a phosphate buffer with added zinc. To grow the crystals large enough, the solution was warmed to 60°, brought to a pH of 6.2, and cooled slowly over three days. Since the original air-dried crystals were brightly birefringent, the crystals were dried and photographed. Though this was formally a step back, it did make it easier to attempt to measure the molecular weight of the protein; it also established that it was possible to get interpretable, if limited, x-ray data from dry crystals (59 reflections in fact), provided the crystals were large enough and the drying slow. The first measurements posed problems that have continued with protein x-ray molecular weights. The unit cell weight of 39,700, with rather wide limits of error on account of the limited dry crystal x-ray data, should crystallographically contain $3n$ protein units and water. The first measured water content, 5.35%, was too low and hence led to rather too high a figure (37,600) for protein in the unit cell compared with 35,100 measured by Svedberg and Sjögren in solution. The correction was made as protein chemical analyses themselves improved. It was Du Vigneaud who first observed during the sulphur analysis of insulin that the water content of crystalline insulin tended to be measured too low unless special precautions were taken: Chibnall provided an accurate measurement, 10.1% of the insulin "dry" crystal water content in 1942, which enabled the protein molecular weight to be corrected to $35,330/n$. By this time it seemed fairly clear from osmotic pressure measurements that n was 3, and the true molecular weight of insulin was ~12,000. But there were indications from the very beginning of a lower figure, about 6000—crystallographically measured as 5,888—though this was not established until Sanger's work on the sequence many years later. A further worry in the early 1940's was that the preferred ultracentrifuge overall molecular weight, from sedimentation-diffusion measurements, was higher than the old Svedberg sedimentation equilibrium value. So that a new discrepancy appeared between the crystallographic and centrifuge figures, in the opposite direction from the old one. This applied also to some other proteins, such as lactoglobulin. (In retrospect, the original crystallographic measurement with insulin can be seen as fitting with the sequence value, $5,776 \times 6$, or 34,656 within the estimated limits of error.)

The next crystal to be x-ray photographed, excelsin, also suffered from drying, at least in part, and this Astbury always thought was very lucky.[48] Kenneth Bailey prepared, by different methods, a number of excelsin crystals, rhombohedral in general habit. One large one was balanced on the tip of a glass capillary tube, partly immersed in water On the oscillation photograph taken with x-rays along the threefold axis. a pattern of sharp crystalline reflections appeared, together with a strong fiber pattern showing reflections at 11.4 and 4.55 Å, repeating around the three-fold axis. This pattern occurred whether the crystal was moving or not. It seemed to Astbury that the peptide chains in the protein molecules must collapse on crystal drying into an arrangement very much like that suggested for fibrous proteins such as silk fibroin.

Lactoglobulin, given to Crowfoot and Riley by R. A. Kekwick, was the first protein to be photographed both wet and dry and in two different crystalline modifications.[49] ¶ Horse methemoglobin, prepared by Adair, and chymotrypsin, prepared by Northrop, were photographed soon after by Bernal, Fankuchen, and Perutz[50]; the three proteins were published together with a little additional note by Crowfoot and Fankuchen[51] on the possibility of interpreting photographs of air-dried cubic crystals of a tobacco seed globulin (contrary to our views of what ought to be done, but we could not get large wet crystals). Zinc insulin crystals were prepared and photographed wet a little later (Crowfoot and Riley,

¶ There is a small error in the first measurements, corrected later.[53]

1939)[44] and added to our list. Very slowly during the war and at a gradually increasing pace afterwards x-ray data on new protein crystals were added: ribonuclease 1941,[52] ferritin and apoferritin 1943,[53] the tobacco necrosis virus derivative 1945,[54] lysozyme,[55] tomato bushy stunt virus,[56] myoglobin,[57] lactoglobulin (new measurements), all 1948,[58] turnip yellow virus with and without nucleic acid (1948),[59] and six new varieties of hemoglobin crystals between 1946–48.[60] John Kendrew's list of 1953[61] includes x-ray data on 55 protein crystals, but one should add perhaps that twelve of these are myoglobins of different species measured by John Kendrew and seventeen are different varieties of hemoglobin, all but two of which[62,63] were measured by Max Perutz and his students.

From the taking of the first pepsin photographs the problem before us was clear. How could the thousands of observable x-ray spectra be used in practice to give us a view of the electron density in protein crystals? And from the beginning this question was posed in the form, how could the appropriate phase constants be directly determined? In Bernal's laboratory in 1934 there were a number of much smaller structures being investigated, particularly the sterols, where exactly the same questions were being asked. In 1932, Bernal obtained crystals of thallium sodium tartrate, isomorphous with Rochelle salt, with a view to getting one of us students to practice the structure determination of isomorphous crystals; the project eventually passed over to Beevers. When Bernal received the news of the x-ray photography of the insulin crystals, he looked up D. A. Scott's papers himself and wrote a quick letter to me (dated 1935 by the address), noting that cadmium as well as zinc could be used in the crystallization of insulin and promising to get cadmium insulin from one of his Cambridge friends (Chibnall?), which he did. The calculations I made on the difference in the scattering contributions of zinc and cadmium and one abortive trial on the photography of cadmium insulin made me pessimistic about the use of this particular pair as isomorphous derivatives.** But I did spend some time trying to obtain iodine-substituted insulin crystals, thinking a heavier atom might be easier to detect, particularly from Dr. L. Reiner of Burroughs Welcome who had made iodo-benzene-azo insulin. None of the iodine-containing crystals proved useful—too little iodine was present distributed over too

---

** In the abortive trial of cadmium and zinc insulin I compared powder photographs of wet crystals (which I realized were inadequate) because I found it impossible for a time to grow large enough single crystals. My first letter to Bernal about my difficulties in growing cadmium insulin crystals is dated March 15, 1935. Probably the preparations were not pure enough. Much later, in 1942, Chibnall offered me new cadmium insulin, but I was beginning work on penicillin and I put it aside for very much later use, as things turned out.

many different sites and the crystals were very small. In the meantime the use of isomorphous derivatives to solve a number of very much simpler structures was investigated; as a result, J. M. Robertson pointed out in 1939 that mercury in an insulin crystal could be as effective in phase determining as nickel in phthalocyanin.[64] My own trials with cholesteryl chloride and bromide were less promising since the crystals proved not sufficiently isomorphous. I tried out the calculations while photographing insulin at the Royal Institution in 1935.

In 1938 a possible new approach to the phase problem was raised by further theoretical ideas about the treatment of x-ray diffraction and the actual properties, as observed, of the wet protein crystals. Formally the scattering factor of a single molecule is a continuous function in which the succeeding changes of amplitude and phase might visually be traced, could it be observed alone. In practice, the study of x-ray diffraction effects from crystals confined the observation of the transform to intervals in diffraction space defined by the crystal lattice constants. Bernal, Fankuchen, and Perutz in their paper about hemoglobin and chymotrypsin in 1938[50] said, "As can be seen from Figure 2 [of Reference 50], the dried crystals of chymotrypsin show not only alterations of spacing but also of relative intensities of reflections. If we assume that drying takes place by the removal of water from between protein molecules, studies of these changes provide an opportunity of separating the effects of inter- and intra-molecular scattering. This may make possible the direct Fourier analysis of the molecular structure, once complete sets of reflections are available in different states of hydration."

The idea that in protein crystals the molecules were rigid and remained internally unchanged, simply moving relatively when the water content of the crystal changed (a necessary condition for the application of this method), was encouraged by the calculation of Patterson vector distributions for wet and dry insulin crystals and for hemoglobin crystals in different shrinkage stages. The maps showed very complicated distributions of peaks that included strong concentrations at 4.5–5 A and 11 Å (FIGURE 4), suggesting some form of chain packing within the molecules as Astbury hoped. but impossible to interpret in detail[53] (though a number of attempts were made). The peak pattern relative to the origin, however, remained unchanged on drying or shrinking the crystals and encouraged the attempt to trace the molecular contribution as such to the x-ray scattering.

It was Max Perutz's major work for the next ten years of his life to test these ideas within the framework of the crystal structure of horse methemoglobin, and particularly in relation to the centrosymmetrical projection based on the h0l reflections where the phase constants re-

136   ANNALS NEW YORK ACADEMY OF SCIENCES

FIGURE 4. Section in three-dimensional Patterson distribution for wet insulin at ∼ 3.0-Å resolution, showing 5-Å vectors (calculated in 1940 by D. Crowfoot).

duced to signs.[65] He was interrupted by internment at the beginning of the war, and later by work on Habbakuk in Canada,†† but by 1942 he had obtained a series of shrinkage states of the crystals that made it possible to narrow down the solution of the sign contributions for the 001 reflections of hemoglobin from 64 to 8. Further studies of salt-free crystals make it possible to limit the alternatives to two, one of which gave a very improbable solution. So he calculated a first one dimensional projection of all the electron density in the hemoglobin molecule. I found a letter from W. L. Bragg in my files, dated 3rd August 1942, trying to stimulate me into further action on insulin in which he says, "I have been very interested in Perutz's latest work, and light seems to be beginning to break in the case of hemoglobin." In the following years Perutz was able to trace the relative sign relations along parallel lines of

†† Habbakuk was a gigantic aircraft carrier, made of ice.

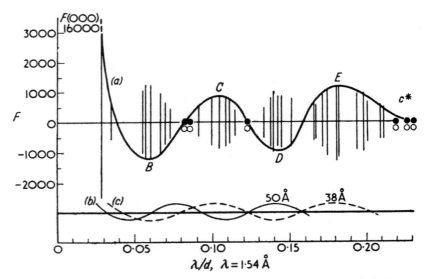

FIGURE 5. The tracing of the molecular transform of hemoglobin in one dimension from the measurement of the (001) reflections (vertical lines) at different shrinkage stages of horse methemoglobin. (After Bragg and Perutz.[65])

h01 reflections but the crystal shrinkage was in one direction only (see excelsin), which made it extremely difficult to interrelate the phase relations of reflections in different row lines.[61] In retrospect, the researches are of great interest both in the light they throw on the behavior of water between protein molecules in various states of humidity and on the possibility of limiting from the observations the overall shape of hemoglobin.[77] But their major importance historically is that they required for their achievement the measurement of the intensities of the x-ray intensities with the highest possible accuracy, preferably on an absolute scale (FIGURES 5–7).

Throughout the 1930s and 1940s there were many attempts to devise model structures for protein molecules with varying degrees of detail and precision as knowledge of the structures of amino acids and peptides increased.[66, 67] In the early years speculation was very free, even the presence of peptide chains was questioned by some. The most complete theory proposed, the cyclol theory of Dorothy Wrinch.[68] in which peptide chains were further condensed to form fabrics had for a time a stimulating effect on both protein and crystallographic research though it soon appeared to be untenable in detail. On the other hand it was also

138 ANNALS NEW YORK ACADEMY OF SCIENCES

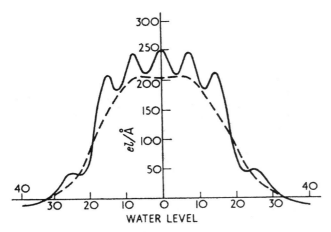

FIGURE 6. Electron density in hemoglobin projected on $\underline{c}^*$ axis. (After Bragg & Perutz.[65])

shown from the general characteristics of the Patterson distributions that protein molecules could not contain even approximately identical peptide chains in simple close packing. A serious reexamination of the probable geometry of peptide chain folding and hydrogen bond formation occurred in a number of laboratories immediately after the war. It led Pauling, Branson, and Corey, using all the accurate x-ray analyses of the Pasadena school to propose precise pleated sheet models for the atomic arrangement in β folded chains and a wholly new helical model, the α helix, for α folded chains in protein molecules.[69] It led also, gradually, to the solution of a number of other fiber structures, collagen among proteins, and DNA.

There were a number of other historically interesting developments in the immediate postwar period. A quite new approach to the problem of phase determination was raised by observations by D. Harker and J. Kasper[70] on mathematical equations relating the magnitudes and phases of the structure amplitudes. Though almost immediately E. W. Hughes[71] (compare Wilson[72]) showed that with many atom molecules as complicated as proteins these relations were unlikely to be useful, they focused everyone's mind again on absolute and accurate intensity measurements and new, more rapid methods of accurate measurement were developed: film scanning and counter diffractometers. There were also very great developments in electronic computing, which made it possible by the early 1950s to handle the enormous calculations required by protein x-ray analysis. And Bokhoven, Schoone, and Bijvoet[73] showed that

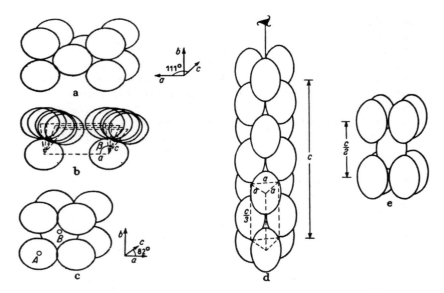

FIGURE 7. Attempts to derive shapes for the hemoglobin molecule by considering packing in different crystals (After Bragg & Perutz.[65])

by accurate measurements on more than two isomorphous derivatives of noncentrosymmetric crystals such as strychnine, general phase angles could be calculated directly from the x-ray data to show atoms in their full three-dimensional relationship.

John Kendrew's review of "Crystalline Proteins; Recent X-Ray Studies and Structural Hypotheses," published early in 1954,[61] gives a picture of the state of our knowledge at the end of this period—measurements, calculations, speculations, good ideas, no certainty, no details. All that was to change as the result of Perutz's experiment in 1953. I give here an account he wrote for me about the sequence of events.

I had been doubtful if any heavy atom would change the intensities of hemoglobin successfully until I asked Bill Cochran if he would let me use his counter spectrometer—consisting of a Unicam oscillation camera with a geiger counter on one arm—to measure the absolute intensity of the hemoglobin reflections. I was surprised how small the absolute Fs were and did some simple calculations which showed that a heavy atom would produce changes that should be easily measured. The original purpose of this experiment had been quite a different one: to put the molecular transform derived from Bragg and my salt-water Fourier on an absolute scale. This happened in 1951 or 1952. [Compare FIGURE 5.]

At that stage I had no idea how I might attach a heavy atom to hemo-

globin. As a side line I had done some work on the crystal structure of
sickle cell hemoglobin. One day I received a set of reprints from the
Journal of General Physiology, a journal that I would not normally have
looked at, from an unknown man at Harvard called Austin Riggs. He
had wondered whether hemoglobin A and hemoglobin S differed in the
number of reactive SH groups and had titrated them with paramercuri-
benzoate. He also examined the effect of PMB on the oxygen equilibrium
curve and found that heme-heme interaction was largely preserved.

I got very excited by this observation, because it suggested that you
can attach molecules of PMB to hemoglobin without changing its structure significantly. I discussed the crystallization of PMB-hemoglobin
with Vernon Ingram who had used this reagent before and kindly made
the compound for me.

When I developed the first precession picture of PMB-hemoglobin
and compared it with that of native hemoglobin, I saw that the two
crystals were isomorphous and that the intensity changes were just of
the magnitude that my measurements of the absolute intensities had led
me to expect. Madly excited, I rushed up to Bragg's room and fetched
him down to the basement dark room. Looking at the two pictures in
the viewing screen, we were confident that the phase problem was
solved. [Compare FIGURE 8.]

Two weeks later, the first two-dimensional electron density projection of hemoglobin was calculated, and six *years* later—strictly out of our
period—the first protein, myoglobin, was seen in a three-dimensional
electron density distribution at sufficient resolution (2.0 Å) to show
clearly the arrangement of the atoms in the molecule.

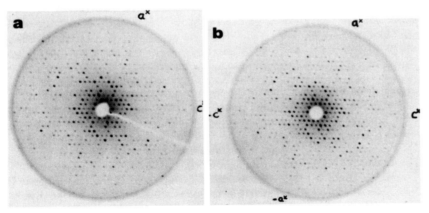

FIGURE 8. X-ray photographs of (h01) reflections of unsubstituted (a) and
p-mercuribenzoyl (b) horse methemoglobin crystals taken by M. F. Perutz.

Today, through the x-ray analysis of isomorphous crystals,[74, 75] we know the atomic arrangement in detail of some fifty protein molecules (I cannot risk an exact figure, the number seems to increase so rapidly).

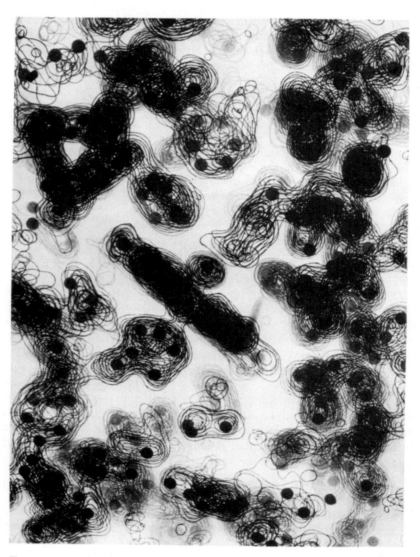

FIGURE 9. Part of the electron density map of myoglobin at 1.4-Å resolution. (Center) The heme group on edge. (Top left) An α helix, end on. (Right) An α helix, on its side. (After J. C. Kendrew, Les Prix Nobel, 1962.)

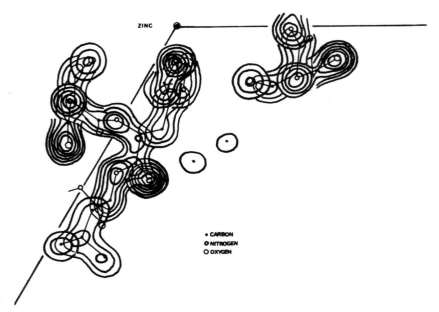

FIGURE 10. The electron density over the residues B9–B11 in 2-zinc insulin crystals showing individual definition of the atoms in serine, histidine, and leucine residues and in the peptide chain (1.5-Å resolution).

In many three-dimensional electron density maps we can observe individually the atoms in the peptide chains and residues (FIGURES 9 & 10). We see in these maps the answers to many of our old controversies.

Though the structure of the pepsin crystal first photographed has not yet been solved, we know that it is very likely to be closely similar to that of the acid protease shown in FIGURE 11. Within this molecule there is a single peptide chain but its course is at first sight rambling; there is very little specific α helical structure and the β pleated sheet structure that occurs is far from regular. It is not surprising that we wondered briefly long ago whether chains existed at all within it; it is reasonable to us now that the β chain character should become rather clearer as Astbury observed when the crystals dry. In protein molecules in general the proportions of α helices and β structures vary widely; both tend to be more distorted than we expected; β sheets are usually much twisted. And within the more irregular strands of chains that run between the specific structures we can trace small stretches of chains in other folds, postulated in various papers and notebooks of those who thought about proteins in the 1930s and 40s. Many protein molecules we find are very

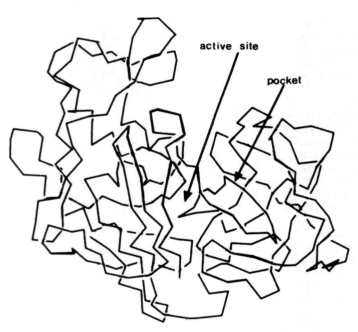

active site

pocket

FIGURE 11. The course of the peptide chain in the acid protease; only α carbon positions are marked. (After T. L. Blundell.)

irregular in shape, and many aggregate as we suspected long ago: insulin into hexamers, lactoglobulin into dimers, hemoglobin into tetramers. We are far still from understanding all the details of the structures we find: protein molecules are marvelously various.

REFERENCES

1. BAUMGÄRTNER, K. H. 1830. Beobachtungen über die Nerven und das Blut. :25 (Table V). Freiburg.
2. HÜNEFELD. 1840. Die Chemismus in der thierische Organisation: 160. Leipzig.
3. REICHERT, K. E. 1849. Arch. P. Anat. u. Physiol. :198.
4. PREYER, W. 1871. Die Blutkrystalle. Jena.
5. REICHERT, E. T. & A. P. BROWN. 1909. The differentiation and specificity of corresponding proteins and other vital substances in relation to biological classification and organic evolution. Carnegie Institute of Washington.
6. COHN, F. 1860. J. Pr. Chem. *80*: 129.
7. SCHIMPER, A. F. W. 1880. Z. Krist. *5*: 131.
8. ZINOFFSKY. 1886. Z. Physiol. Chem. *10*: 16.
9. FRIEDRICH, W., P. KNIPPING & M. v. LAUE. 1912. Sitz. Ber. Math. Phys. Klasse. Bayer. Akad. Wiss. München. :303.

144   ANNALS NEW YORK ACADEMY OF SCIENCES

10. EWALD, P. P. Ed. 1962. Fifty Years of X-Ray Diffraction. International Union of Crystallography. Oosthoek, Utrecht.
11. FRIEDRICH, W. 1913. Phys. Z. *14*: 317.
12. BRAGG, W. L. 1913. Proc. R. Soc. A *89*: 248.
13. NISHIKAWA, S. & S. ONO. 1913. Proc. Math. Phys. Soc. Tokyo 7: 113.
14. HERZOG, R. D. & W. JANKE. 1920. Berichte *53*: 2062.
15. POLANYI, M. 1921. Naturwissenschaften *9*: 288.
16. BRILL, R. 1923. Annalen *434*: 204.
17. MEYER, K. H. & H. MARK. 1928. Berichte *6*: 1932.
18. WEISSENBERG, K. 1926. Berichte *59*: 1535.
19. GONELL, H. W. & H. MARK. 1923. Z. Phys. Chem. *107*: 181.
20. DICKINSON, R. G. & A. L. RAYMOND. 1923. J. Am. Chem. Soc. *45*: 22.
21. HENDRICKS, S. B. 1928. J. Am. Chem. Soc. *50*: 2455.
22. ASTBURY, W. T. & H. J. WOODS. 1930. Nature. *126*: 913.
23. ASTBURY, W. T. & A. STREET. 1931. Phil. Trans. R. Soc. *230*: 75.
24. ASTBURY, W. T. & T. C. MARWICK. 1932. Nature *130*: 309.
25. ASTBURY, W. T., T. C. MARWICK & J. D. BERNAL. 1932. Proc. R. Soc. B. *109*: 443.
26. BERNAL, J. D. 1931. Z. Krist. *78*: 363.
27. BERNAL, J. D. 1932. Biochem. J. *26*: 75.
28. WARREN, B. & W. L. BRAGG. 1928. Z. Krist. *69*: 467.
29. BRAGG, W. H. 1915. Phil. Trans. R. Soc. *215*: 253.
30. DUANE, W. 1925. Proc. Nat. Acad. Sci. U.S.A. *11*: 489.
31. HAVIGHURST, R. J. 1925. Proc. Nat. Acad. Sci. U.S.A. *11*: 502.
32. CORK, J. M. 1927. Phil. Mag. *4*: 688.
33. LONSDALE, K. 1931. Proc. R. Soc. *133*A: 536.
34. BEEVERS, C. A. & H. LIPSON. 1934. Proc. R. Soc. *146*A: 570.
35. GEORGE, W. H. 1929. Proc. Leeds Phil. Lit. Soc. *1*: 412.
36. CLARK, G. L. & K. E. CORRIGAN. 1932. Phys. Rev. *40*(ii): 639.
37. BERNAL, J. D. & D. CROWFOOT. 1934. Nature. *133*: 794.
38. PERUTZ, M. F. 1949. Research *2*: 52.
39. ASTBURY, W. T. & R. LOMAX. 1934. Nature *133*: 795.
40. PATTERSON, A. L. 1935. Z. Krist. *90*: 517, 543.
41. COREY, R. B. & R. W. G. WYCKOFF. 1936. J. Biol. Chem. *114*: 407.
42. WYCKOFF, R. W. G. & R. B. COREY. 1936. J. Biol. Chem. *116*: 51.
43. BAWDEN, F. C.. N. W. PIRIE, J. D. BERNAL & I. FANKUCHEN. 1936. Nature *138*: 1051.
44. BERNAL, J. D. & I. FANKUCHEN. 1941. J. Gen. Physiol. *25*: 111.
45. CROWFOOT, D. 1935. Nature *135*: 591.
46. CROWFOOT, D. 1938. Proc. R. Soc. A *164*: 580.
47. CROWFOOT, D. & D. P. RILEY. 1939. Nature *144*: 1011.
48. ASTBURY, W. T., S. DICKINSON & K. BAILEY. 1935. Biochem. J. *29*: 2351.
49. CROWFOOT, D. & D. P. RILEY. 1938. Nature *141*: 521.
50. BERNAL, J. D., I. FANKUCHEN & M. PERUTZ. 1938. *Ibid.* 523.
51. CROWFOOT, D. & I. FANKUCHEN. 1938. *Ibid.* 522.
52. FANKUCHEN, I. 1943. J. Biol. Chem. *150*: 57.
53. CROWFOOT HODGKIN, D. 1950. Cold Spring Harbor Symp. Quant. Biol. *14*: 65.
54. CROWFOOT, D. & G. M. J. SCHMIDT. 1945. Nature *155*: 504.
55. PALMER, K. J., M. BALLANTYNE & J. A. GALVIN. 1948. J. Am. Chem. Soc. 70: 906.
56. CARLISLE, C. H. & K. DORNBERGER. 1948. Acta Crystallogr. *1*: 194.

57. KENDREW, J. C. 1948. Acta Crystallogr. *1*: 366.
58. SENTI, F. R. & R. C. WARNER. 1948. J. Am. Chem. Soc. *70*: 3319.
59. BERNAL, J. D. & C. H. CARLISLE. 1948. Nature *162*: 139.
60. KENDREW, J. C. & M. F. PERUTZ. 1948. Proc. R. Soc. A *194*: 375.
61. KENDREW, J. C. 1954. Prog. Biophys. Biophys. Chem. *4*: 244.
62. ZINSSER, H. H. & Y. C. TANG. 1951. Arch. Biochem. *34*: 81.
63. TANG, Y. C. 1951. Acta Crystallogr. *4*: 564.
64. ROBERTSON, J. M. 1939. Nature *143*: 75.
65. BRAGG, W. L. & M. F. PERUTZ. 1952. Proc. R. Soc. A *213*: 425.
66. ASTBURY, W. T. 1939. Ann. Rev. Biochem. *8*: 113.
67. HUGGINS, M. L. 1942. Ann. Rev. Biochem. *11*: 27.
68. WRINCH, D. 1936. Nature *137*: 411.
69. PAULING, L., R. B. COREY & H. R. BRANSON. 1951. Proc. Nat. Acad. Sci. U.S.A. *37*: 235.
70. HARKER, D. & J. S. KASPER. 1948. Acta Crystallogr. *1*: 70.
71. HUGHES, E. W. 1949. Acta Crystallogr. *2*: 34.
72. WILSON, A. J. C. 1942. Nature *150*: 152.
73. BOKHOVEN, C., J. C. SCHOONE & J. M. BIJVOET. 1951. Acta Crystallogr. *4*: 275.
74. GREEN, D. W., V. M. INGRAM & M. F. PERUTZ. 1954. Proc. R. Soc. A *225*: 287.
75. KENDREW, J. C., R. E. DICKERSON, B. E. STRANDBERG, R. G. HART, D. R. DAVIES, D. C. PHILLIPS & V. C. SHORE. 1960. Nature *185*: 422.
76. JENKINS, J., I. TICKLE, T. SEWELL, L. UNGARETTI, A. WOLLMER & T. BLUNDELL. 1977. Acid Proteases. :43. Plenum Publishing Corporation. New York, N.Y.
77. BRAGG, W. L. & M. F. PERUTZ. 1952. Acta Kristallogr. *5*: 323.

DISCUSSION OF THE PAPER

GORDON: I got the impression that the pepsin crystal when it was wet was so very much better than when it was dry as tried first. Now is it a fortunate thing that Bernal happened to find such an enormous difference in the sharpness of the spots for that one? Or, if he had used another crystalline protein and got some sort of pattern from the dried crystal might he not have bothered to go on to the wet one?

HODGKIN: I am always glad that Bernal took the first x-ray photographs of pepsin himself a little hurriedly, and saw almost no diffraction effects as others had observed before him, who all had tried taking x-ray photographs of dried crystals. We now know that it is possible to obtain extensive x-ray reflections from most wet protein crystals and more limited data from most dried protein crystals, including pepsin, provided they are allowed to dry sufficiently slowly. In the wet crystals the molecules are in regular contact with one another, with water filling the spaces between them. If you let the water out, the molecules sag irregularly, order diminishes, and most of the x-ray reflections fade or become diffuse. I think that Bernal even if he had obtained such a limited diffraction pattern would have realized it could not correspond with the beautiful appearance of the wet pepsin crystals.

W. DUAX (*Medical Foundation of Buffalo, Buffalo, N.Y.*): I found a letter from Max Perutz about his excitement when he realized that he was on his way to the answer concerning the hemoglobin structure. Do you have any similar feeling in regard to either insulin or vitamin B_{12}? Was there a time when you suddenly realized it was finally going to be downhill the rest of the way, or was it just a sort of excitement of discovery?

HODGKIN: Yes, there are always these moments of discovery. I should also add that I was so deeply involved in the protein story, that I shared Max's excitement. I remember how Max rang me up the day he got the first projection of the hemoglobin structure drawn out and I just got in the car and drove straight over to Cambridge to see it, and I knew I would see essentially nothing because the projection was not interpretable down 63 angstroms. But I could not help going to look at it.

E. PATTERSON (*Institute for Cancer Research, Philadelphia, Pa.*): My husband would have been so excited had he lived to know what the method did. I think the reason he felt that it never would be so useful was because he could not conceive of the development of computers, which it seems to me have been exceedingly helpful in solving protein structures. Again the intellectual development and what you might call a mechanical development have gone hand in hand with the solution of the structures.

HODGKIN: Of course, the very interesting thing is that the first steps, the calculation of the one-dimensional projection, the calculation of the two-dimensional projections, and the first Patterson projections, were done by hand computation on adding machines. However, the first three-dimensional map at 2 angstroms that showed where the atoms really were, the map of myoglobin, depended upon computers.

KARLSON: I have a question regarding model building, we all know that the detection of the alpha helix and also of the pleated sheet relied very much from the first on Corey's measurement of peptides and, secondly, on the model building with the exact angles and distances of the atoms in Pauling's laboratory. What was the attitude in Astburys's laboratory and in Cambridge to this kind of model building? Did Astbury ever try such a thing and what did they do at Cambridge?

HODGKIN: Yes, actually there was a history of this going on all the time. The actual determination of crystal structures gave more and more exact ideas of the size of atoms and how they were arranged. The very first step, the old silk fibroin model, really depended upon structure analyses of urea by Hendricks of hexamethylenetetramine in Pasadena and in

Berlin. Then later more structures showed that you not only had to have the sizes right but that you had to have the angles and conformations right. Just before the helix model was produced there was a very long model-building paper by Bragg, Perutz, and Kendrew in the *Proceedings of the Royal Society*, in which a variety of structures were proposed for peptide chains. Astbury certainly also took new evidence into account and revised earlier proposals in his later model building. The alpha helix was just a stroke of genius on the part of Linus Pauling himself. Again I know because I was there. He was having a cold or flu in Oxford the year he was there and was rolling bits of paper around his fingers, which took him away from the crystallographic repeats and allowed the helix to go its own way. Before that, structures very close to the alpha helix had been built by several people, particularly H. S. Taylor and by Maurice Huggins, following the idea that a helix was a likely kind of repeating pattern; but it was the nonintegral character of the alpha helix repeat that really produced a change of outlook, and that was due to Pauling.

One of the stories is that when John Kendrew's group calculated the first three-dimensional electron density map of myoglobin in Cambridge, at 2.5 angstroms with all the terms included, all of the group stayed up until the alpha helix came through. When they could see that there were definitely alpha helices in the structure most of them went home to bed.

T. SEJNOWSKI (*Princeton University, Princeton, N.J.*): I have two questions. The first is a matter of strategy. I wonder why the decision was made to put all the effort into hemoglobin first, rather than, for example, myoglobin, which is closely related but much simpler and easy to work with.

And the second question concerns remarkable persistence over many years. There must have been moments when the group was discouraged, or at least it may not have been clear that the end was in sight. And I wonder what was the driving force that kept you on that particular problem rather than going on to some simpler one first?

HODGKIN: If you look back at the whole history of protein crystals, which goes back to the 1830s, hemoglobin is the first protein observed to crystallize. Different hemoglobins give very beautiful crystals and Max Perutz was keen on working on hemoglobin as a problem in biochemistry before he even came to Cambridge in 1936. Myoglobin was not isolated until sometime later, I think during the war or soon after. When its characteristics were realized, Bragg and Perutz decided to concentrate work on myoglobin with a new very good research student, John Kendrew, who came to work with them directly after the war. However, Max Perutz still had his old love of hemoglobin and had already put in an enormous amount of his own time measuring hemoglobin reflections.

So the work on myoglobin ran ahead once three dimensional calculations became possible, first at low resolution, the 6-angstrom map, and then at high resolution, the 2-angstrom map—because, as you say, fewer x-ray reflections had to be measured to get the answers. But hemoglobin, partly for the reasons given in Max Perutz's letter, partly because he had measured hemoglobin crystals so often already, was the first protein for which heavy atom derivatives that were interpretable were obtained. These led to making similar derivatives for myoglobin.

Actually I myself did put protein x-ray analysis largely aside and worked on simpler compounds for many years before beginning to concentrate again on insulin. And even Max Perutz did some pieces of research over the years, not connected with hemoglobin, to sustain his morale.

SMITH: Myoglobin was actually first isolated from horse heart in the 1930s. It was very difficult to obtain it in pure form and it was very difficult to obtain in any quantity. Kendrew's x-ray work was done on whale myoglobin; that became available in large quantities immediately after the war because people thought that whale meat was going to be a good substitute for the meat shortage in Britain.

HODGKIN: Thank you very much, yes. I should say that the immediate next step, which I skipped over, was that after the set of photographs of protein crystals that were measured up to 1950, there was a fantastic thrust in which a large number of crystals were studied, but nearly all of these are different species of myoglobin and different species of hemoglobin.

PNINA ABIR-AM (*Université de Montréal, Montréal, Québec*): It was mentioned that a gentlemen's agreement existed between Bernal and Astbury with regard to crystalline versus fibrous material for x-ray work. Bernal and Astbury agreed that Bernal would work on crystalline material and Astbury on fibers. Bernal said later that this division, by a gentlemen's agreement, led to his having reached DNA later on. I wonder what your opinion on this is.

HODGKIN: I do not think they took this agreement quite that seriously. Bernal certainly became most interested in solving the DNA structure just after the war and his line of study was the same as that adopted for the α helix. That is to say, he planned to have studies made of the structures of crystals of the nucleotides and nucleosides of which DNA is composed and one such study was done at Birkbeck by S. Fürberg. Fürberg certainly built what he hoped might be models of DNA as a consequence of his work. Of course, they had only enough diffraction data to do rough calculations. He just did not happen to find the right structure.

REFERENCES

Alfrey, T., Agron, P., Bohner, J., Haas, H. & Wechsler, H. 1948 *J. Poly. Sci.* **3**, 157.
Alfrey, T. & Goldfinger, G. 1944 *J. Chem. Phys.* **12**, 205, 322.
Alfrey, T. & Price, C. C. 1947 *J. Poly. Sci.* **2**, 101.
Arlman, E. J., Melville, H. W. & Valentine, L. 1949 *Rec. Trav. chim. Pays-Bas*, **68**, 945.
Bickel, A. F. & Melville, H. W. 1949 *Trans. Faraday Soc.* **45**, 1049.
Burnett, G. M., Valentine, L. & Melville, H. W. 1949 *Trans. Faraday Soc.* **45**, 960.
Cohen, S. G., Ostberg, B. E., Sparrow, D. B. & Blout, E. R. 1948 *J. Poly. Sci.* **3**, 269.
Degens, P. N. & Gouverneur, P. 1950 *Anal. Chim. Act.* (in the Press).
Grassie, N. & Melville, H. W. 1950 *J. Poly. Sci.* (in the Press).
Masson, C. R. & Melville, H. W. 1949 *J. Poly. Sci.* **4**, 337.
Mayo, F. R. & Lewis, F. M. 1944 *J. Amer. Chem. Soc.* **66**, 1594.
Mayo, F. R., Lewis, F. M. & Walling, C. 1947 *Disc. Faraday Soc.* No. 2, 287.
Melville, H. W., Noble, B. & Watson, W. F. 1947 *J. Poly. Sci.* **2**, 229.
Melville, H. W. & Valentine, L. 1950*a* *Proc. Roy. Soc. A*, **200**, 353.
Melville, H. W. & Valentine, L. 1950*b* *Trans. Faraday Soc.* **46**, 210.
Price, C. C. 1947 *Disc. Faraday Soc.* No. 2, 307.
Rodebush, W. H. & Feldman, I. 1946 *J. Amer. Chem. Soc.* **68**, 897.
Schütze, M. 1939 *Z. anal. Chem.* **118**, 241.
Simha, R. & Wall, L. A. 1948 *Bur. Stand. J. Res., Wash.*, **41**, 521.
Smakula, A. 1934 *Z. angew. Chem.* **47**, 777.
Unterzaucher, J. 1940 *Ber. dtsch. Chem. Ges.* **73** B, 391.
Walling, C. 1949 *J. Amer. Chem. Soc.* **71**, 1930.

Polypeptide chain configurations in crystalline proteins

BY SIR LAWRENCE BRAGG, F.R.S., J. C. KENDREW AND M. F. PERUTZ

Cavendish Laboratory, University of Cambridge

(*Received* 31 *March* 1950)

Astbury's studies of α-keratin, and X-ray studies of crystalline haemoglobin and myoglobin by Perutz and Kendrew, agree in indicating some form of folded polypeptide chain which has a repeat distance of about 5·1 Å, with three amino-acid residues per repeat. In this paper a systematic survey has been made of chain models which conform to established bond lengths and angles, and which are held in a folded form by N—H—O bonds. After excluding the models which depart widely from the observed repeat distance and number of residues per repeat, an attempt is made to reduce the number of possibilities still further by comparing vector diagrams of the models with Patterson projections based on the X-ray data. When this comparison is made for two-dimensional Patterson projections on a plane at right angles to the chain, the evidence favours chains of the general type proposed for α-keratin by Astbury. These chains have a dyad axis with six residues in a repeat distance of 10·2 Å, and are composed of approximately coplanar folds. As a further test, these chains are placed in the myoglobin structure, and a comparison is made between calculated and observed F values for a zone parallel to the chains; the agreement is remarkably close taking into account the omission from the calculations of the unknown effect of the side-chains. On the other hand, a study of the three-dimensional Patterson of haemoglobin shows how cautious one must be in accepting this agreement as significant. Successive portions of the rod of high vector density which has been supposed to represent the chains give widely different projections and show no evidence of a dyad axis.

The evidence is still too slender for definite conclusions to be drawn, but it indicates that a further intensive study of these proteins, and in particular of myoglobin which has promising features of simplicity, may lead to a determination of the chain structure.

322   Sir Lawrence Bragg, J. C. Kendrew and M. F. Perutz

1. Introduction

Proteins are built of long chains of amino-acid residues. Amino-acids unite to form a chain (figure 1) in which the R groups are of some twenty-three different kinds, varying from a hydrogen atom in glycine to moderately complex groups (e.g. the linked five- and six-membered rings of tryptophan). As examples, insulin of molecular weight 12,000 has, according to Chibnall (1945), 106 residues, in four chains linked by six disulphide bridges of cystine (Sanger 1948). Myoglobin, 17,000, has about 146 residues; haemoglobin, 67,000, has about 580 residues.

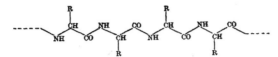

Figure 1. Polypeptide chain.

In this paper an attempt is made to glean as much information as possible about the nature of the chains from X-ray studies of crystalline proteins, and to survey the possible types of chain which are consistent with such evidence as is available. The configuration and arrangement of the R groups is not discussed; we are merely concerned with the configuration of the —CO—$\overset{|}{C}$H—NH— chain to which they are attached.

Certain features of all protein structures deserve special mention.

(*a*) The mean molecular weight of the residue $\overset{\text{R}}{\underset{\text{NH}\quad\text{CO}}{\text{CH}}}$ is much the same in most proteins, ranging between 110 and 120 (e.g. haemoglobin 112·5). Since the molecular weight of the chain element —CO—$\overset{|}{C}$H—NH— is 56, one-half the weight of the protein is in the chain and one-half in the side groups, excluding associated water. To put this in another way, the average side-chain contains about four atoms other than hydrogen.

(*b*) All the amino-acids occurring in proteins (except glycine which is non-enantiomorphous, and a very few amino-acids found in primitive organisms such as bacteria) have the steric configuration about the central carbon atom conventionally termed 'laevo'.

(*c*) X-ray determinations of the structures of simple amino-acids or dipeptides yield consistent information about the interatomic distances and bond angles in their crystals, and presumably these angles and distances will not be very different in a long polypeptide chain.

2. Previous speculations about the configuration of the polypeptide chain

Astbury and his co-workers in their pioneer investigations have made an exhaustive study of the fibrous proteins such as the keratin of hair and wool. Their most important result, in the present connexion, is their inference that the marked

Polypeptide chain configurations in proteins   323

5·1 Å repeat along the fibre axis which is shown prominently by X-ray photographs of α-keratin and its analogues corresponds to an element of folded chain containing three amino-acid residues. Briefly, Astbury (private communication, 1949) summarizes the evidence as follows:

(*a*) The β-keratin X-ray diagram with its strong meridional 3·4 Å reflexion represents a system of extended polypeptide chains as in figure 2(*b*), for which the average length per residue must be about 3·4 Å to correspond to the accepted bond lengths and angles.

(*b*) The β configuration is approximately twice as long as the folded α configuration, since the reversible extensibility of wool and hair is approximately to twice the normal length. The average length per residue in the α chain is therefore 1·7 Å, giving three residues per repeat of 5·1 Å.

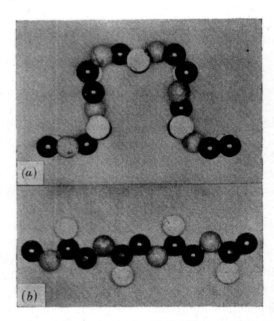

FIGURE 2. Chain configurations proposed by Astbury (1949*a*) for (*a*) α-keratin, (*b*) β-keratin.

The model of the folded chain in figure 2 (*a*) is proposed for α-keratin by Astbury as that which at present seems to fit the facts best, other indications in favour of this structure being obtained from considerations of side chain packing. The complete repeat is at distances of 10·2 Å, each loop occupying 5·1 Å and containing three residues as indicated in the diagram. It will be referred to again below.

Huggins (1943) has made an extensive review of possible types of polypeptide chains. He gives numerous examples of chains in both extended and shrunk forms, and of ways in which these may be linked together, showing how many the possibilities are until further evidence is obtained which restricts them.

Recently a type of chain has been proposed by Ambrose & Hanby (1949; see also Ambrose, Elliott & Temple 1949) based on measurements of the dichroism of

324   Sir Lawrence Bragg, J. C. Kendrew and M. F. Perutz

α-keratin, myosin and tropomyosin in the infra-red. This is a folded chain with two residues in a repeat of 5·1 Å, and Ambrose and his collaborators picture these chains as occurring in pairs so that there are four residues in the α-keratin unit.

3. POLYPEPTIDE CHAINS IN CRYSTALLINE HAEMOGLOBIN AND MYOGLOBIN

The X-ray diffraction pictures given by haemoglobin, in particular horse methaemoglobin, have been studied in detail by Perutz and his collaborators (Boyes-Watson, Davidson & Perutz 1947; Perutz 1949), and Kendrew (1950) has published an account of work with myoglobin. The evidence for the existence and features of chains in haemoglobin and myoglobin has been very fully discussed in these papers, but it may be useful to recapitulate here the nature of the problem and the way in which an attack on it has been made. Protein crystals give a wealth of diffracted beams, extending out to angles which correspond to spacings as small as 2 Å, but the interpretation of these photographs as a complete picture of the arrangement of the thousands of atoms in the unit cell is a task greater by several orders of complexity than the most complex crystals yet successfully analyzed. It is always possible, however, to represent the experimental observations as a Patterson or 'vector' map. Broadly speaking, if there are in the actual crystal two atoms a and b with 'weights' m_a and m_b, the Patterson map has a peak or lump of density proportional to the product $m_a m_b$ at a point 'ab' such that the line joining 'ab' to the origin of the map is equal and parallel to the line drawn between a and b in the crystal. If the crystal contains n atoms, there are n^2 such points in the Patterson map corresponding to the product $(a+b+c+d+...)(a+b+c+d+...)$. The n terms a^2, b^2, c^2, d^2, ... have zero vectors and superimpose at the origin to form a large concentration. The remainder are distributed throughout the cell. (The meaning of Patterson maps is explained in greater detail by Kendrew & Perutz (1949).) It is common practice in crystal analysis to start by forming two- or three-dimensional Patterson summations and then to seek to decipher the significance of their more prominent peaks in terms of the atomic arrangement. However, the Patterson projection of a protein represents some millions of vector peaks. It can only be deciphered if there is some regular underlying arrangement throughout the crystal structure, such as would be the case if the polypeptide chains are regular in form, straight, and parallel to each other in the molecule. It has already been remarked that half the atoms (hydrogen can be neglected on account of its low scattering power) are in the chain element CO—C—N. Further, if some model of the chain is adopted, the first carbon atom of each side-chain is also in a definite position, so that some five-eighths of all the atoms form part of a regularly recurring pattern. The remainder are in the side groups and will have a wide variety of configurations. The vectors drawn between atoms inside any one chain, and those between its atoms and the atoms of neighbouring parallel chains, will form a large proportion of all the vectors in the Patterson and will be of a relatively few constantly repeated types. Owing to this regularity they may be prominent in the Patterson projection in spite of the irregular welter of other vectors on which they are superimposed. This possibility is the greater because the parallel straight chains would be regions of relatively high density. The atoms in any one

328   Sir Lawrence Bragg, J. C. Kendrew and M. F. Perutz

4. CLASSIFICATION OF CHAIN STRUCTURES

In this and the following sections we attempt to survey systematically all those types of folded polypeptide chain configurations which satisfy certain conditions, established by experiment or plausible on general grounds.

It cannot be assumed as certain that the polypeptide chain has the same configuration in all crystalline proteins, or that a similar configuration occurs in fibrous proteins such as α-keratin. It is, however, not unreasonable to expect that haemoglobin and myoglobin contain chains of the same type, because these proteins appear to be closely related in several ways; and furthermore, the repeat distance, the interchain distance, and the number of residues per repeat, are similar in these two proteins to the corresponding features of α-keratin. It will therefore be assumed as a working hypothesis that the chain configurations in large classes of proteins resemble one another closely, while bearing in mind that this hypothesis is based on slender evidence and may have to be abandoned when further experimental data are available.

In our survey of chain configurations we have adopted the following conditions:

(a) *Interatomic distances and bond angles*

Some relevant data have been obtained by X-ray analysis of the structures of amino-acids and small peptides. Glycine has been analyzed by Albrecht & Corey (1939), DL-alanine by Levy & Corey (1941) and β-glycylglycine by Hughes & Moore (1949). An addition compound of cysteyl-glycine has recently been analyzed in the Cavendish Laboratory by Dyer, and progress has been made with the analysis of tripeptides and tetrapeptides.

In an earlier paper Huggins (1943) has given the following summary of the distances and angles to be expected in the chain unit $-\mathrm{N-\overset{\overset{\textstyle H}{|}}{C}-\overset{\overset{\textstyle O}{\|}}{C'}-}$ (with H, R below C):

	(Å)		(°)
N—C	1·41	$\widehat{NCC'}$	112
C—C′	1·52	$\widehat{CC'N}$	118
C′—O	1·25	$\widehat{C'NC}$	118
C′—N	1·33		

The experimental results, including the most recent, agree generally with these figures, though with some individual variations (for summary see Corey 1948). For example, in cysteyl-glycine Dyer has found $\widehat{C'NC}$ considerably greater than the figure given. All the structures contain hydrogen bonds N—H...O which vary in length between 2·6 and 2·85 Å.

For the construction of models we have adopted the following values, which agree with those given by Huggins or with generally recognized standards:

Polypeptide chain configurations in proteins   329

(i) Covalent distances: C—C, 1·52 Å; C—N, 1·36 Å; C—O, 1·24 Å.
(ii) Hydrogen bond distances: N—O (in N—H...O) 2·85 Å.
(iii) Bond angles: C: tetrahedral distribution.

N: tetrahedral (interbond angle 109° 28') or planar (120°).*

N—H...O═C$\big\langle$: we have placed no restriction on $\widehat{\mathrm{NHO}}$ or on $\widehat{\mathrm{HOC}}$, though we have generally attempted to make NHOC as nearly collinear as possible.

(b) Optical configuration of amino-acids

We have assumed that all the amino-acids (except glycine which is optically inactive) belong to the *laevo* series.

(c) Symmetry of the chain

Huggins (1943) has made the following two general points about the symmetry of a stable chain configuration:

(i) 'Polypeptide chains extending through the crystalline regions must each have a screw axis of symmetry, or else two or more chains must be grouped around screw axes or other symmetry elements. The unbalanced forces on opposite sides of a chain which has no screw axis, e.g. any of the earlier chain structures advocated for α-keratin by Astbury or the one that he has most recently proposed for collagen, would tend to bend it continuously in the same direction.'

(ii) 'In general, a structural pattern for a protein in which like groups are all surrounded in a like manner, except for differences between the R groups, is more probable than one in which this is not the case.'

We have accepted Huggins's first criterion throughout; indeed, we have used it as a basis for classifying types of configuration. In other words, all the structures examined possess a screw axis of symmetry (not necessarily restricted to crystallographic types of screw symmetry, e.g. a fivefold axis would be permissible).

Huggins's second criterion, that each element of the chain should be in a similar relation to neighbouring elements, we have not regarded as essential, and some of the chains described below do not obey this rule.

(d) The role of hydrogen bonding

We have made the plausible (but still unproved) assumption that the chain is held in a folded condition by hydrogen bonds between $\rangle$NH and $\rangle$CO groups of nearby amino-acid residues. In other words, the folded chain is thrown into a series of rings which must be ruptured at the hydrogen bond before unfolding can take place. Formally speaking, these rings may be of two types, illustrated diagrammatically in figure 4 (*a*).

* No complete structure analysis of any compound containing nitrogen bound analogously to nitrogen in proteins of the α-keratin type is yet available. We have therefore thought it better to leave this question open for the present, and to examine, in each type of structure, the effect of the two configurations given.

330   Sir Lawrence Bragg, J. C. Kendrew and M. F. Perutz

We have used criteria (c) and (d) as a basis for classifying possible types of chain configuration:

(i) *By symmetry*. The chains may possess a twofold, threefold, fourfold or higher screw axis of symmetry.

(ii) *By ring size*. The ring, being formed by elements of the chain and the hydrogen atom between O and N, must be of one of the following types:

Type A. Number of atoms in ring, $R = 3n + 4 = 7, 10, 13, \ldots$.

Type B. Number of atoms in ring, $R = 3n + 5 = 8, 11, 14, \ldots$.

These rings are shown schematically in figure 4 (b).

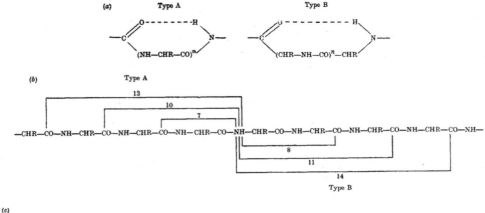

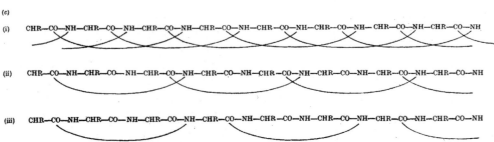

FIGURE 4. (a) Possible types of hydrogen-bonded ring formed from folded polypeptide chain. (b) Scheme showing the numbers of atoms in the various possible ring structures. (c) Families of structures of type S_{10}: (i) $1/q = 1$, (ii) $1/q = \frac{1}{2}$, (iii) $1/q = \frac{1}{3}$.

Each type of structure may be designated by a symbol S_R, where S denotes the screw symmetry and R the number of members in the hydrogen-bonded rings; e.g. 2_7 is a structure having a dyad axis of symmetry with seven-membered rings. Formally speaking each of these structural types may be further subdivided into families according as all, or only a fraction $1/q$, of the NH (or CO) groups form hydrogen bonds. For example, the structure type S_{10} might form families of the kind shown in figure 4 (c); in (i) every possible hydrogen bond is made, in (ii) $\frac{1}{2}$ and in (iii) $\frac{1}{3}$ of the possible bonds are made. Similar families can be described for each of the permitted values of R. These may be designated by the symbols $S_R(1/q)$, e.g. $2_7 \cdot \frac{1}{2}$ would be a structure with a seven-membered ring, a dyad screw axis, and one-half of the possible hydrogen bonds made.

Polypeptide chain configurations in proteins   331

It will be evident that the number of configurations formally possible is rather large, and even though many of the formal schemes are found on examination not to be sterically possible it is still true that the number of configurations which should be examined in a comprehensive survey is considerable. We regard those structures in which all NH and CO groups are hydrogen bonded ($q = 1$) as inherently the more probable, because their free energy is presumably lower. We have therefore examined structures of this type in considerable detail and have made efforts to build all possible models provided they conformed to the conditions outlined above.

We have also tried to build all the possible structures having only a fraction of the NH and CO groups hydrogen bonded, determining at least the repeat distance of the molecular pattern and the number of amino-acid residues which that pattern contains. This survey was less thorough, so that some possible structures may have been overlooked, and only a few of the structures were examined in detail. Among the latter was $2_{13} \cdot \frac{1}{3}$ because of the evidence which Astbury produced in its favour, and $2_{14} \cdot \frac{1}{3}$ because of its close similarity with $2_{13} \cdot \frac{1}{3}$.

5. STRUCTURES EXAMINED IN DETAIL

Within the limits discussed in the last two paragraphs we have attempted to make models of as many types of structure as possible, considering in turn each type of chain symmetry and each ring size. In some instances no structure could be devised for steric reasons. In general, very small or very large rings may be excluded; small rings on account of the strains involved in closing them, large ones because they involve either excessively long hydrogen bonds or else chains of very large cross-sectional area containing many residues per repeating unit.

These structures examined in detail are listed in table 1, classified in accordance with their symmetry and ring size. It is to be noted that a given chain configuration can in general give rise to two types of structure depending on the way the side groups are attached, since the chain itself is enantiomorphous; broadly speaking, these may be distinguished as *laevo* groups in a left-handed chain, or *laevo* groups in a right-handed chain—structures which are *not* mirror images of each other and which are designated by suffixes '*a*' and '*b*'. In some cases steric considerations rule out one of these configurations.

We now proceed to discuss briefly each of the structures listed in table 1, leaving over for more detailed treatment later those which we regard as displaying the most promising agreement with the experimental data. Those not further considered have generally been excluded because they do not satisfy one or both of the following criteria:

(*a*) *Repeat distance*. The structure must account for an apparent 'repeat' distance of just over 5 Å. This 'repeat' may be the true repeat distance between identical points, or a fractional translation produced by the action of the screw axis.

(*b*) *Number of residues per repeat*. We have preferred structures in which the repeat (or pseudo-repeat) contains three amino-acid residues, for the reasons given above (p. 325). We do not feel, however, that we can entirely exclude the possibility that the repeat contains four residues, and two such structures are considered in

332   Sir Lawrence Bragg, J. C. Kendrew and M. F. Perutz

TABLE 1

screw axis of symmetry	no. of atoms in ring	repeat distances (Å)	no. of residues per repeat	illustrations (figure no.)	comments
twofold	7 a	5–5·6	2	5	structure proposed by Huggins (1943)
	7 b	5–5·6	2	6	structure proposed by Zahn (1947) and Ambrose *et al.* (1949); readily folds in pairs; see §7, 8
	8	4·6–4·8	2	7	structure proposed by Huggins (1943); only one configuration possible; repeat distance too short
	13	10·2	6	8	structure proposed by Astbury & Bell (1941); see §7, 8
	14	10·2	6	9	see §7, 8
threefold	7	7·5	3	—	repeat distance too short
	8	5·4	3	10	hydrogen bonds mutually perpendicular; see §7
	10	5·2	3	11	structure proposed by Taylor (1941) and Huggins (1943); hydrogen bonds oriented nearly parallel to the chain direction; see §8
	11				
	13	—	—	—	no possible structures
	14				
fourfold	7				
	8	—	—	—	no possible structures
	10				
	11	5·4	4	—	rings somewhat strained; similar to 4_{13}; a possible structure
	13	5·6	4	12	a possible structure
	14 or greater	—	—	—	no possible structures
fivefold and higher symmetries	—	—	—	—	all such structures contain more than four amino-acid residues per repeat unit

detail (2_7a and 4_{13}). It appears highly improbable that the number of residues in the 5 Å repeat is greater than four, and this criterion excludes symmetries which are more than fourfold; a fivefold chain repeating at 5 Å, for example, would necessarily contain at least five amino-acid residues per repeat.

The following diagrams illustrating the types of chain were drawn from optical projections of models, and are only intended to illustrate the structure of the chain. The co-ordinates of the atoms given in the accompanying tables are accurate to about 0·1 Å.

(i) *Twofold structures*

Seven-atom ring (2_7a, 2_7b)

One form of this ring (2_7a) was proposed by Huggins (1943), whose drawings of it are reproduced in figure 5. The chain itself forms a more or less flat ribbon while R groups project alternately up and down from the plane of the ribbon. The repeat distance is 5 to 5·6 Å and contains two residues.*

* The precise repeat distances in 2_7 and 2_8 cannot be predicted since the N—H—O—C bonds deviate widely from a straight line, and precise experimental evidence is not yet available for the N—O distance in such a case.

Polypeptide chain configurations in proteins   333

The other form (2_7b), differing only in that the R groups and H atoms on the α-carbon atoms are interchanged, was suggested by Zahn (1947) and described in more detail by Ambrose & Hanby (1949) and Ambrose *et al.* (1949). As shown in figure 6 the R groups now lie approximately in the plane of the ribbon, instead of normal to it. This is an attractive feature of the structure, since it enables the chain

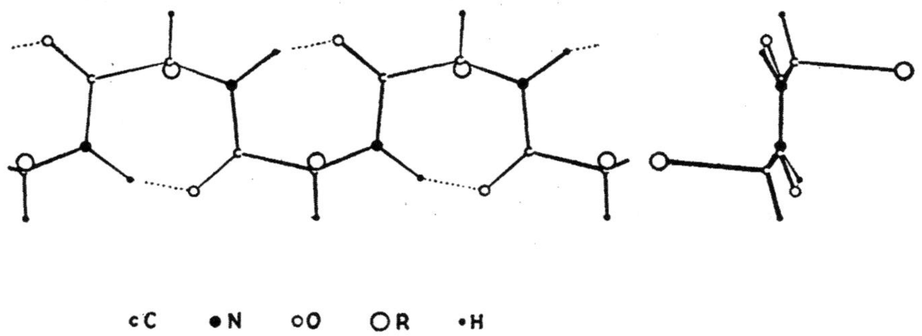

cC   ●N   oO   OR   •H

FIGURE 5. Projections of 2_7a chain, after Huggins (1943).   Repeat distance about 5 Å

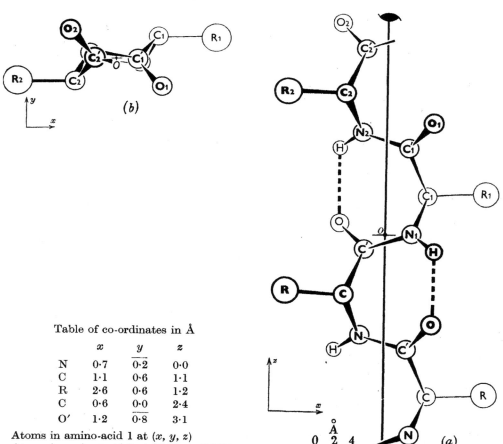

Table of co-ordinates in Å

	x	y	z
N	0·7	$\overline{0·2}$	0·0
C	1·1	0·6	1·1
R	2·6	0·6	1·2
C	0·6	0·0	2·4
O′	1·2	$\overline{0·8}$	3·1

Atoms in amino-acid 1 at (x, y, z)
Atoms in amino-acid 2 at $(\overline{x}, \overline{y}, z + 2·85)$

FIGURE 6. Projections of 2_7b chain.

334   Sir Lawrence Bragg, J. C. Kendrew and M. F. Perutz

to bend back on itself about an axis in the ribbon plane, without mutual interference of R groups or rupture of hydrogen bonds; the two limbs then lie about 4 Å apart.

Astbury (1949*b*) has criticized 2_7b on several grounds, one being that a chain having a true repeat of 5 Å and a screw dyad axis would give a weak 010 (5 Å) and a strong 020 (2·5 Å) reflexion, whereas in fact in α-keratin it is the 5 Å meridional reflexion which is overwhelmingly strong. This objection would apply to both forms of 2_7, and to overcome it Huggins (1943) introduced 'the additional assumption that alternate R groups are much more potent X-ray scatterers than the intermediate ones'. There is, however, no other evidence that such an arrangement exists.

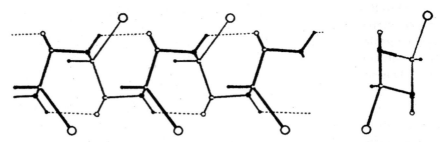

FIGURE 7.   Projections of 2_8 chain after Huggins (1943).   Repeat distance about 5 Å.
Symbols as for figure 5.

2_7b has also been criticized by Darmon & Sutherland (1949) on grounds connected with the interpretation of infra-red absorption spectra given by Ambrose *et al.* (1949).

Another serious objection to both forms of 2_7 is that it contains only two amino-acid residues per repeat of 5·5 Å. Alternatively, 2_7b can be arranged so that two close-spaced chains (see above) run through a single subcell of $5 \times 10 \times 10$ Å, giving *four* residues per subcell. We have given reasons for preferring models containing *three* residues per subcell; but these reasons are not conclusive ones, and so we have given more detailed consideration (see below) to 2_7b, in particular in relation to haemoglobin and myoglobin.

Both structures can be built using either a planar or a tetrahedral configuration of bonds about the nitrogen atom. We have preferred a planar configuration, since this gives a more nearly linear arrangement of the hydrogen bond N—H—O, and because the ring is flatter than with the tetrahedral configuration, which prevents the packing of neighbouring chains to as close a distance as the myoglobin data would appear to demand (see p. 350).

Eight-atom ring (2_8)

This model was first suggested by Huggins (1943), whose projections of it are reproduced in figure 7. In the configuration of its R groups it resembles 2_7b rather than 2_7a; no second form, corresponding to 2_7a, is possible for steric reasons; in such a structure R would lie far too close to C'.

The 2_8 structure is not immediately attractive since, like 2_7, it contains only two residues in the repeat distance and has other disadvantages not shared by 2_7:

(*a*) The repeat distance is distinctly less than 5 Å; 4·6 with planar, and 4·7 to 4·8 Å with tetrahedral nitrogen atoms.

Polypeptide chain configurations in proteins   335

(b) The angle HOC in the hydrogen bonds is only about 100°, which seems unlikely on general grounds, since it brings the N and O atoms very close to one another.

(c) The whole chain is much more rigid than $2_7 b$, leaving no possibility of folding about an axis in the plane of the ribbon. In any case the side chains would interfere with such a fold.

For all these reasons we regard 2_8 as an unlikely structure, and do not consider it further.

Thirteen-atom ring $(2_{13} \cdot \frac{1}{3})$

This is the well-known structure proposed for α-keratin by Astbury & Bell (1941). We have constructed models in conformity with the diagrams published by these authors; our projections are illustrated in figure 8. The repeat distance is just more than 10 Å, and contains six amino-acid residues; side chains project alternately up and down. The structure satisfies all the experimental data and is discussed in more detail below.

We found it impossible to construct a $2_{13} \cdot \frac{1}{3} \cdot b$ configuration. On the other hand, we obtained a quite distinct version of $2_{13} \cdot \frac{1}{3}$ in which the side-chains project alternately up and down as in the Astbury version, and the repeat distance is 9 Å, containing six amino-acid residues. The ring is somewhat strained and the structure seems in general an unlikely one; we mention it merely as an illustration of the fact, which we have observed here and in $2_{14} \cdot \frac{1}{3}$, that these large ring structures have considerable flexibility and their detailed configurations can be altered within wide limits. At the present stage there is clearly nothing to be gained by studying all such minor variations in detail.

Fourteen-atom ring $(2_{14} \cdot \frac{1}{3})$

This structure, which we believe has not been previously described, is illustrated in figure 9. It possesses all the attractive features of $2_{13} \cdot \frac{1}{3}$: the ring is not strained and is almost planar; R groups project alternately up and down from the ring plane (though they are not close-packed, their distances apart varying from 4 to 6 Å); the repeat distance is 10·2 Å and contains six residues. Accordingly, we discuss the structure in detail below.

Both $2_{14} \cdot \frac{1}{3} \cdot a$ and $2_{14} \cdot \frac{1}{3} \cdot b$ are possible, and each is susceptible of minor variations, but the ring is so flexible that we have only thought it worth while to discuss one typical, and on general grounds highly probable, structure.

(ii) *Threefold structures*

In all these structures (and those of higher symmetry) the R group and H atom on the α-carbon atoms can easily be interchanged, with little effect on the general properties of the structure. We have in each case considered the version in which the side-chains would protrude most nearly normal to the axis of symmetry.

Seven-atom ring (3_7)

This structure would appear to be possible sterically, and we constructed a model without difficulty. We have not considered it further, since its repeat distance is 7·5 Å (containing three amino-acid residues).

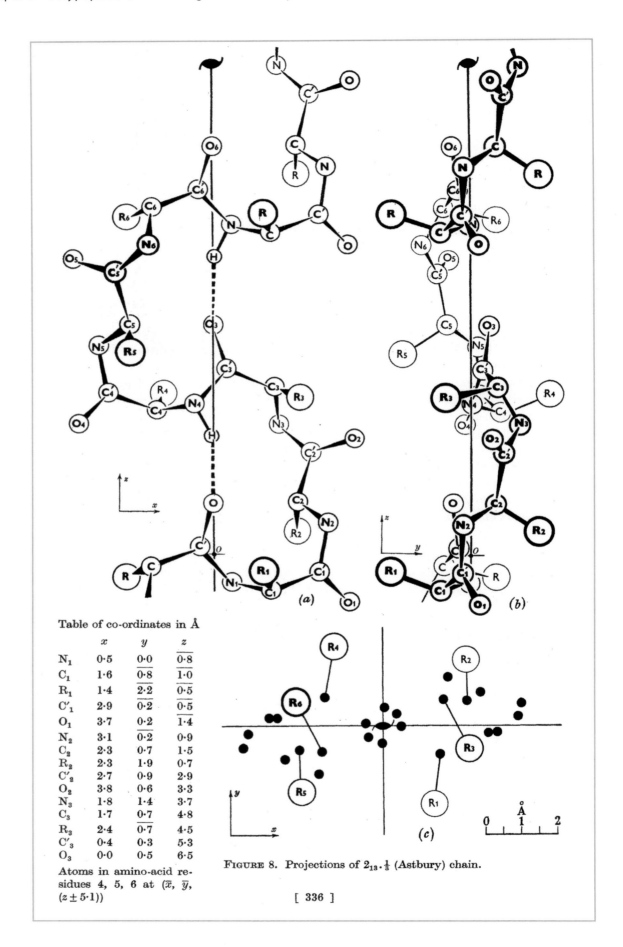

Table of co-ordinates in Å

	x	y	z
N_1	0·5	0·0	0·8
C_1	1·6	$\overline{0·8}$	$\overline{1·0}$
R_1	1·4	$\overline{2·2}$	0·5
C'_1	2·9	$\overline{0·2}$	$\overline{0·5}$
O_1	3·7	0·2	$\overline{1·4}$
N_2	3·1	$\overline{0·2}$	0·9
C_2	2·3	0·7	1·5
R_2	2·3	1·9	0·7
C'_2	2·7	0·9	2·9
O_2	3·8	0·6	3·3
N_3	1·8	1·4	3·7
C_3	1·7	0·7	4·8
R_3	2·4	$\overline{0·7}$	4·5
C'_3	0·4	0·3	5·3
O_3	0·0	0·5	6·5

Atoms in amino-acid residues 4, 5, 6 at $(\overline{x}, \overline{y}, (z \pm 5·1))$

FIGURE 8. Projections of $2_{13} \cdot \frac{1}{3}$ (Astbury) chain.

[336]

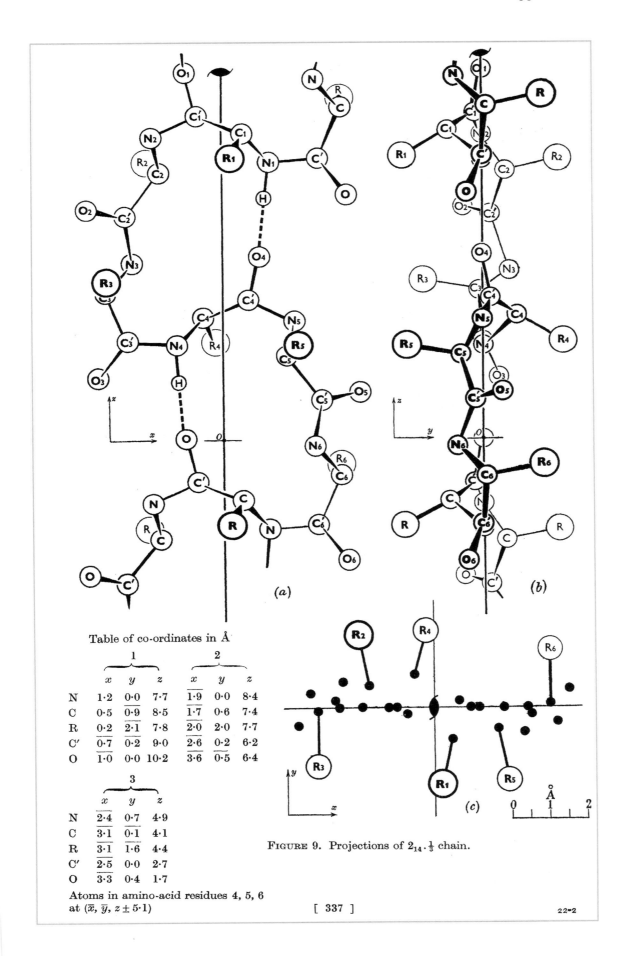

Table of co-ordinates in Å

		1			2	
	x	y	z	x	y	z
N	1·2	0·0	7·7	$\overline{1·9}$	0·0	8·4
C	0·5	$\overline{0·9}$	8·5	$\overline{1·7}$	0·6	7·4
R	0·2	$\overline{2·1}$	7·8	$\overline{2·0}$	2·0	7·7
C′	$\overline{0·7}$	0·2	9·0	$\overline{2·6}$	0·2	6·2
O	$\overline{1·0}$	0·0	10·2	$\overline{3·6}$	0·5	6·4

		3	
	x	y	z
N	$\overline{2·4}$	0·7	4·9
C	$\overline{3·1}$	0·1	4·1
R	$\overline{3·1}$	$\overline{1·6}$	4·4
C′	$\overline{2·5}$	0·0	2·7
O	$\overline{3·3}$	0·4	1·7

Atoms in amino-acid residues 4, 5, 6
at $(\overline{x}, \overline{y}, z \pm 5·1)$

FIGURE 9. Projections of $2_{14}.\frac{1}{3}$ chain.

338   Sir Lawrence Bragg, J. C. Kendrew and M. F. Perutz

Eight-atom ring (3_8)

We have not encountered any earlier description of this structure. Our model of it is illustrated in figure 10. The repeat distance is about 5·4 Å, and of course comprises three amino-acid residues. It is a feature of this structure that the three

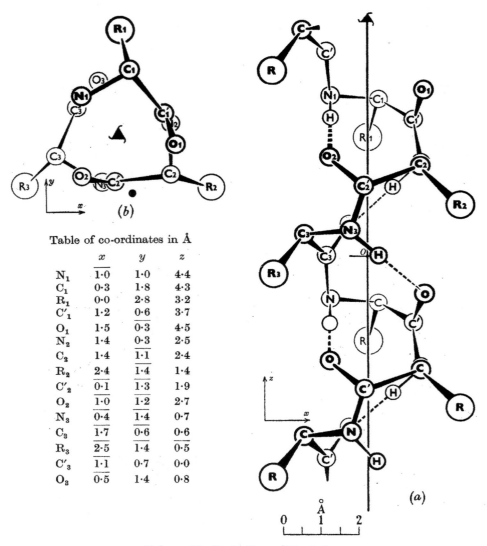

Table of co-ordinates in Å

	x	y	z
N_1	1·0	1·0	4·4
C_1	0·3	1·8	4·3
R_1	0·0	2·8	3·2
C'_1	1·2	0·6	3·7
O_1	1·5	0·3	4·5
N_2	1·4	0·3	2·5
C_3	1·4	$\overline{1·1}$	2·4
R_2	2·4	1·4	1·4
C'_2	$\overline{0·1}$	1·3	1·9
O_2	$\overline{1·0}$	1·2	2·7
N_3	$\overline{0·4}$	1·4	0·7
C_3	$\overline{1·7}$	0·6	0·6
R_3	$\overline{2·5}$	$\overline{1·4}$	$\overline{0·5}$
C'_3	$\overline{1·1}$	0·7	0·0
O_3	$\overline{0·5}$	1·4	0·8

FIGURE 10. Projections of 3_8 chain.

N—H...O bonds in each repeating unit are mutually perpendicular and therefore could not possibly be responsible for any infra-red pleochroic effects which might be observed in the crystal or fibre. However, we regard the whole interpretation of these effects as still equivocal, and hence, though we shall not discuss this model further (owing to its close similarity to 3_{10}), we cannot definitely reject it at this stage.

Polypeptide chain configurations in proteins   339

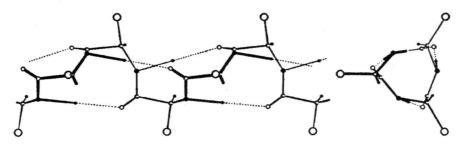

FIGURE 11. Projections of 3_{10} chain, after Huggins (1943). Repeat distance about 5·1 Å. Symbols as for figure 5.

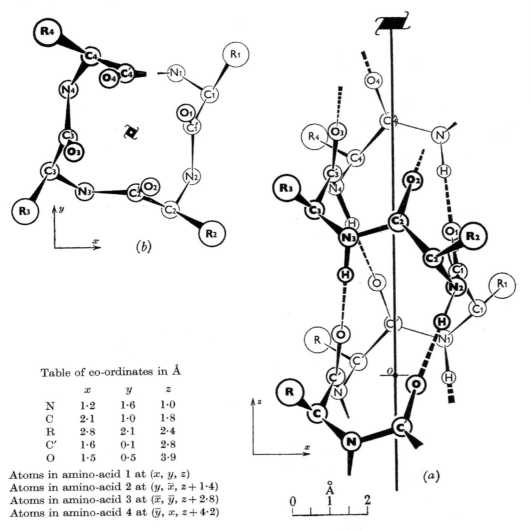

Table of co-ordinates in Å

	x	y	z
N	1·2	1·6	1·0
C	2·1	1·0	1·8
R	2·8	2·1	2·4
C'	1·6	0·1	2·8
O	1·5	0·5	3·9

Atoms in amino-acid 1 at (x, y, z)
Atoms in amino-acid 2 at $(y, \bar{x}, z+1\cdot4)$
Atoms in amino-acid 3 at $(\bar{x}, \bar{y}, z+2\cdot8)$
Atoms in amino-acid 4 at $(\bar{y}, x, z+4\cdot2)$

FIGURE 12. Projections of 4_{13} chain.

340   Sir Lawrence Bragg, J. C. Kendrew and M. F. Perutz

Ten-atom ring (3_{10})

This structure was first proposed by Taylor (1941); it was later discussed in detail by Huggins (1943), whose projections of it are reproduced in figure 11. Like 3_8 its repeat distance is rather over 5 Å, and contains three residues; it has the further attractive feature that the hydrogen bonds are oriented nearly parallel to the triad axis and could therefore contribute to the infra-red pleochroism of the molecule as a whole, granted a suitable orientation of the chains.

We discuss this structure in detail below.

Other threefold structures

We have found it impossible to build threefold structures containing rings of 11, 13 or more members; structures with these large rings invariably assume a symmetry higher than threefold.

(iii) Fourfold structures

We were able to build no fourfold structures having rings of fewer than eleven or more than thirteen members.

Eleven-atom ring (4_{11})

A structure of this type can be built, though neighbouring pairs of oxygen atoms fall abnormally close to one another (2·2 Å instead of the normal minimum of 2·8 Å). Because of this feature, and because the repeat unit (of 5·4 Å) contains *four* amino-acid residues, the structure is regarded as improbable.

Thirteen-atom ring (4_{13})

This version seems to be more plausible than 4_{11} since its rings are less strained; projections are given in figure 12. The repeat distance is again rather over 5 Å and, of course, contains four amino-acid residues. This structure will be further referred to in §7.

6. STRUCTURES NOT EXAMINED IN DETAIL

(a) Structures of higher symmetry

We have not examined in detail structures with fivefold, sixfold or higher symmetry since they would inevitably contain more than four residues per repeat unit. There is, nevertheless, no difficulty in building models of such structures. For example, there is a hexad structure 6_{20}, with a repeat distance of about 5 Å and containing six residues per repeat; it is an open helix whose internal diameter is between 7 and 8 Å.

(b) Structures in which only part of the CO and NH groups are hydrogen bonded

We mentioned in §4 above that structures of this type were considered comparatively unlikely, because they would be expected to have a higher free energy than those in which all CO and NH groups are bonded. Nevertheless, a general survey has been made. The list of possible structures is not as formidable as might appear at first sight, because the repeat distances along the fibre axis become too long when the fraction of bonded NH or CO groups falls below $\frac{1}{3}$, and often already when it falls

Polypeptide chain configurations in proteins 341

below $\frac{1}{2}$. Some structures were found to be sterically impossible, and others again, though capable of shortening to give the desired repeat of $n \times 5$ Å, contained no bond that would keep them in this shortened configuration.

The structures are described in table 2, which includes only three models having the correct repeat distance. Though these three, as well as some of the other structures listed in table 2, may have any type of screw symmetry owing to the rotational freedom of the individual rings, we decided for simplicity's sake to describe them all in terms of twofold screw axes. We append some comments on the three structures having approximately correct repeat distances.

TABLE 2

screw axis of symmetry	no. of atoms in ring	fraction of NH (and CO) groups which are hydrogen bonded	true repeat	no. of residues per repeat	illustrations (figure no.)	comments
twofold	7	$\frac{1}{2}$	11·6	4	13	two side-chains above and two below plane of main chain; free rotation between rings
	7	$\frac{1}{3}, \frac{1}{4}$, etc.	$\geqslant 10$	6, 8, etc.	—	'repeat' too long
	8	$\frac{1}{2}$	9·8	4	14	free rotation between rings
	8	$\frac{1}{3}, \frac{1}{4}$, etc	$\geqslant 10$	6, 8, etc.	—	'repeat' too long
	10	$\frac{1}{2}$	8·2	4	—	'repeat' too short
	10	$\frac{1}{3}$	12	6	—	'repeat' too long
	11	$\frac{1}{2}$	8·0	4	—	'repeat' too short
	11	$\frac{1}{3}$	9·6	6	15	possible structure; properties similar to 2_{13} and 2_{14}, though rings are not so flat; side chains point alternately up and down from the plane of the rings; free rotation between rings
	11	$\frac{1}{4}$	16·5	8	—	'repeat' too long
	13	$\frac{1}{2}$	—	—	—	no possible structure
	13	$\frac{1}{3}$	10·2	6	8	described in table 1
	13	$\frac{1}{4}$	$\geqslant 10$	8	—	'repeat' too long
	14	$\frac{1}{2}$	—	—	—	no possible structure
	14	$\frac{1}{3}$	10·2	6	9	described in table 1
	14	$\frac{1}{4}$	$\geqslant 10$	8	—	'repeat' too long
threefold	10	$\frac{1}{2}$	13·4	6	—	'repeat' too long
	11	$\frac{1}{2}$	13·0	6	—	'repeat' too long

(i) *Seven-atom ring. One-half of CO and NH groups hydrogen bonded* $(2_7 . \frac{1}{2})$

In its fully extended state this structure has a repeat distance of 11·6 Å, with four amino-acid residues per repeat, and pairs of side-chains protruding alternately above and below the plane of the main chains in the *a* form. In the *b* form pairs of side-chains protrude alternately right and left within the plane of the main chain. Neighbouring side-chains are thus brought improbably close together. This proximity can be avoided in a variety of ways, however, by departing from the twofold screw symmetry. Such departure produces a slight shortening of the repeat distance. In end-on projection a structure of the type $S_7 . \frac{1}{2}$ with high-order screw symmetry is likely to have an appearance similar to that of 3_8, 3_{10} or 4_{13}. The structure with twofold screw symmetry is illustrated in figure 13, which shows that neighbouring rings are free to rotate about the C'—C—N bonds.

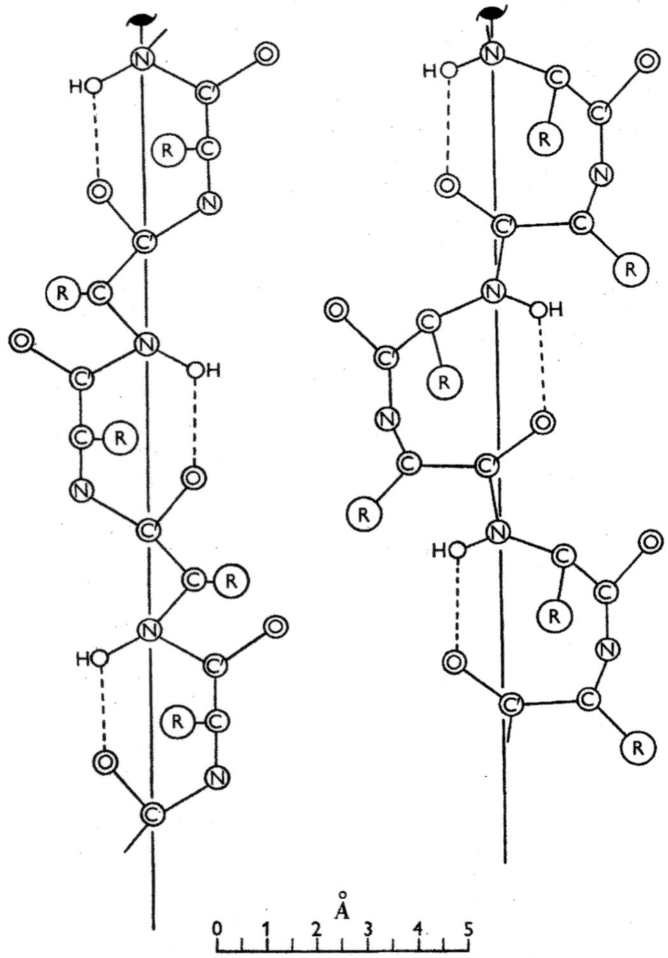

FIGURE 13   FIGURE 14

FIGURE 13. Sketch of the $2_7 \cdot \frac{1}{2}$ chain.
FIGURE 14. Sketch of the $2_8 \cdot \frac{1}{2}$ chain.

(ii) *Eight-atom ring. One-half of CO and NH groups hydrogen bonded* ($2_8 \cdot \frac{1}{2}$)

This structure has a repeat distance of 9.8 Å with four amino-acid residues and is shown in figure 14. Neighbouring rings are free to rotate about the N—C' bond, without appreciably affecting the length of the repeat. Side-chains protrude alternately parallel and perpendicular to the plane of the rings.

(iii) *Eleven-atom ring. One-third of NH and CO groups hydrogen bonded* ($2_{11} \cdot \frac{1}{3}$)

This structure has a repeat distance of 9.6 Å with six amino-acid residues, and is shown in figure 15. Side-chains protrude alternately up and down from the plane

Polypeptide chain configurations in proteins   343

of the rings. The properties of this model are very similar to those of 2_{13} and 2_{14} except that the individual rings are free to rotate about the C'—N bond. Viewed in end-on projection with the resolving power of the Pattersons of haemoglobin and myoglobin this structure would be indistinguishable from 2_{13} or 2_{14}.

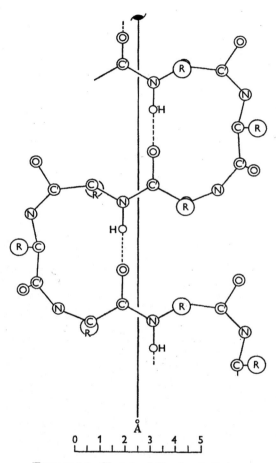

0 1 2 3 4 5 Å

FIGURE 15.   Sketch of the $2_{11} \cdot \frac{1}{3}$ chain.

It is seen that in each of the three structures listed above the hydrogen bonded rings are free to rotate relative to each other. None of the models can be tied down to one specific configuration, so that definite parameters, on which a comparison of observed and calculated intensities might be based, cannot be assigned. Depending on the nature of the screw symmetry, end-on projections of these models would resemble one or the other of the types of structure described in detail in § 5. With the resolving power available no distinction can be made between models whose end-on projections are roughly similar.

(c) Structures with secondary folds

We may distinguish structures of a higher degree of complexity in which the 'asymmetric unit' which the screw axis repeats consists not of one hydrogen-

344   Sir Lawrence Bragg, J. C. Kendrew and M. F. Perutz

bridged ring alone, but of more than one ring or of some combination of rings and chain. Some of the structures in this category are similar to those described in the preceding paragraph and suffer from the same drawbacks. Examples of such structures are those recently discussed by Mizushima, Simanouti, Tsuboi, Sugita & Kato (1949); for example, these authors illustrate one in which the 'asymmetric unit' is formed of two $2_7 b$ rings and one amino-acid residue which is not part of a ring; this asymmetric unit is operated upon by a screw dyad to form the chain.

Another structure of this type was proposed by Huggins (1943) and is illustrated in figure 16; here the asymmetric unit consists of three linked $2_7 a$ rings, and the whole chain has screw dyad symmetry. Using space-filling models we have found the repeat distance of this chain to be 8 Å, not 10·2 Å as stated by Huggins. We ourselves have found it possible to coil a $2_7 b$ structure into helical forms, with, for example, a screw hexad symmetry. The structure has an asymmetric unit consisting of a single seven-membered ring; the internal diameter of the helix is about 5 Å and its pattern repeats at intervals of 6 Å.

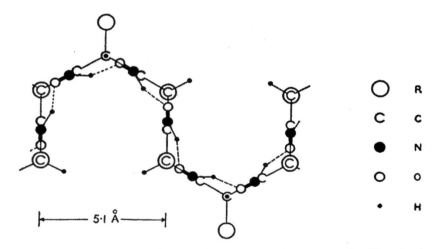

○	R
Ⓒ	C
●	N
○	O
·	H

FIGURE 16. Structure with secondary folds built up of $2_7 a$ rings, after Huggins (1943).

We have not made an exhaustive study of structures of this general type, because they all suffer from what we believe is a serious objection, namely, that, although the rings are held together by hydrogen bonds as in all the configurations mentioned earlier, the structure as a whole is not so held together. For example, there seems no obvious reason why Huggins's structure referred to in the last paragraph should not stretch to the ordinary $2_7 b$ structure, since such stretching would involve only the overcoming of van der Waals attractions, and not the rupture of hydrogen bonds. It seems to be generally accepted that the α-β transformation in keratin involves the breaking of hydrogen bonds at least, though it must be admitted that up to now studies of the energetics of this transformation have not given any clear indication of the nature of the processes involved.

To sum up, after excluding the types of chain which are incompatible with three or four residues in a sub-cell of $10 \times 10 \times 5$ Å, or are improbable for other reasons,

Polypeptide chain configurations in proteins   345

we are left with $2_7 b$, $2_{13} \cdot \frac{1}{3}$, $2_{14} \cdot \frac{1}{3}$, 3_8, 3_{10} and 4_{13}*. In the next section an attempt is made to narrow the possibilities still further by comparing the vector projections of these chains with the Patterson projections of haemoglobin and myoglobin. If, as is believed, the chains are parallel in their crystals, the vectors between atoms of the same chain should have a deciding influence on the form of the Patterson around the origin when the projection is upon a plane at right angles to the chain direction. This is a more favourable basis of comparison than the Patterson projections shown in figure 3 (*a*) and (*d*), because the chain vectors are crowded into a small area and their effects are superimposed.

7. Vector projections of chain models

(*a*) *End-on projections of the chains*

A series of vector projections on a plane at right angles to the chain are shown in figure 17. The atoms C, N, O have been assigned equal weights, and vectors have been drawn between every pair of atoms in one repeat of the chain. These are transferred to a common origin, and their ends are plotted as projected on a plane. For instance, in the $2_{14} \cdot \frac{1}{3}$ chain which has six amino-acid residues in a repeat distance of 10·2 Å or 30 atoms per repeat, there are 900 vectors, 30 of which constitute the peak at the origin. The projection is divided into squares of 1 Å side, and the number of vector ends in each square is counted. The density distribution can then be represented by contours at convenient levels. The vectors are distributed irregularly, and considerable smoothing is necessary in drawing the contours, but these contours serve to represent sufficiently accurately the type of Patterson diagram which the chains would give if arranged regularly in a simple unit cell with a wide spacing between the chains so that there is little overlap of the origin peaks. The correct representation would be attained by calculating the F's of this arrangement, and constructing a Patterson projection with them. Such a calculation has been carried out for certain chains in the myoglobin cell in §8 below.

Figure 17 (*a*) is the vector projection of the $2_{14} \cdot \frac{1}{3}$ chain illustrated in figure 9. The atoms lie very nearly in a plane, with the exception of the first atoms R of the side-chains, and the Patterson projection is correspondingly elongated in one direction and narrow in the other. In comparing this projection with the origin peak of myoglobin, it must be remembered that the crystal has a dyad screw axis perpendicular to the chains, so that the part of the Patterson distribution around the origin due to the chains is a combination of (*a*) with its counterpart formed by rotation about the axis. Assuming the axis to be nearly parallel to the plane of the chain, the result would be as shown in figure 17 (*b*), which is formed by superposing (*a*) and its mirror image. Figure 17 (*e*) shows the distribution around the origin in a Patterson projection of myoglobin on a plane perpendicular to the chains, and it will be noted that it has a general similarity to (*b*). Figure 17 (*f*) is a similar projection for horse methaemoglobin. Its form suggests that it may be due to the superposition of

* The chains $2_7 \cdot \frac{1}{2}$, $2_8 \cdot \frac{1}{2}$ and $2_{11} \cdot \frac{1}{3}$ are also possible, but these chains are so ill-defined by the limiting conditions hitherto available that detailed tests of their conformity with the observations cannot be made at the present stage.

356   Sir Lawrence Bragg, J. C. Kendrew and M. F. Perutz

Assuming an arrangement of this kind, pairs of chains have been inserted into the projection in the way indicated in figure 21 (e). The resulting values of F^2 are compared in table 6 with those observed, and it will be noticed again that the agreement is not so good as in the case of the $2_{13}.\frac{1}{3}$ and $2_{14}.\frac{1}{3}$ configurations. The calculated vector projection (figure 22 (d)) fails in the following respects:

(i) the origin peak is too thick and is insufficiently elongated; there is no sign of resolution into three peaks;

(ii) in the middle layer the four peaks are unresolved. The same difficulties are encountered in the 3_8 and 4_{13} structures.

9. Discussion of results

If we base our conclusions only on the projections on a plane at right angles to the chain direction, undoubtedly the best agreement with the experimentally determined projections and intensities of diffraction is given by the $2_{13}.\frac{1}{3}$ and $2_{14}.\frac{1}{3}$ chains. Considering that the side-chains and intermolecular liquid have been left completely out of account, the resemblance between the projections (figures 22 (a), (b), and 20 (b)) is very striking. In contrast to this agreement, in the structures shown in figures 22 (c) and (d) the lack of resolution of peaks in the middle layer would appear to be a decisive criterion for rejecting these structures for myoglobin. The nearest peaks in the middle layer of the observed vector projection are only 4·2 Å apart; it can be shown by drawing cross-sections of the vector equivalent of the various chains (as illustrated in figure 17) that resolution could be expected only for chains about as narrow, in the x direction, as the $2_{13}.\frac{1}{3}$ or $2_{14}.\frac{1}{3}$ configurations.

The extension of the comparison to the three-dimensional Patterson projection of haemoglobin in §7, however, shows how misleading it may be to base conclusions on reflexions around a single zone. It will be necessary to get a three-dimensional Patterson for myoglobin in order to assess whether these resemblances are significant or fortuitous.

The problem is very complex, and one is forced to rely on a number of items of evidence each of which is very slight. The conclusion that the chains are of a folded coplanar form resembling the $2_{13}.\frac{1}{3}$ or $2_{14}.\frac{1}{3}$ type would be more convincing if there were any indication that the form has obvious advantages over others. In X-ray analysis in general, when a crystal structure has been successfully analyzed and a model of it is built, it presents so neat a solution of the requirements of packing and interplay of atomic forces that it carries conviction as to its essential correctness. In the present case the models to which we have been led have no obvious advantages over their alternatives. Much more evidence must be accumulated before conclusions can be safely drawn. There are certain hopeful features, however, at the present stage of the investigation. Assuming that criteria of the type adopted in this paper are justified, a survey shows that the number of possible forms of chain is very restricted. There appears to be a real simplicity of chain structure in myoglobin, which will perhaps be shown by other favourably built proteins, and which makes it particularly suitable for intensive X-ray investigation. There is hope that the study of such proteins may lead to a reliable determination of the chain structure.

Polypeptide chain configurations in proteins 357

The authors wish to thank Mr C. W. Bunn for his helpful advice on the technique of model building, Mr F. H. C. Crick for his valuable criticisms of a number of points raised in §7, and Miss V. E. Marting for calculating the Fourier transform of the 3_{10} helix. They also thank the Medical Research Council for financial support of two of them (J. C. K. and M.F.P.)

REFERENCES

Albrecht, G. & Corey, R. B. 1939 *J. Amer. Chem. Soc.* **61**, 1087.
Ambrose, E. J., Elliott, A. & Temple, R. B. 1949 *Nature*, **163**, 859.
Ambrose, E. J. & Hanby, W. E. 1949 *Nature*, **163**, 483.
Astbury, W. T. 1949a *Brit. J. Radiol.* **22**, no. 259.
Astbury, W. T. 1949b *Nature*, **164**, 439.
Astbury, W. T. & Bell, F. O. 1941 *Nature*, **147**, 696.
Boyes-Watson, J., Davidson, E. & Perutz, M. F. 1947 *Proc. Roy. Soc.* A, **191**, 83.
Bragg, W. L. 1949 *Nature*, **164**, 7.
Chibnall, A. C. 1945 2nd Procter Memorial Lecture. Croydon: Int. Soc. of Leather Trades' Chemists.
Corey, R. B. 1948 *Advances in protein chemistry*, **4**, 385. New York: Academic Press Inc.
Darmon, S. E. & Sutherland, G. B. B. M. 1949 *Nature*, **164**, 440.
Huggins, M. L. 1943 *Chem. Rev.* **32**, 195.
Hughes, E. W. & Moore, W. J. 1949 *J. Amer. Chem. Soc.* **71**, 2618.
Kendrew, J. C. 1950 *Proc. Roy. Soc.* A, **201**, 62.
Kendrew, J. C. & Perutz, M. F. 1949 *Haemoglobin*, p. 161. London: Butterworths.
Levy, H. A. & Corey, R. B. 1941 *J. Amer. Chem. Soc.* **63**, 2095.
Mizushima, S., Simanouti, T., Tsuboi, M., Sugita, T. & Kato, E. 1949 *Nature*, **164**, 918.
Perutz, M. F. 1949 *Proc. Roy. Soc.* A, **195**, 474.
Sanger, F. 1948 *Nature*, **162**, 491.
Taylor, H. S. 1941 *Proc. Amer. Phil. Soc.* **85**, 1.
Zahn, H. 1947 *Z. Naturforschung*, **2b**, 104.

by an increase in protein content, while the amount of desoxyribonucleic acid remains unchanged.

Acknowledgments.—This work was supported by research grants from the University of California Board of Research.   We are greatly indebted to Professor A. W. Pollister, Dept. of Zoology, Columbia University, for allowing the senior author use of his laboratory facilities to conduct the measurements described herein.

[1] Salvatore, C. A., *Biol. Bull.*, **99**, 112–119 (1950).

[2] Caspersson, T., *Skand. Arch. Physiol.*, **73**, Suppl. 8 (1936).

[3] Pollister, A. W., and Ris, H., *Cold Spring Harbor Symp. Quant. Biol.*, **12**, 147–157 (1947).

[4] Swift, H. H., *Physiol. Zool.*, **23**, 169–198 (1950).

[5] Swift, H. H., these PROCEEDINGS, **36**, 643–654 (1950).

[6] Ris, H., and Mirsky, A. E., *J. Gen. Physiol.*, **33**, 125–146 (1949).

[7] Leuchtenberger, C., Vendrely, R., and Vendrely, C., these PROCEEDINGS, **37**, 33–37 (1951).

[8] Alfert, M., *J. Cell. Comp. Physiol.*, **36**, 381–410 (1950).

[9] Schrader, F., and Leuchtenberger, C., *Exp. Cell Res.*, **1**, 421–452 (1950).

[10] Pollister, A. W., and Leuchtenberger, C., these PROCEEDINGS, **35**, 66–71 (1949).

[11] Leuchtenberger, C., *Chromosoma*, **3**, 449–473 (1950).

[12] Mirsky, A. E., and Ris, H., *Nature*, **163**, 666–667 (1949).

THE STRUCTURE OF PROTEINS: TWO HYDROGEN-BONDED HELICAL CONFIGURATIONS OF THE POLYPEPTIDE CHAIN

BY LINUS PAULING, ROBERT B. COREY, AND H. R. BRANSON*

GATES AND CRELLIN LABORATORIES OF CHEMISTRY,
CALIFORNIA INSTITUTE OF TECHNOLOGY, PASADENA, CALIFORNIA†

Communicated February 28, 1951

During the past fifteen years we have been attacking the problem of the structure of proteins in several ways. One of these ways is the complete and accurate determination of the crystal structure of amino acids, peptides, and other simple substances related to proteins, in order that information about interatomic distances, bond angles, and other configurational parameters might be obtained that would permit the reliable prediction of reasonable configurations for the polypeptide chain. We have now used this information to construct two reasonable hydrogen-bonded helical configurations for the polypeptide chain; we think that it is likely that these configurations constitute an important part of the structure of both fibrous and globular proteins, as well as of synthetic polypeptides. A letter announcing their discovery was published last year.[1]

The problem that we have set ourselves is that of finding all hydrogen-bonded structures for a single polypeptide chain, in which the residues are equivalent (except for the differences in the side chain R).   An amino acid residue (other than glycine) has no symmetry elements.   The general operation of conversion of one residue of a single chain into a second residue equivalent to the first is accordingly a rotation about an axis accompanied by translation along the axis.   Hence the only configurations for a chain compatible with our postulate of equivalence of the residues are helical configurations.   For rotational angle 180° the helical configurations may degenerate to a simple chain with all of the principal atoms, C, C′ (the carbonyl carbon), N, and O, in the same plane.

We assume that, because of the resonance of the double bond between the carbon-oxygen and carbon-nitrogen positions, the configuration of each residue

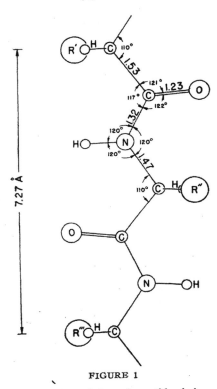

FIGURE 1

Dimensions of the polypeptide chain.

is planar.

This structural feature has been verified for each of the amides that we have studied.   Moreover, the resonance theory is now so well grounded and its experimental substantiation so extensive that there can be no doubt whatever about its application to the amide group.   The observed C—N distance, 1.32 Å, corresponds to nearly 50 per cent double-bond character, and we may conclude that rotation by as much as 10° from the planar configuration would result in instability by about 1 kcal. mole^{-1}.   The interatomic distances and bond angles within the residue are assumed to have the values shown in figure 1.   These values have been formulated[2] by consideration of the experimental values found in the crystal structure studies of DL-alanine,[3] L-threonine,[4] N-acetylglycine[5], and β-glycylglycine[6] that have been made in our

Laboratories.   It is further assumed that each nitrogen atom forms a hydrogen bond with an oxygen atom of another residue, with the nitrogen-oxygen distance equal to 2.72 Å, and that the vector from the nitrogen atom to the hydrogen-bonded oxygen atom lies not more than 30° from the N—H direction.   The energy of an N—H · · · O=C hydrogen bond is of the order

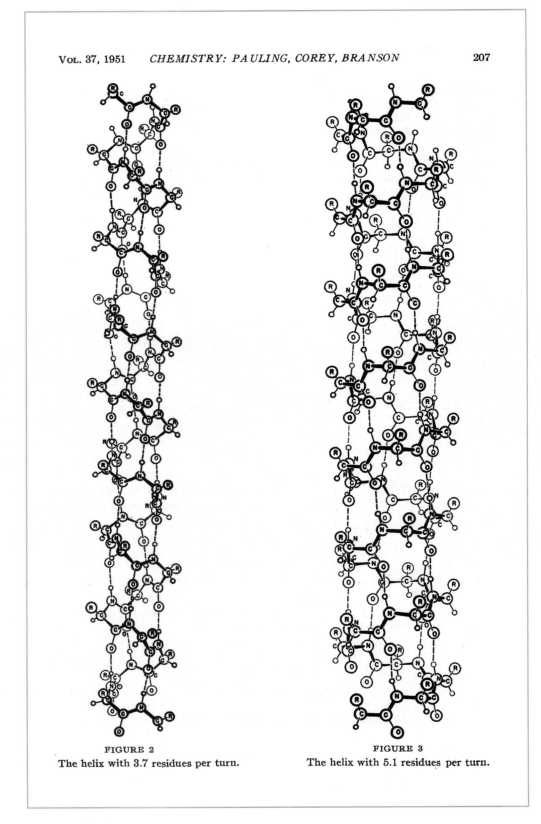

VOL. 37, 1951 *CHEMISTRY: PAULING, COREY, BRANSON* 207

FIGURE 2
The helix with 3.7 residues per turn.

FIGURE 3
The helix with 5.1 residues per turn.

of 8 kcal. mole^{-1}, and such great instability would result from the failure to form these bonds that we may be confident of their presence.   The N—H $\cdots$ O distance cannot be expected to be exactly 2.72 Å, but might deviate somewhat from this value.

Solution of this problem shows that there are five and only five configurations for the chain that satisfy the conditions other than that of direction of the hydrogen bond relative to the N—H direction.   These correspond to the values 165°, 120°, 108°, 97.2° and 70.1° for the rotational angle.   In the first, third, and fifth of these structures the $\diagdown$CO group is negatively and the $\diagdown$N—H group positively directed along the helical axis, taken as the direction corresponding to the sequence—CHR—CO—NH—CHR— of atoms in the peptide chain, and in the other two their directions are reversed.   The first three of the structures are unsatisfactory, in that the

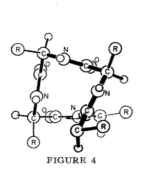

FIGURE 4

Plan of the 3.7-residue
helix.

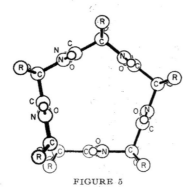

FIGURE 5

Plan of the 5.1-residue helix.

N—H group does not extend in the direction of the oxygen atom at 2.72 Å; the fourth and fifth are satisfactory, the angle between the N—H vector and N—O vector being about 10° and 25° for these two structures respectively.   The fourth structure has 3.69 amino acid residues per turn in the helix, and the fifth structure has 5.13 residues per turn.   In the fourth structure each amide group is hydrogen-bonded to the third amide group beyond it along the helix, and in the fifth structure each is bonded to the fifth amide group beyond it; we shall call these structures either the 3.7-residue structure and the 5.1-residue structure, respectively, or the third-amide hydrogen-bonded structure and the fifth-amide hydrogen-bonded structure.

Drawings of the two structures are shown in figures 2, 3, 4, and 5.

For glycine both the 3.7-residue helix and the 5.1-residue helix could occur with either a positive or a negative rotational translation; that is, as either a positive or a negative helix, relative to the positive direction of the helical axis given by the sequence of atoms in the peptide chain. For other amino acids with the L configuration, however, the positive helix and the negative helix would differ in the position of the side chains, and it might well be expected that in each case one sense of the helix would be more stable than the other. An arbitrary assignment of the R groups has been made in the figures.

The translation along the helical axis in the 3.7-residue helix is 1.47 Å, and that in the 5.1-residue helix is 0.99 Å. The values for one complete turn are 5.44 Å and 5.03 Å, respectively. These values are calculated for the hydrogen-bond distance 2.72 Å; they would have to be increased by a few per cent, in case that a larger hydrogen-bond distance (2.80 Å, say) were present.

The stability of our helical structures in a non-crystalline phase depends solely on interactions between adjacent residues, and does not require that the number of residues per turn be a ratio of small integers. The value 3.69 residues per turn, for the third-amide hydrogen-bonded helix, is most closely approximated by 48 residues in thirteen turns (3.693 residues per turn), and the value 5.13 for the other helix is most closely approximated by 41 residues in eight turns. It is to be expected that the number of residues per turn would be affected somewhat by change in the hydrogen-bond distance, and also that the interaction of helical molecules with neighboring similar molecules in a crystal would cause small torques in the helixes, deforming them slightly into configurations with a rational number of residues per turn. For the third-amide hydrogen-bonded helix the simplest structures of this sort that we would predict are the 11-residue, 3-turn helix (3.67 residues per turn), the 15-residue, 4-turn helix (3.75), and the 18-residue, 5-turn helix (3.60). We have found some evidence indicating that the first and third of these slight variants of this helix exist in crystalline polypeptides.

These helical structures have not previously been described. In addition to the extended polypeptide chain configuration, which for nearly thirty years has been assumed to be present in stretched hair and other proteins with the β-keratin structure, configurations for the polypeptide chain have been proposed by Astbury and Bell,[7] and especially by Huggins[8] and by Bragg, Kendrew, and Perutz.[9] Huggins discussed a number of structures involving intramolecular hydrogen bonds, and Bragg, Kendrew, and Perutz extended the discussion to include additional structures, and investigated the compatibility of the structures with x-ray diffraction data for hemoglobin and myoglobin. None of these authors proposed either our 3.7-residue helix or our 5.1-residue helix. On the other hand, we would eliminate, by our basic postulates, all of the structures proposed by them. The reason for the difference in results obtained by other investigators and by us through essentially similar arguments is that both Bragg and his collaborators and Huggins discussed in detail only helical structures with an integral number of residues per turn, and moreover assumed only a rough approximation to the requirements about interatomic distances, bond angles, and planarity of the conjugated amide group, as given by our investigations of simpler substances. We contend that these stereochemical features must be very closely retained in stable configurations of polypeptide chains in proteins, and that there is no special stability associated with an integral number of residues per turn in the helical molecule. Bragg, Kendrew, and Perutz have described a structure topologically similar to our 3.7-residue helix as a hydrogen-bonded helix with 4 residues per turn. In their thorough comparison of their models with Patterson projections for hemoglobin and myoglobin they eliminated this structure, and drew the cautious conclusion that the evidence favors the non-helical 3-residue folded α-keratin configuration of Astbury and Bell, in which only one-third of the carbonyl and amino groups are involved in intramolecular hydrogen-bond formation.

It is our opinion that the structure of α-keratin, α-myosin, and similar fibrous proteins is closely represented by our 3.7-residue helix, and that this helix also constitutes an important structural feature in hemoglobin, myoglobin, and other globular proteins, as well as of synthetic polypeptides. We think that the 5.1-residue helix may be represented in nature by supercontracted keratin and supercontracted myosin. The evidence leading us to these conclusions will be presented in later papers.

Our work has been aided by grants from The Rockefeller Foundation, The National Foundation for Infantile Paralysis, and The U. S. Public Health Service. Many calculations were carried out by Dr. S. Weinbaum.

Summary.—Two hydrogen-bonded helical structures for a polypeptide chain have been found in which the residues are stereochemically equivalent, the interatomic distances and bond angles have values found in amino acids, peptides, and other simple substances related to proteins, and the conjugated amide system is planar. In one structure, with 3.7 residues per turn, each carbonyl and imino group is attached by a hydrogen bond to the complementary group in the third amide group removed from it in the polypeptide chain, and in the other structure, with 5.1 residues per turn, each is bonded to the fifth amide group.

* Present address, Howard University, Washington, D. C.
† Contribution No. 1538.
[1] Pauling, L., and Corey, R. B., *J. Am. Chem. Soc.*, **72**, 5349 (1950).
[2] Corey, R. B., and Donohue, J., *Ibid.*, **72**, 2899 (1950).

   [3] Lévy, H. A., and Corey, R. B., *Ibid.*, **63**, 2095 (1941).   Donohue, J., *Ibid.*, **72**, 949 (1950).

   [4] Shoemaker, D. P., Donohue, J., Schomaker, V., and Corey, R. B., *Ibid.*, **72**, 2328 (1950).

   [5] Carpenter, G. B., and Donohue, J., *Ibid.*, **72**, 2315 (1950).

   [6] Hughes, E. W., and Moore, W. J., *Ibid.*, **71**, 2618 (1949).

   [7] Astbury, W. T., and Bell, F. O., *Nature*, **147**, 696 (1941).

   [8] Huggins, M. L., *Chem. Rev.*, **32**, 195 (1943).

   [9] Bragg, L., Kendrew, J. C., and Perutz, M. F., *Proc. Roy. Soc.*, **A203**, 321 (1950).

CONCERNING NON-CONTINUABLE, TRANSCENDENTALLY TRANSCENDENTAL POWER SERIES

By F. Bagemihl

Department of Mathematics, University of Rochester

Communicated by J. L. Walsh, February 23, 1951

The main purpose of this note is to show that power series of the kind described in the title can be obtained from a given power series by simply multiplying certain of its coefficients by -1.

Consider the class $\mathcal{K}$ of power series of the form $\sum_{\nu=0}^{\infty} a_\nu z^\nu$ whose circle of convergence is the unit circle.   There are $\mathfrak{c}$ elements in $\mathcal{K}$ (where $\mathfrak{c}$ denotes the power of the continuum).   Let $\mathcal{C}$ be the class of those series in $\mathcal{K}$ which can be continued beyond the unit circle, and let $\mathcal{A}$ be the class of those series in $\mathcal{K}$ which satisfy an algebraic differential equation.   Denote by $\mathcal{C}'$, $\mathcal{A}'$, the respective complements of $\mathcal{C}$, $\mathcal{A}$, with respect to $\mathcal{K}$.

There are the following sufficient conditions for a series in $\mathcal{K}$ to belong to $\mathcal{C}'$, $\mathcal{A}'$, respectively:

$(A)^1$ *Let* $\{\lambda_\nu\}$ $(\nu = 1, 2, 3, \ldots)$ *be an increasing sequence of non-negative integers such that* $\lambda_\nu/\nu \to \infty$ *as* $\nu \to \infty$. *If* $\sum_{\nu=1}^{\infty} a_\nu z^{\lambda_\nu}$ *belongs to* $\mathcal{K}$, *then it also belongs to* $\mathcal{C}'$.

$(B)^2$ *Let* $\{\lambda_\nu\}$ $(\nu = 1, 2, 3, \ldots)$ *be a sequence of non-negative integers such that* $\lambda_{\nu+1} > \nu\lambda_\nu$ *for every* ν. *If* $\sum_{\nu=1}^{\infty} a_\nu z^{\lambda_\nu}$ *belongs to* $\mathcal{K}$, *then it also belongs to* $\mathcal{A}'$.

The series $\sum_{\nu=0}^{\infty} z^\nu$, which represents $(1 - z)^{-1}$ for $|z| < 1$, belongs to $\mathcal{C}\mathcal{A}$ (i.e., to both $\mathcal{C}$ and $\mathcal{A}$).   The series $\sum_{\nu=0}^{\infty} b_\nu z^\nu$, which represents the meromorphic function $\Gamma(z + 1)$ for $|z| < 1$, belongs to $\mathcal{C}$ and^3 to $\mathcal{A}'$.   According to (A), $\sum_{\nu=0}^{\infty} z^{\nu^2}$ belongs to $\mathcal{C}'$, and it is known^4 that this series belongs to $\mathcal{A}$.   Finally, it follows from (A) and (B) that $\sum_{\nu=0}^{\infty} z^{\nu!}$ belongs to $\mathcal{C}'\mathcal{A}'$.   Thus,

I Wish I'd Made You Angry Earlier

Fifty years ago, the great unsolved problem of biology seemed to be the structure of proteins.

Bill Astbury, a physicist and X-ray crystallographer working for the Wool Research Association in Leeds (United Kingdom), discovered that the fibrous protein keratin, found in wool, horn, nails, and muscle, gave a common X-ray diffraction pattern consisting of just two reflections, a meridional one at 5.1 Å and an equatorial one at 9.8 Å.

Astbury called this the α-keratin pattern. When these fibers were stretched under steam, a new pattern appeared with meridional reflection at 3.4 Å and two equatorial ones at 4.5 and 9.7 Å. Astbury called this the β-keratin pattern. He concluded that it arose from the regular repeat of amino acid residues along straight polypeptide chains, whereas the chains in α-keratin must be folded or coiled such that several amino acid residues form a pattern repeating every 5.1 Å along the fiber axis.

It appeared that the key to understanding the structure of proteins lay in the elucidation of this ubiquitous fold, but the meager information contained in the X-ray diffraction patterns did not provide enough clues to uncover it.

By 1950, J.C. Kendrew and I had evidence that the same fold of the polypeptide chain occurred also in the two globular proteins myoglobin and hemoglobin. W.L. Bragg, the pioneer of X-ray crystallography who was our professor at the Cavendish Laboratory in Cambridge, stimulated us to attack the problem by building molecular models.

189

To start us off, he hammered nails representing amino acid residues into a broomstick in a helical pattern with an axial distance between successive turns (or pitch) of 5.1 centimeters. Kendrew and I had great difficulty in building real models of helical polypeptide chains with the right pitch; no matter whether we built them with two, three or four amino acid residues per turn, their bond angles were always strained. After some months we published our work jointly with Bragg in the *Proceedings of the Royal Society*, but without any firm conclusions about the correct fold.

One Saturday morning shortly after our paper had appeared, I went to the Cavendish library, and in the latest issue of the *Proceedings of the National Academy of Sciences of the United States of America* found a series of papers by Linus Pauling together with the crystallographer R.B. Corey. In their first paper they proposed an answer to the long-standing riddle of the structure of α-keratin, suggesting that it consisted of helical polypeptide chains with a non-integral repeat of 3.6 amino acid residues per turn. Their helix had a pitch not of 5.1 Å, as Astbury's X-ray picture seemed to demand, but of 5.4 Å, which was consistent with the repeat found in fibers of certain synthetic polypeptides by C.H. Bamford and A. Elliott and their colleagues at the Courtaulds Research Laboratories.

I was thunderstruck by Pauling and Corey's paper. In contrast to Kendrew's and my helices, theirs was free of strain; all the amide groups were planar and every carbonyl group formed a perfect hydrogen bond with an imino group four residues further along the chain.

The structure looked dead right. How could I have missed it? Why had I not kept the amide groups planar? Why had I stuck blindly to Astbury's 5.1 Å repeat? On the other hand, how could Pauling and Corey's helix be right, however nice it looked, if it had the wrong repeat? My mind was in a turmoil. I cycled home to lunch and ate it oblivious of my children's chatter and unresponsive to my wife's inquiries as to what the matter was with me today.

Suddenly I had an idea. Pauling and Corey's α-helix was like a spiral staircase in which the amino acid residues formed the steps and the height of each step was 1.5 Å. According to diffraction theory, this regular repeat should give rise to a strong X-ray reflection of 1.5 Å spacing from planes perpendicular to the fiber axis. As far as I knew, such a reflection had never been reported, either from "natural" proteins like hair and muscle or from synthetic polypeptides. Hence, I concluded, the α-helix must be wrong.

But wait! Suddenly I remembered a visit to Astbury's laboratory and realized that the geometry of his X-ray setup would have precluded observation of the 1.5 Å reflection because he oriented his fibers with their long axes perpendicular to the X-ray beam, while observation of the 1.5 Å reflection would have required inclining them at the Bragg angle of 31°. Furthermore, Astbury used a flat plate camera that was too narrow to record a reflection deflected from the incident X-ray beam by 2 x 31°.

In mad excitement, I cycled back to the lab and looked for a horse hair that I had kept tucked away in a drawer. I stuck it on a goniometer head at an angle of 31° to the incident X-ray beam; instead of Astbury's flat plate camera I put a cylindrical film around it that would catch all reflection with Bragg angles of up to 85°.

After a couple of hours, I developed the film, my heart in my mouth. As soon as I put the light on I found a strong reflection at 1.5 Å spacing, exactly as demanded by Pauling and Corey's α-helix. The reflection did not by itself prove anything, but it excluded all alternative models that had been put forward by ourselves and others and was consistent only with the α-helix.

On Monday morning I stormed into Bragg's office to show him my X-ray diffraction picture. When he asked me what made me think of this crucial experiment, I told him that the idea was sparked off by my fury over having missed building that beautiful structure myself. Bragg's prompt reply was, "I wish I had made you angry earlier!" because discovery of the 1.5 Å reflection would have led us straight to the α-helix.

The inconsistency between the 5.4 and 5.1 Å continued to worry me until one morning about two years later when Francis Crick arrived at the lab with two rubber tubes around which he had pinned corks with a helical repeat of 3.6 corks per turn and a pitch of 5.4 centimeters. He showed me that the two tubes could be wound around each other to make a double helix such that the corks neatly interlocked. This shortened the pitch of the individual chains, when projected onto the fiber axis, from 5.4 to 5.1 centimeters, as required by the X-ray pattern of α-keratin.

Such a double helix was eventually found by my colleagues A.D. McLachlan and J. Karn in the muscle protein myosin.

No. 4261   June 30, 1951   NATURE   1053

NEW X-RAY EVIDENCE ON THE CONFIGURATION OF POLYPEPTIDE CHAINS

Polypeptide Chains in Poly-γ-benzyl-L-glutamate, Keratin and Hæmoglobin

POLYPEPTIDE chains in certain synthetic polymers, in fibrous proteins of the keratin–myosin–fibrinogen group, and also in hæmoglobin, appear to be coiled or folded to about half the length of a fully stretched chain. Many different chain configurations have been proposed to account for the X-ray diffraction data[1], the latest being those of Pauling, Corey and Branson[2]. Until now, however, the lack of any simple and decisive criterion in the X-ray diffraction pattern has made it difficult to test the validity of proposed models. This communication describes a new reflexion, not hitherto observed, which is given by the proteins mentioned above. The spacing at which this reflexion appears excludes all models except the 3·7 residue helix of Pauling, Corey and Branson, with which it is in perfect concord.

This model has two types of repeat: (a) the distance between successive turns of the spiral (5·55 A.), and (b) the spacing along the chain of successive amino-acid residues (1·5 A.). Thus there are in this model 3·7 residues per turn. Pauling and Corey[3] quote a calculation by V. Shoemaker showing that the 5·1-A. reflexion can be explained by the turn of the spiral. I have found a new reflexion from planes perpendicular to the fibre axis at a spacing of 1·50 A. which corresponds to the repeat of the amino-acid residues along the chain. This reflexion was discovered by oscillating the specimens about a direction normal to the fibre axis, so as to satisfy Bragg angles for planes perpendicular to that axis, and by taking photographs on cylindrical films of 3-cm. radius instead of the flat plates normally used.

Structure proposed by	Illustration	Repeat of pattern	Screw axis	Spacing of first 0k0 reflexion
Ambrose, Elliott and Temple (ref. 4)	Fig. 6, ref. 1	5·0–5·6	2	2·5–2·8
Bragg, Kendrew and Perutz (ref. 1)	Fig. 10, ref. 1	5·4	3	1·8
Huggins (ref. 5)	Fig. 11, ref. 1	5·2	3	1·7
Bragg, Kendrew and Perutz (ref. 1)	Fig. 12, ref. 1	5·6	4	1·4
Astbury and Bell (ref. 6)	Fig. 8, ref. 1	10·2	2	5·1 ; 1·7
Pauling, Corey and Branson	Fig. 2, ref. 2	$X \times 1·5$	–	1·5

The appearance of a 1·50-A. reflexion from planes perpendicular to the fibre axis is incompatible with any other model so far proposed, for the following reason. All these models possess screw axes of symmetry. Thus if the chain axis Y is an n-fold screw, $0k0$ should be absent for $k = nm$, where m is any integer. The first $0k0$ reflexion should occur at a spacing of $d = b/n$, where b is the repeat of pattern along the chain. The table lists the structures so far proposed, together with their repeat distances, their types of screw axes and the spacing of the first $0k0$ reflexion to be expected. It is seen that the only structure to give a 1·5-A. reflexion is that of Pauling, Corey and Branson; the one that comes closest to it (1·4 A.) is a topographically similar model with a fourfold screw axis proposed by Bragg, Kendrew and Perutz.

Bamford, Hanby and Happey[7] attempted to interpret their X-ray photograph of poly-γ-benzyl-L-

glutamate in terms of a structure possessing a repeat of 5·0–5·6 A. and a screw dyad, which should give a strong reflexion at 2·5–2·8 A. Oscillation photographs of an oriented film of the polymer, kindly lent to me by Dr. A. Elliott, were taken about the normal to the fibre axis, but showed no such reflexion. On the other hand, an extremely powerful reflexion was discovered at a spacing of 1·50 A. (Fig. 1), in agreement with the structure postulated by Pauling and Corey[8]. The presence of this reflexion does not in itself prove the correctness of Pauling and Corey's structure, but taken in conjunction with the favourable agreement of observed and calculated intensities of other reflexions already obtained by these authors, it leaves little doubt about their structure being right.

The reflexion at 1·50 A. was also discovered on oscillation photographs of horse hair and of porcupine quill tip. (It is actually listed as one of the meridional reflexions of porcupine quill tip by MacArthur[9], but its significance does not seem to have been realized.) Fig. 2 is a Geiger counter spectrometer record of horse hair kindly placed at my disposal by Mr. Andrew Lang, which shows the relative intensities of the 5·1-A. and the 1·5-A. reflexions. When hair is stretched to the β-form the 1·5-A. reflexion vanishes.

Pauling and Corey also propose a structure of feather rachis keratin[10]. So far as can be seen from their illustration, this consists to about three-quarters of its volume of 3·7 residue helices, the remaining quarter being made up of 'pleated sheets' of extended chains. If this structure were correct, feather rachis keratin should show the 1·5-A. reflexion. Prolonged exposure of seagull's feather rachis set at the appro-

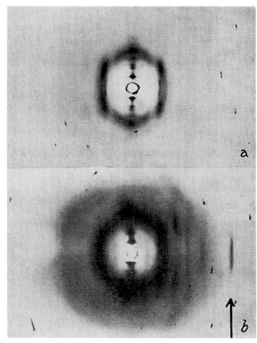

Fig. 1.   X-ray photographs of poly-γ-benzyl-L-glutamate. (a) Conventional picture with fibre axis horizontal and normal to X-ray beam. (b) Fibre axis oscillated between angles of 55° and 85° from X-ray beam. Arrow indicates 1·50-A. reflexion. 3-cm. cylindrical camera, copper Kα radiation

1054   N A T U R E   June 30, 1951 VOL. 167

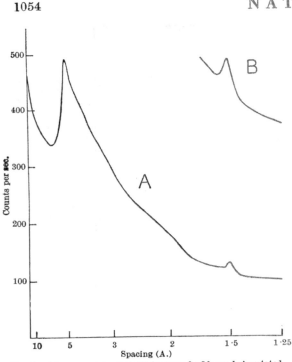

Fig. 2. Geiger counter spectrometer record of horse hair rotated about a normal to the fibre axis. (*A*) Bragg angle for 0*k*0 planes $2\theta = 7°-75°$, taken with reduced aperture of beam. (*B*) Repeat of $2\theta = 55°-75°$ with full aperture of beam

priate Bragg angle revealed no trace of such a reflexion, which shows that the structure proposed by Pauling and Corey must be wrong.

The three-dimensional Patterson synthesis of horse methæmoglobin shows rod-like concentrations of high vector density parallel to the crystallographic X-axis[11]. My interpretation of that synthesis led me to the conclusion that the hæmoglobin molecule consists of a compact bundle of close-packed chains running parallel to the X-axis and having the α-keratin configuration. The X-ray diffraction pattern of hæmoglobin fades out at a spacing of about 2 A., and no diffraction pattern at smaller spacings had hitherto been observed. Taking the X-axis as a possible chain axis, a search was made for reflexions in the 1·5-A. region by taking a 5° oscillation photograph in the appropriate orientation of the crystal. A picture was obtained in which the bulk of the reflexions fade out at a spacing of 2 A. as usual, but protruding from this is a faint bulge with a distinct maximum of intensity at 1·50 A. A preliminary search has not revealed any such effect in other crystallographic directions.

The three-dimensional Patterson synthesis of hæmoglobin shows neighbouring chains to be 10·5 A. apart and arranged in cylindrical close-packing. Taking the density of the protein as 1·30 and the mean residue weight as 112·5[11], the number of residues in a 'sub-cell', $10·5 \times 10·5 \times 1·5 \times \sin 60° = 143$ A.[3], can be calculated :

$$\frac{143 \times 1·3 \times N}{112·5} = 1·00 \text{ residue,}$$

which is the number to be expected for the 3·7 residue helix. These results add to the evidence in favour of the hæmoglobin structure proposed by me[11] and indicate that the chains are coiled to form 3·7 residue helices in poly-γ-benzyl-L-glutamate, in α-keratin and in hæmoglobin.

The discovery of the 1·5-A. reflexion shows that even relatively disordered substances like hair may contain an atomic pattern of such high intrinsic regularity that it gives rise to X-ray diffraction effects at spacings where they had never before been suspected.

M. F. PERUTZ

Cavendish Laboratory,
University of Cambridge.

[1] Bragg, W. L., Kendrew, J. C., and Perutz, M. F., *Proc. Roy. Soc.*, A, **203**, 321 (1950).
[2] Pauling, L., Corey, R. B., and Branson, H. R., *Proc. U.S. Nat. Acad. Sci.*, **37**, 205 (1951).
[3] Pauling, L., and Corey, R. B., *Proc. U.S. Nat. Acad. Sci.*, **37**, 261 (1951).
[4] Ambrose, E. J., Elliott, A., and Temple, R. B., *Nature*, **163**, 859 (1949).
[5] Huggins, M. L., *Chem. Rev.*, **32**, 195 (1943).
[6] Astbury, W. T., and Bell, F. O., *Nature*, **147**, 696 (1941).
[7] Bamford, C. H., Hanby, W. E., and Happey, F., *Proc. Roy. Soc.*, A, **205**, 30 (1951).
[8] Pauling, L., and Corey, R. B., *Proc. U.S. Nat. Acad. Sci.*, **37**, 241 (1951).
[9] MacArthur, I., *Nature*, **152**, 38 (1943).
[10] Pauling, L., and Corey, R. B., *Proc. U.S. Nat. Acad. Sci.*, **37**, 256 (1951).
[11] Perutz, M. F., *Proc. Roy. Soc.*, A, **195**, 474 (1949).
[12] Tristram, G. R., "Hæmoglobin", 109 (London, 1949).

Polypeptide Chains in Frog Sartorius Muscle

Oscillation photographs of dried frog sartorius muscle were taken about a direction normal to the fibre axis using a 3-cm. cylindrical camera and copper radiation. The specimens included muscle dried in the stretched, relaxed and contracted state (contraction by electrical stimulus). All photographs show the 1·5-A. reflexion from planes normal to the fibre axis, as described in the preceding communication. The reflexion is most intense on photographs of stretched muscle, slightly weaker on pictures of relaxed muscle, and very faint on photographs of contracted muscle. Thus both stretched and relaxed muscle seem to contain polypeptide chains coiled in the 3·7 residue helix and running parallel to the fibre axis. It is too early to say whether the weakening of the 1·5-A. reflexion on contraction is due to a change in chain configuration or to the disorientation of larger units. In any event, clear evidence on the mechanism of contraction cannot be expected unless X-ray photographs are taken during an actual twitch. In addition to the 1·5-A. reflexion, which can be recognized as an 0*k*0 reflexion by its small angular spread, the photographs also show a wide arc at 2·9–3·0 A. which is about as intense as the 5·1-A. reflexion. This arc is strongest in stretched and weakest in contracted muscle, and was also observed in oscillation photographs of hair.

Our results are incompatible with the mechanism of muscle contraction proposed by Pauling and Corey[1], who suggest that the chains in extended muscle are almost fully stretched, and that they coil up to form 3·7 residue helixes on contraction. On the other hand, our findings are in accord with those of Astbury and Dickinson[2], who showed both extended and relaxed muscle to have the α-keratin structure which becomes disorientated on contraction.

We are grateful to the Medical Research Council for financing the work described above.

H. E. HUXLEY
M. F. PERUTZ

Cavendish Laboratory, Cambridge.

[1] Pauling, L., and Corey, R. B., *Proc. U.S. Nat. Acad. Sci.*, **37**, 261 (1951).
[2] Astbury, W. T., and Dickinson, S., *Proc. Roy. Soc.*, B, **129**, 307 (1940).

Vol. 37, 1951   *CHEMISTRY: PAULING AND COREY*   729

with positive slope but their salts with small anions yield linear plots with negative slope.

[1] Onsager, L., *Physik. Z.*, **28**, 277 (1927); Onsager, L., and Fuoss, R. M., *J. Phys. Chem.*, **36**, 2689 (1932); Fuoss, R. M., *Physik. Z.*, **35**, 59 (1935).

[2] Shedlovsky, T., *J. Am. Chem. Soc.*, **54**, 405 (1932).

[3] Shedlovsky, T., Brown, A. S., and MacInnes, D. A., *Trans. Electrochem. Soc.*, **66**, 165 (1934); Benson, G. C., and Gordon, A. R., *J. Chem. Phys.*, **13**, 473 (1948); Monk, C. B., *J. Am. Chem. Soc.*, **70**, 328 (1948); Owens, B. B., *Ibid.*, **61**, 1393 (1939).

CONFIGURATIONS OF POLYPEPTIDE CHAINS WITH FAVORED ORIENTATIONS AROUND SINGLE BONDS: TWO NEW PLEATED SHEETS

By Linus Pauling and Robert B. Corey

Gates and Crellin Laboratories of Chemistry,* California Institute of Technology, Pasadena, California

Communicated September 4, 1951

In recent papers we have described several configurations of polypeptide chains with interatomic distances, bond angles, and other structural features as indicated by the studies in these Laboratories of the structure of crystals of amino acids, simple peptides, and related substances, and have presented evidence for their presence in synthetic polypeptides, fibrous proteins, and globular proteins.[1-9] The requirements that we have imposed for a satisfactory polypeptide configuration, in addition to the correct bond distances and bond angles, are that each amide group be planar, with either the cis configuration or the trans configuration about the C'—N bond (which has nearly 50 per cent double-bond character), and that each carbonyl and imino group (except for proline or hydroxyproline residues) be involved in the formation of a hydrogen bond with N—H···O distance approximately 2.8 A and with the oxygen atom nearly on the N—H axis. The only other structural parameters involved in a configuration of polypeptide chains are the orientations around the N—C and C—C' single bonds. In the following paragraphs we discuss the question of the relative stability of structures with different values of these orientational parameters.

We are interested in the potential function for orientation around a single bond between the α carbon atom, which forms three single bonds in addition to the bond under consideration, and either the nitrogen atom, which forms a single bond to its hydrogen atom and a bond with somewhat less than 50 per cent double-bond character to the amide carbon atom, or the carbon atom C', which forms a bond with oxygen which has somewhat more than 50 per cent double-bond character and a bond with nitrogen with

The remaining configuration, 1, is of great interest—it seems likely that it is present in many proteins with the β-keratin configuration. The predicted identity distance along the chain axis is 6.68 A, corresponding to 3.34

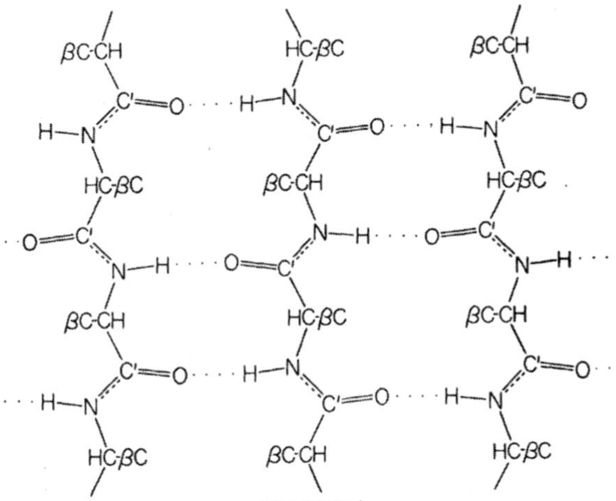

FIGURE 4

A diagrammatic representation of the antiparallel-chain pleated sheet structure.

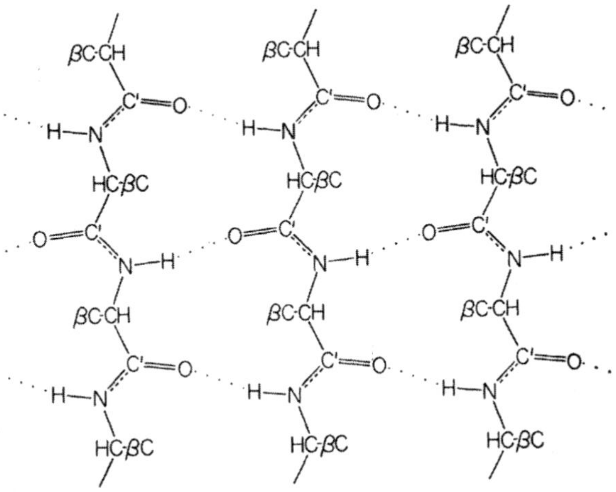

FIGURE 5

A diagrammatic representation of the parallel-chain pleated sheet structure.

A per residue. This agrees very well with the residue length reported by Astbury and Street for proteins with the β-keratin structure, 3.32 A.[22]

We have previously suggested for β keratin a pleated-sheet configuration

738   *CHEMISTRY: PAULING AND COREY*   Proc. N. A. S.

antiparallel-chain pleated sheet, considered as made of L-amino acid residues, has orthorhombic symmetry, its symmetry elements being a twofold screw axis in the direction of the chains, a twofold screw axis normal to this direction and in the plane of the sheet, and a twofold axis normal to the plane of the sheet. The parallel-chain pleated sheet has only a twofold screw axis in the direction of the chains.

In the parallel-chain pleated sheet the oxygen atom lies nearly in the N—H direction. The deviation from the N—H axis would be zero in case that the bond angle C—N—H were equal to 125°. It is likely that this bond angle is about 116°, and that there is accordingly a bend in the hydrogen bond of about 10°. This is not so great as to lead to significant instability.

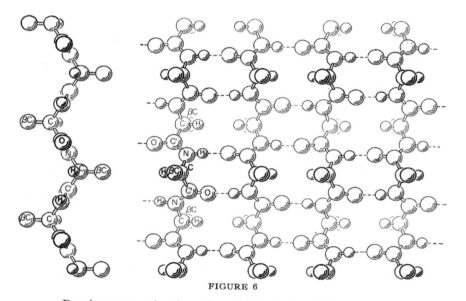

FIGURE 6

Drawing representing the anti-parallel-chain pleated sheet structure.

It is interesting to note that the ability of polypeptide chains with configuration 1 to form lateral hydrogen bonds with either a parallel or an antiparallel adjacent chain strongly suggests that proteins with the β-keratin structure may exist in which there is a randomness in sequence of parallel and antiparallel chains. We plan to discuss this question later, in connection with a comparison of observed and calculated intensities of x-ray reflections for synthetic polypeptides and proteins of the β-keratin type.

It is to be noted that our 5.1-residue helix[1-3] is not represented in table 1, although it satisfies the conditions of containing trans amide groups, and of having all amide groups equivalent. The reason for its failure to appear as a result of the considerations of the present paper is that it represents unfavorable orientation about both the N—C bond and the C—C' bond; the orientation about each bond is such as to correspond to a rotation of about 30° for both C—H and C—R from the position of coplanarity with each of the two amide groups adjacent to the α carbon atom. If we accept the assumption expressed above that these orientations represent an instability of about 1 kcal. mole⁻¹ relative to the favored orientations, then the 5.1-residue helix would be predicted to be less stable than the 3.7-residue helix by 2 kcal. mole⁻¹ per residue. There is no indication from the structures that a corresponding extra stability would be conferred on the 5.1-residue helix by, for example, the presence of N—H···O hydrogen bonds of abnormally great stability, the absence of steric hindrance between side

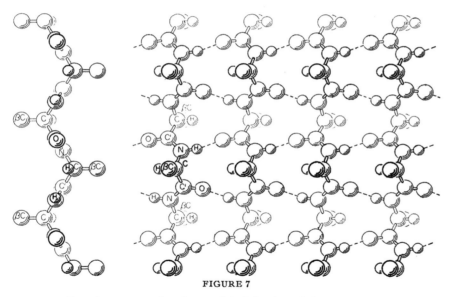

FIGURE 7

Drawing representing the parallel-chain pleated sheet structure.

chains of adjacent turns of the helix which might be present in the 3.7-residue configuration and make it less stable, or any other structural feature—indeed, as was pointed out in an earlier paper,[23] the cylindrical cavity down the center of the 5.1-residue helix would be expected to lead to an extra instability for this configuration by decreasing the intramolecular van der Waals attraction. We thus are led to the conclusion that the 5.1-residue helix is a less stable configuration of polypeptide chains than the 3.7-residue helix, and that it would not be expected to occur very often as an important feature of the structure of proteins. In particular, we are provided with grounds for withdrawing our tentative suggestion[7] that proteins of the α-keratin type undergo transformation to the 5.1 residue configuration on supercontraction. This suggestion was made because the amount of contraction involved in the change from the 3.7-residue helix to the 5.1-residue helix, about 35 per cent, is just that reported to occur in the process of supercontraction. We now think that it is likely that, as suggested by Astbury, the process of supercontraction involves merely a disorientation of α-keratin molecules, without significant change in the configuration of the individual polypeptide chains.

This investigation was aided by grants from The Rockefeller Foundation, The National Foundation for Infantile Paralysis, and The National Institutes of Health, Public Health Service.

* Contribution No. 1629.

[1] Pauling, L., and Corey, R. B., *J. Am. Chem. Soc.*, **72**, 5349 (1950).
[2] Pauling, L., Corey, R. B., and Branson, H. R., these Proceedings, **37**, 205 (1951).
[3] Pauling, L., and Corey, R. B., *Ibid.*, **37**, 235 (1951).
[4] Pauling, L., and Corey, R. B., *Ibid.*, **37**, 241 (1951).
[5] Pauling, L., and Corey, R. B., *Ibid.*, **37**, 251 (1951).
[6] Pauling, L., and Corey, R. B., *Ibid.*, **37**, 256 (1951).
[7] Pauling, L., and Corey, R. B., *Ibid.*, **37**, 261 (1951).
[8] Pauling, L., and Corey, R. B., *Ibid.*, **37**, 272 (1951).
[9] Pauling, L., and Corey, R. B., *Ibid.*, **37**, 282 (1951).
[10] Pitzer, K. S., and Gwinn, W. D., *J. Am. Chem. Soc.*, **63**, 3313 (1941).
[11] DeVries, T., and Collins, B. T., *Ibid.*, **64**, 1224 (1942).
[12] Jones, W. N., and Giauque, W. F., *Ibid.*, **69**, 983 (1947).
[13] Pitzer, K. S., *J. Chem. Phys.*, **5**, 473 (1937).
[14] Crawford, B. L., Jr., Kistiakowsky, G. B., Rice, W. W., Wells, A. J., and Wilson, E. B., Jr., *J. Am. Chem. Soc.*, **61**, 2980 (1939).
[15] Wilson, E. B., Jr., and Wells, A. J., *J. Chem. Phys.*, **9**, 319 (1941).
[16] Telfair, D., and Pielemeier, W. H., *Ibid.*, **9**, 571 (1941).
[17] Kistiakowsky, G. B., *Ibid.*, **10**, 78 (1942).
[18] Carpenter, G. B., and Donohue, J., *J. Am. Chem. Soc.*, **72**, 2315 (1950).
[19] Hughes, E. W., and Moore, W. J., *Ibid.*, **71**, 2618 (1949).
[20] Crowfoot, D., Dunn, C. W., Rogers-Low, B. W., and Turner-Jones, A., The X-ray Crystallographic Investigation of the Structure of Penicillin, in *The Chemistry of Penicillin*, Princeton University Press, **1949**.
[21] Perutz, M. F., *Nature*, **167**, 1053 (1951); see also Pauling, L., and Corey, R. B., *Ibid.*, **168**, 550 (1951).
[22] Astbury, W. T., and Street, A., *Phil. Trans. Roy. Soc.*, A230, 75 (1931).
[23] Pauling, L., "The Structure of Proteins," Phi Lambda Upsilon Second Annual Lecture Series, Ohio State University, February, 1951.

The discovery of the α-helix and β-sheet, the principal structural features of proteins

David Eisenberg*

Howard Hughes Medical Institute and University of California–Department of Energy Institute of Genomics and Proteomics, University of California, Los Angeles, CA 90095-1570

PNAS papers by Linus Pauling, Robert Corey, and Herman Branson in the spring of 1951 proposed the α-helix and the β-sheet, now known to form the backbones of tens of thousands of proteins. They deduced these fundamental building blocks from properties of small molecules, known both from crystal structures and from Pauling's resonance theory of chemical bonding that predicted planar peptide groups. Earlier attempts by others to build models for protein helices had failed both by including nonplanar peptides and by insisting on helices with an integral number of units per turn. In major respects, the Pauling–Corey–Branson models were astoundingly correct, including bond lengths that were not surpassed in accuracy for >40 years. However, they did not consider the hand of the helix or the possibility of bent sheets. They also proposed structures and functions that have not been found, including the γ-helix.

A decade before the structures of entire proteins were first revealed by x-ray crystallography, Linus Pauling and Robert Corey of the California Institute of Technology (Fig. 1) deduced the two main structural features of proteins: the α-helix and β-sheet, now known to form the backbones of tens of thousands of proteins. Their deductions, triumphs in building models of large molecules based on features of smaller molecules, were published in a series of eight articles, communicated to PNAS in February and March 1951. Their work had a significance for proteins comparable to that 2 years later of the Watson–Crick paper for DNA, which adopted the Pauling–Corey model-building approach. Here I summarize the main points of

these historic articles, and then mention some surprising omissions from them.

The most revolutionary of these articles is the first, submitted to PNAS on Pauling's 50th birthday, February 28th, 1951. It is *The Structure of Proteins: Two Hydrogen-Bonded Helical Configurations of the Polypeptide Chain* (1), in which Pauling and Corey are joined by a third coauthor, H. R. Branson, an African-American physicist, then on leave from his faculty position at Howard University (Fig. 1). In the opening paragraph, the authors state that "we have been attacking the problem of the structure of proteins in several ways. One of these ways is the complete and accurate determination of the crystal structure of amino acids, peptides, and other simple substances related to proteins, in order

that information about interatomic distances, bond angles, and other configurational parameters might be obtained that would permit the reliable prediction of reasonable configurations of the polypeptide chain." In other words, the structural chemist Pauling believed that with an accurate parts list for proteins in hand he would be able to infer major aspects of their overall architecture, and this proved to be so.

The next two paragraphs concisely set out the method: "The problem we have set ourselves is that of finding all hydrogen-bonded structures for a single polypeptide chain, in which the residues are equivalent (except for the differences in the side chain R)." That is, the authors sought all possible repeating structures (helices) in which the carbonyl C=O group of each amino acid residue accepts an N—H hydrogen bond from another residue. Why did they believe that there would be only a small number of types of helices? This was because of the constraints on structure imposed by the precise bond lengths and bond angles they had found from their past studies of crystal structures of amino acids and peptides, the components from which proteins are built up. These constraints are summarized in the third paragraph of their paper, which specifies to three significant figures the bond lengths and bond angles that they had found.[†] The most important constraint was that all six atoms of the amide (or peptide) group, which joins each amino acid residue to the next in the protein chain, lie in a single plane.

Fig. 1. Linus Pauling and Robert Corey (*A*) and Herman Branson (*B*). Pauling's deep understanding of chemical structure and bonding, his retentive memory for details, and his creative flair were all factors in in the discovery of the α-helix. Robert Corey was a dignified and shy x-ray crystallographer with the know-how and patience to work out difficult structures, providing Pauling with the fundamental information he needed. Herman Branson was a physicist on leave at the California Institute of Technology, who was directed by Pauling to find all helices consistent with the rules of structural chemistry that he and Corey had determined. The wooden helix between Pauling and Corey has a scale of 1 inch per Å, an enlargement of 254,000,000 times. (*A*) Courtesy of the Archives, California Institute of Technology. (*B*) Courtesy of the Lincoln University of Pennsylvania Archives.

This perspective is published as part of a series highlighting landmark papers published in PNAS. Read more about this classic PNAS article online at www.pnas.org/misc/classics.shtml.

*E-mail: david@mbi.ucla.edu.

[†]The bond lengths are all within 1 standard deviation of those determined 40 years later (15).

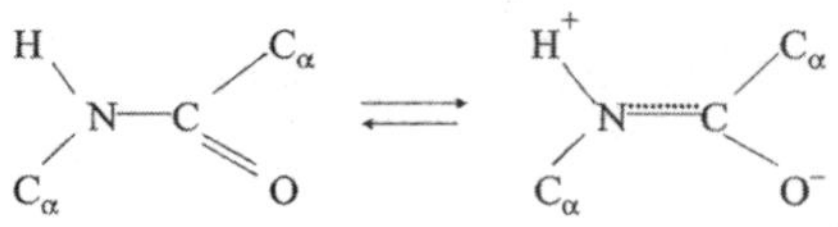

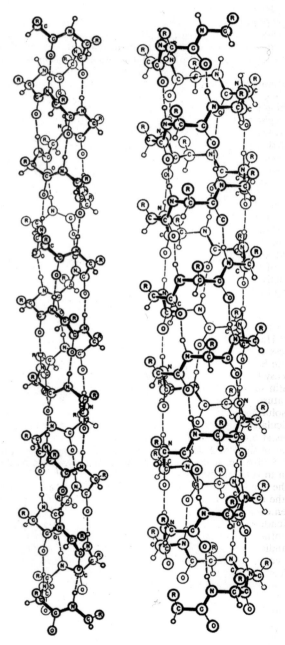

Scheme 1.

Pauling had predicted planar peptide groups because of resonance of electrons between the double bond of the carbonyl group and the amide C—N bond of the peptide group (Scheme 1).

In fact, such planar peptide groups had been observed in the crystal structures of *N*-acetylglycine and β-glycylglycine. As the authors put it: "This structural feature has been verified for each of the amides that we have studied. Moreover, the resonance theory is now so well grounded and its experimental substantiation so extensive that there can be no doubt whatever about its application to the amide group."

When Pauling, Corey, and Branson constructed helices with planar amide groups, with the precise bond dimensions they had observed in crystal structures, and with linear hydrogen bonds of length 2.72 Å, they found there were only two possibilities. These two they called the helix with 3.7 residues per turn and the helix with 5.1 residues per turn (Fig. 2), soon to be called the α-helix and the γ-helix.

Much of the rest of this short, brilliant paper is taken up with a comparison of these two helices with helices proposed earlier by others, most notably Bragg, Kendrew, and Perutz (2) in a paper the year before, that attempted to enumerate all possible protein helices, but missed these two. In their α-helix paper, Pauling *et al.* take a tone of triumph: "None of these authors propose either our 3.7-residue helix or our 5.1-residue helix. On the other hand, we would eliminate by our basic postulates all of the structures proposed by them. The reason for the difference in results obtained by other investigators and by us through essentially similar arguments is that both Bragg and his collaborators . . . discussed in detail only helical structures with an integral number of residues per turn, and moreover assume only a rough approximation to the requirements about interatomic distances bond angles, and planarity of the conjugated amide group, as given by our investigations of simpler substances. We contend that these stereochemical features must be very closely retained in stable configurations of polypeptide chains in proteins, and that there is no special stability associated with an integral number of residues per turn in the helical molecule." In short, stereochem-

istry is important in determining which helices are possible, and integral symmetry has no role whatever.

Today, we accept without a second thought that helices do not need to have an integral number of monomer units per turn. But in 1950, the crystallographic backgrounds of Bragg, Kendrew, and Perutz, three of the greatest structural scientists of the 20th century, saddled them with the notion of integral numbers of units per unit cell. They also missed the necessity of planar peptide groups. Working in the physics department at Cambridge University (Cambridge, U.K.), they were unaware of conjugation with nearby double bonds. The professor of organic chemistry at Cambridge at that time was Alexander Todd, who worked across the courtyard

Fig. 2. The α-helix (*Left*) and the γ-helix (*Right*), as depicted in the 1951 paper by Pauling, Corey, and Branson (1). Biochemists will note that the C=O groups of the α-helix point in the direction of its C terminus, whereas those of the γ-helix point toward its N terminus, and, further, that the α-helix shown is left-handed and made up of D-amino acids. (Reproduced with permission from Linda Pauling Kamb.)

from Bragg and his team. Todd recalled (3) that "despite the proximity, Bragg never, to my knowledge, set foot in the chemical laboratory . . . until one day . . . he came to my room in a somewhat agitated state of mind, bearing a bunch of papers in his hand," including the Pauling–Corey–Branson paper and his own on helices. Bragg asked Todd whether he preferred the α-helix over the helices that Bragg and his coworkers had invented. Todd responded, "I think that, given the evidence, any organic chemist would accept Pauling's view. Indeed, if at any time since I have been in Cambridge you had come over to the chemical laboratory, I . . . would have told you that."

The idea of the nonintegral α-helix had come to Pauling 3 years before, when he was visiting professor at Oxford University. He caught cold in the damp weather and spent several days in bed. He recalled (4) that he was soon bored with detective novels and "I didn't have any molecular models with me in Oxford but I took a sheet of paper and sketched the atoms with the bonds between them and then folded the paper to bend one bond at the right angle, what I thought it should be relative to the other, and kept doing this, making a helix, until I could form hydrogen bonds between one turn of the helix and the next turn of the helix, and it only took a few hours of doing that to discover the α-helix."

Why did Pauling delay 3 years in publishing this finding that came to him in only a few hours? He gave the answer in his banquet address at the third symposium of the Protein Society in Seattle in 1989. He was uneasy that the diffraction pattern of α-keratin shows as its principal meridional feature a strong reflection at 5.15-Å resolution, whereas the α-helix repeat calculated from his models with Corey was at 5.4 Å. As he says in his fourth paper of the PNAS series with Corey: "The 5.15-Å arc seems on first consideration to rule out the α-helix, for which the c-axis period must be a multiple of the axis distance per turn . . . " But then came the paper in 1950 by Bragg, Kendrew, and Perutz enumerating potential protein helices. Pauling told his audience in 1989: "I knew that if they could come up with all of the wrong helices, they would soon come up with the one right one, so I felt the need to publish it."

The origin of the discrepancy between the repeat of the α-helix and the x-ray reflection of α-keratin was hit on a year later by Francis Crick (5), then a graduate student with Perutz, and also by Pauling. It is that keratin is a coiled-coil, with α-helices winding around each other. The wider excursion of the α-helix in the coiled-coil reduces its repeat distance to 5.1 Å. This knack of knowing which contradictory fact to ignore was one of Pauling's great abilities as a creative scientist.

The β-Sheets

The second paper of the series appeared as one of a group of seven in a single issue of PNAS. It was: *The Pleated Sheet, A New Layer Configuration of Polypeptide Chains* (6). In this article, Pauling and Corey report that they have discovered a hydrogen-bonded layer configuration of polypeptide chains, in which the planar peptide groups lie in the plane of the sheet, and successive protein chains can run in opposite directions, giving an antiparallel sheet, as well as a parallel sheet. In both, linear H-bonds are again formed, but between protein chains rather than within a single chain. This results in protein chains that are not fully extended: the rise per residue is 3.3 Å, a spacing seen in x-ray diffraction patterns of β-keratin, rather than 3.6 Å, expected for a fully extended protein chain.

Confirmation of the α-Helical and β-Sheet Models

Confirmation of the α-helix came from Max Perutz, one of the three authors of the 1950 article that had enumerated the wrong helices. One Saturday morning in spring 1951, he came across the PNAS paper (7). "I was thunderstruck by Pauling and Corey's paper. In contrast to Kendrew's and my helices, theirs was free of strain; all of the amide groups were planar and every carbonyl group formed a perfect hydrogen bond with an imino group four residues further along the chain. The structure looked dead right. How could I have missed it ? . . . I cycled home to lunch and ate it oblivious of my children's chatter and unresponsive to my wife's inquiries as to what the matter was with me today."

Suddenly Perutz had an idea: "Pauling and Corey's α-helix was like a spiral staircase in which the amino acid residues formed the steps and the height of each step was 1.5 Å. According to diffraction theory, this regular repeat should give rise to a strong x-ray reflection of 1.5 Å spacing from planes perpendicular to the fiber axis . . . In mad excitement, I cycled back to the lab and looked for a horse hair that I had kept tucked away in a drawer . . . " and put it in the x-ray beam at an angle of 31° to the beam to bring the 1.5-Å repeat into the reflecting position. "After a couple of hours, I developed the film, my heart in my mouth. As soon as I put the light on I found a strong reflection at 1.5-Å spacing, exactly as demanded by Pauling and Corey's α-helix."

On Monday morning, Perutz showed his x-ray diffraction picture to Bragg. "When he asked me what made me think of this crucial experiment, I told him that the idea was sparked off by my fury over having missed building that beautiful structure myself. Bragg's prompt reply was, 'I wish I had made you angry earlier!' because discovery of the 1.5-Å reflection would have led us straight to the α-helix." Perutz also found the 1.5-Å reflection in diffraction from hemoglobin. He wrote to Pauling (8), "The fulfillment of this prediction and, finally, the discovery of this reflection in hemoglobin has been the most thrilling discovery of my life." Perutz, along with his coworkers Dickerson, Kendrew, Strandberg, and Davies, was to make even more thrilling discoveries later, including seeing direct pictures of α-helices in myoglobin and hemoglobin.

β-Sheets and single-stranded β-ribbons were first seen in globular proteins as in the structure of egg white lysozyme in 1965 (9). An initial surprise was that both the strands and the sheets are twisted, unlike the straight strands and pleated sheets of Pauling and Corey. In 1989 Pauling recalled that as soon as he saw the structure of lysozyme with its twisted sheet he realized he should have incorporated the twist in the original model. More recently there have been thorough analyses of twist and shear in β-structures (10, 11).

Some Surprising Omissions from the 1951 Papers

Chemists who take a careful look at the α-helix of Fig. 2 will notice two surprising features: (*i*) It is a left-handed helix, unlike α-helices of biological proteins, which are now known to be right-handed. That is, if your left thumb points along the helix axis, the helix turns in the direction of the fingers of your left hand. (*ii*) The configuration of chemical groups around each α-carbon atom have the D-configuration, rather than the naturally occurring L-configuration of amino acid residues in proteins. That is, this model of Pauling *et al.* is the mirror image of an α-helix in a natural protein. In contrast, the γ-helix in Fig. 2 is a right-handed helix made up of D-amino acid residues. Why did the authors choose to draw the α-helix as left-handed, with D-amino acids?

The basis for this choice has recently been analyzed by Dunitz (12), who had been a postdoctoral fellow at the California Institute of Technology at the time of the Pauling–Corey research. In fact, it was Dunitz who persuaded Pauling to change his terminology from "spiral" to "helix" in describing the new protein structures. In his analysis, Dunitz notes that 1951, the year of the α-helix, was also the year in which J. M. Bijvoet established the absolute configuration of molecules by the anomalous scattering of x-rays. After recalling discussions of handedness at the California Institute of Technology in that year, Dunitz concludes: "Either Pauling was unaware of these developments when he wrote the α-helix paper, or he knew about them but was uninterested. . . I tend to believe that when they wrote the paper, or quite possibly even when they made the models, Pauling (or his colleague Robert B. Corey) simply picked one of the two amino acid configurations (as it happened, the wrong one) to illustrate the helical structures and did not give the problem of absolute configuration much thought. . . Problems of absolute configuration received little or no attention because there seemed to be no need for them then. Perhaps they were even regarded as a distraction from the task at hand. Sometimes one can focus more clearly by closing one eye."

Also missing from the first paper is anything more than passing mention of the 3_{10} helix, a component of globular proteins found rarely in short segments, but more common than the Pauling–Corey–Branson γ-helix, which is virtually never seen. The H-bonds of the 3_{10} helix are somewhat too long and bent to have been acceptable by the stringent thresholds set by the authors. Their intuition about bent and long hydrogen bonds destabilizing structures was basically correct, but the thresholds they set are

more stringent than those used today (13), now that we know nature accepts the 3_{10} helix.

One other omission from the set of 1951 papers is the Ramachandran diagram. This is a 2D plot of the allowed values of rotation about the N—Cα and Cα—C≡O bonds in the protein backbone, introduced by Ramachandran and others in 1964 (14). This diagram shows that most values of rotation about these two bonds are forbidden by collisions of protein atoms. Only two major regions of the diagram are allowed: one corresponds to the α-helix, and one to the nearly extended chains of the β-sheets. Today the Ramachandran diagram is taught in all classes on protein structure and is featured in every textbook to give insight into the forces that determine the structures of proteins. But there is nothing in this diagram beyond what Pauling and Corey knew well: they built models of their proposed structures that embodied all features of the Ramachandran diagram. Apparently they understood the principles so well that they felt no need to explain them by a diagram of this sort. Another factor may have been that Pauling and Corey focused more on the stability provided by hydrogen bonds and less on the restrictions on possible structures dictated by collisions between nonbonded atoms.

The Other Six PNAS Articles by Pauling and Corey and the Wider Context

The remaining six articles in PNAS give the atomic coordinates of the models and interpret the diffraction patterns of fibrous proteins in terms of the models. There is much in these papers than has not been borne out, including a proposal

that muscle contraction is a transition from extended β-strands to compact α-helices. Nevertheless, the breathtaking correctness of the α-helix and β-sheets and the bold approach of modeling biological structures from chemical principles overshadow the rest.

These papers are all the more remarkable when we consider the political context in which they were written. During this period, Pauling was also heavily involved in defending academics, including himself, against charges of disloyalty to the United States, brought about by the pressures of the Cold War and what became known as McCarthyism. He was subpoenaed to appear before various anticommunist investigating committees, he received hate mail for his work on liberal causes, and he faced cancellation of his major consulting contract and coolness from some California Institute of Technology colleagues. On the day after Pauling and Corey submitted their seven protein papers for publication, the House Un-American Activities Committee named Pauling one of the foremost Americans involved in a "Campaign to Disarm and Defeat the United States" (8). The press release read, "His whole record. . . indicates that Dr. Linus Pauling is primarily engrossed in placing his scientific attainments at the service of a host of organizations which have in common their complete subservience to the Communist Party of the USA, and the Soviet Union." Somehow, even in the face of such false invective and multiple distractions, Pauling could maintain his focus as a top creative scientist.

I thank David R. Davies, Richard E. Dickerson, Jack Dunitz, Richard E. Marsh, and Doug Rees for discussion.

1. Pauling, L., Corey, R. B. & Branson, H. R. (1951) *Proc. Natl. Acad. Sci. USA* **37**, 205–211.
2. Bragg, L., Kendrew, J. C. & Perutz, M. F. (1950) *Proc. R. Soc. London Ser. A* **203**, 321–357.
3. Todd, Lord Alexander (1990) in *The Legacy of Sir Lawrence Bragg*, eds. Thomas, J. M. & Phillips, D. (Science Reviews Limited, Northwood, U.K.), p. 95.
4. Serafini, A. (1989) *Linus Pauling: A Man and His Science* (Paragon House, New York), p. 131.
5. Crick, F. (1952) *Nature* **170**, 882–883.

6. Pauling, L. & Corey, R. B. (1951) *Proc. Natl. Acad. Sci. USA* **37**, 251–256.
7. Perutz, M. (1998) *I Wish I'd Made You Angry Earlier* (Cold Spring Harbor Lab. Press, Plainview, NY), pp. 173–175.
8. Hager, T. (1995) *Force of Nature: The Life of Linus Pauling* (Simon & Schuster, New York), pp. 379–380.
9. Blake, C. C. F., Koenig, G. A., Mair, G. A., North, A. C. T., Phillips, D. C. & Sarma, V. R. (1965) *Nature* **206**, 759–761.

10. Chothia, C. (1973) *J. Mol. Biol.* **75**, 295–302.
11. Bosco, K. H. & Curmi, P. M. G. (2002) *J. Mol. Biol.* **317**, 291–308.
12. Dunitz, J. D. (2001) *Angew. Chem. Int. Ed.* **40**, 4167–4173.
13. Kortemme, T., Morozov, A. V. & Baker, D. (2003) *J. Mol. Biol.* **326**, 1239–1259.
14. Ramachandran, G. N., Sasisekharan, B. & Ramakrishnan, C. (1963) *J. Mol. Biol.* **7**, 95–99.
15. Engh, R. A. & Huber, R. (1991) *Acta Crystallogr. A* **47**, 392–400.

-5-

THE RACE FOR THE DNA DOUBLE HELIX

"Regrettably, the future is not always Rosy;
you never can tell Watson the agenda."

KEY PAPERS

(A) 1950 E. Chargaff, *Experientia* VI, 201–209. "Chemical Specificity of Nucleic Acids" [3 pages of 9].

(B) 1951 E. Chargaff, *Fed. Proc.* 10, 654–659. "Structure and Function of Nucleic Acid as Cell Constituents" [2 pages of 6].

(C) 2004 R. D. B. Fraser, *J. Struct. Biol.* 145, 184–6. "The Sructure of Deoxyribose Nucleic Acid" (written in 1951 but not published then).

(D) 1953 L. Pauling and R. B. Corey, *Proc. Nat. Acad. Sci. USA* 39, 84–97. "A Proposed Structure for the Nucleic Acids."

1953 R. E. Franklin and R. G. Gosling, *Acta Cryst.* 6, 673–677. "The Structure of Sodium Thymonucleate Fibers. I. The Influence of Water Content."

1953 R. E. Franklin and R. G. Gosling, *Acta Cryst.* 6, 678–685. "The Structure of Sodium Thymonucleate Fibers. II. The Cylindrically Symmetrical Patterson Function."

Molecular Structure of Nucleic Acids, *Nature* 171, 737–741

(E) 1953 J. D. Watson and F. H. C. Crick, pp. 737–738. "A Structure for Deoxyribose Nucleic Acid."

(F) 1953 M. H. F. Wilkins, A. R. Stokes and H. R. Wilson, pp. 738–740. "Molecular Structure of Deoxypentose Nucleic Acids."

(G) 1953 R. E. Franklin and R. G. Gosling, pp. 740–741. "Molecular Configuration in Sodium Thymonucleate."

Subsequent DNA Structure Papers

1953 J. D. Watson and F. H. C. Crick, *Cold Spring Har. Symp.* 18, 123–131. "The Structure of DNA."

1953 J. D. Watson and F. H. C. Crick, *Nature* 171, 964–967. "Genetical Implications of the Structure of Deoxyribonucleic Acid."

1953 R. E. Franklin and R. G. Gosling, *Nature* 172, 156–157. "Evidence for 2-Chain Helix in Crystalline Structure of Sodium Deoxyribonucleate."

1953 M. H. F. Wilkins, W. E. Seeds, A. R. Stokes and H. R. Wilson, *Nature* 172, 759–762. "Helical Structure of Crystalline Deoxypentose Nucleic Acid."

1954 F. H. C. Crick and J. D. Watson, *Proc. Roy. Soc. London* A223, 80–96. "The Complementary Structure of Deoxyribonucleic Acid."

1956 J. Donohue, *Proc. Nat. Acad. Sci. USA* 4 2, 60–65. "Hydrogen-Bonded Helical Configurations of Polynucleotides."

1956 J. Donohue and G. S. Stent, *Proc. Nat. Acad. Sci. USA* 42, 734–736. "An Identical Duplex Structure for Polynucleotides."

The Issue of Rosalind Franklin's Contributions

    1968   A. Klug, *Nature* 219, 808–810, 843–844. "Rosalind Franklin and the Discovery of the Structure of DNA." See also pp. 879 and 1192.

(H)   1974   **A. Klug, *Nature* 248, 787–788. "Rosalind Franklin and the Double Helix."**

    2004   A. Klug, *J. Mol. Biol.* 335, 3–26. "The Discovery of the DNA Double Helix."

Postscripts

(I)   1968   E. Chargaff, *Science* 159, 1448–1449. "A Quick Climb up Mount Olympus."

(J)   1969   M. F. Perutz; M. H. F. Wilkins, J. D. Watson, *Science* 164, 1537–1539. "DNA Helix."

(K)   1978   E. Chargaff, in *Heraclitean Fire*, Rockefeller U. Press, pp. 100–103. "Gullible's Troubles."

The drama of the discovery of the structure of DNA has been presented so often, in reminiscences, semi-novels, histories, TV documentaries and docu-dramas, that yet another presentation scarcely seems necessary. But we shall focus more on some of the lesser-known aspects of the science, as well as the drama. To understand and appreciate what happened and how it happened, we shall need a little of what can be termed "crystallography without mathematics."

Crystallography Without Mathematics

When a crystal or a fiber is irradiated by a finely collimated beam of x-rays of a particular wavelength, the x-rays are scattered in different directions and can be collected with photographic film, scintillation detectors, or other devices. It is the job of the crystallographer to measure this diffraction pattern, and use it to figure out the structure of the objects doing the scattering. The task is not trivial.

The diffraction pattern is produced by the addition of waves of x-rays scattered by all the atoms (actually, by the electrons) of the subject molecules. The way that two waves add depends not only on their magnitudes, but on the phase shift between them. Figure 5.1 shows at left two sine waves of the same magnitude that are completely in phase, or with relative phase angle 0°. The summed wave is everywhere twice as large as the individual components. In the center of Figure 5.1, the lower wave has been shifted right by half a wavelength (phase shift 180°), so it is completely out of phase with the wave above it. Their sum is zero everywhere; the two waves cancel. At the far right the lower wave has been shifted right by yet another 180°, to give a total phase shift between waves of 360°. Now the waves are in phase again, and their sum is identical to the starting situation with phase shift 0°.

Figure 5.2 shows the simplest possible diffraction experiment, using light waves and two pinholes punched in a cardboard mask, rather than x-rays and a crystal. As

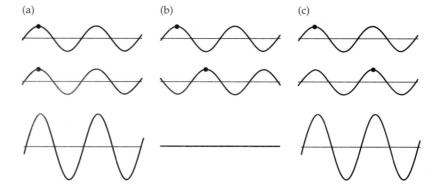

(a)   (b)   (c)

FIGURE 5.1   Addition of two sine waves with various relative phase shifts, (a) 0°, (b) 180° and (c) 360°, to yield the sum functions below. Waves with phase shifts 0° or any integral number of wavelengths add in phase. Two waves with 180° phase shifts cancel one another.

FIGURE 5.2   Diffraction experiment involving scattering of
light through two pinholes in a mask, A and B. The diffraction
pattern is observed on a screen at right. The path difference
between the two rays is BC = δ = **a** sin α, where α is the angle of
deflection of the rays. The two waves are in phase as they pass
through the pinholes, but waves deflected through an angle α
are not necessarily still in phase when they reach the screen; this
depends on α. In this example the two scattered waves are very
nearly 180° out of phase, and the screen would be nearly dark at
point D. The resulting diffraction pattern is a series of bright horizontal parallel bands across the screen, with spacing inversely
proportional to α. This is a one-dimensional analogue of the diffraction in three dimensions of x-rays by a crystal.

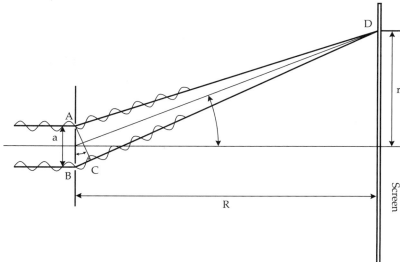

the beam of light approaches the two pinholes, labeled A and B, all elements of the
beam are in phase. At a distance R to the right is an observing screen. What will one
see at D, a distance **r** up from the center of the screen?

While the ray passing through hole A traverses a distance AD to the observing
screen, the ray through hole B travels a longer distance BD. How will the two
waves add together at their destination? The answer, as Figure 5.1 told us, is that
maximum light intensity occurs at those values of D for which the path difference,
(BD – AD), is an integral number of wavelengths of light; that is, for which the
phase shift is **n** × 360°, where **n** is an integer. To find out the conditions for which
this is satisfied, drop a perpendicular from point A to line BD, and call the intersection C. We will make the simplifying assumption that the distance from pinholes to
screen, R, is so much greater than the separation between pinholes, AB = **a**, that the
two scattered rays AD and BD can be regarded as effectively parallel. This is not
quite true for the experiment shown here, but is certainly valid for diffraction of x-
rays, where the distance to the screen is measured in centimeters while the spacings
between repeats in the crystal are measured in Ångstroms (1 cm = 10^8 Å).

The path difference between waves scattered by pinholes A and B is BC = δ.
Maximum intensity on the screen occurs whenever δ is an integral number of
wavelengths of light, λ, or when: δ h = λ = **a** sin α where h is an integer. In the
large-R approximation, sin α is essentially the same as tan α = r/R. Hence:

$$\delta = h\lambda = \mathbf{a} \sin \alpha \approx \mathbf{a} \tan \alpha = \mathbf{a}(\mathbf{r}/R) \text{ or: } \mathbf{r} = (\lambda R)h(1/\mathbf{a})$$

Hence if the spacing between the scattering pinholes is **a**, then the diffraction pattern on the screen will be a set of horizontal bright lines, of spacing proportional to
1/**a**. (The term λR is merely an experimental scale factor.) The best way of thinking
of it is that the observed pattern of bright lines marks those positions at which the
rays from the two pinholes do NOT cancel one another out. If the spacing of the
scatterers is made smaller, then the spacing between diffraction maxima becomes
larger. This is the reason that a diffraction pattern is said to be in "reciprocal space."

Diffraction is extended to two dimensions in Figure 5.3. With light, this could be
a lattice of pinholes punched in a mask, or the wires of a fine-mesh sieve. If the
spacing between pinholes (or wires) in the vertical direction is **a**, and in the horizontal direction is **b**, then the diffraction pattern will be a pattern of spots, with vertical
coordinates of h(1/**a**), and horizontal coordinates of k(1/**b**), where h and k are positive or negative integers. One can think of the pattern as the intersections of two
sets of fringes, one from the vertical repeat in the scattering mask and the other
from the horizontal.

FIGURE 5.3 Perspective view of a two-dimensional diffraction experiment involving the scattering of light through a rectangular grid of holes in a mask. The vertical repetition along **a** produces cancellation of light rays at the screen except on a series of horizontal bands of spacing 1/**a**. At the same time, the horizontal repetition along **b** produces cancellation except on a series of vertical bands of spacing 1/**b**. The product of these two sets of bands is a lattice of discrete spots of light with spacings 1/**a** and 1/**b**. Spots are labeled (h,k) as shown. The strong beam in this drawing leads to reflection (2,2).

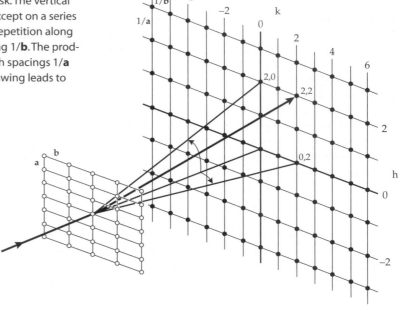

The results of an actual experiment with fine-mesh wire sieves are shown in Figure 5.4. The lens of an ordinary slide projector has been covered by cardboard, leaving only a small central hole to simulate a fine x-ray beam. When you look directly toward the projector through a sieve, the diffraction pattern from the wires appears to hang in space between you and the sieve. Figure 5.4a shows the results with a 60-micron sieve. Because horizontal and vertical spacings of the wires are the same, the diffraction pattern is a square array. Figure 5.4b uses a finer 30-micron sieve, with half the distance between adjacent wires. As expected, the diffraction pattern in

(a)
(b)
(c)

FIGURE 5.4 Actual diffraction experiment analogous to Figure 5.3, using a pinhole light source from a slide projector, and geologists' fine-mesh wire screens. The bright spots that you see are the diffraction pattern; the wire mesh itself is not visible in these photographs. Faint spots on lattice lines between the larger ones are artifacts derived from the small size of the sieve (the "crystal"), and can be ignored here. (a) 60-micron wire grid spacings, (b) 30-micron spacings, (c) the 30-micron sieve tilted in order to foreshorten the vertical repeat of the wires, and hence to elongate the vertical repeat of the diffraction pattern. From (1).

"reciprocal space" has spots twice as far apart. One ordinarily cannot find a sieve with unequal wire spacings in the two directions. But the same practical result can be obtained by tilting the sieve, thereby shortening the effective repeat distance in the vertical direction. This was done in Figure 5.4c, and one can see that the effect is to *increase* the vertical spacings between spots in the diffraction pattern.

At this point we must deal in a completely nonmathematical way with two important concepts: that of a Fourier Transform, T(A), of some function A, and the Convolution Product, A*B, of two functions A and B. (As in the standard cinema disclaimer, no numbers will be injured during the making of this presentation.) The first point is that *the diffraction pattern from a group of scattering objects is just the Fourier transform of the objects*. Figure 5.5 depicts several useful examples. We have already seen case 1: the transform of a row of scattering objects with spacing **a** is simply a set of parallel fringes or diffraction maxima of spacing $1/$**a**. We also have seen case 2: the transform of a two-dimensional lattice of scatterers of spacings **a** and **b** is a grid of spots with spacings proportional to $1/$**a** and $1/$**b**. It should not be hard to extend these ideas in your mind to the three-dimensional array of a crystal: given unit cell dimensions **a**, **b** and **c**, the diffraction pattern is a three-dimensional array with spacings $1/$**a**, $1/$**b** and $1/$**c**. (I will assume throughout that all axes are mutually perpendicular, making angles of 90° with one another. Of course this is a simplification, but is good enough for our present purposes.)

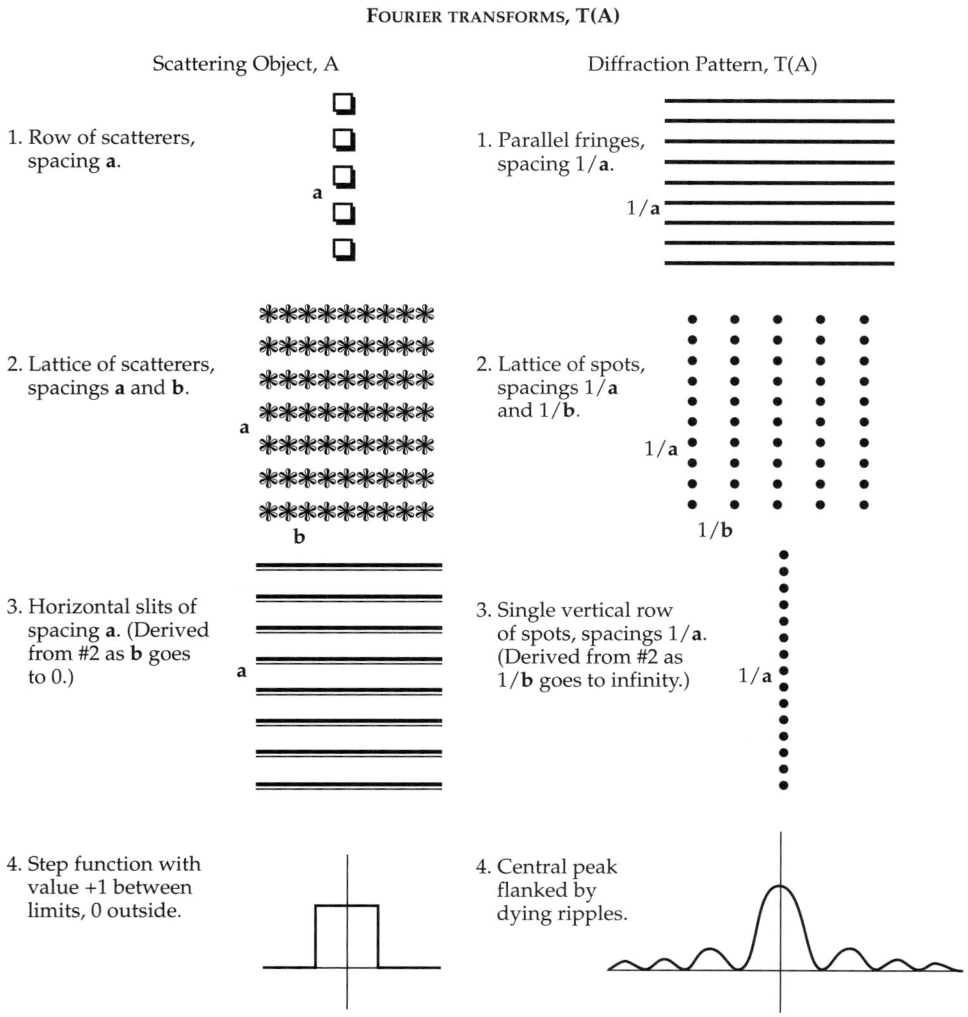

FOURIER TRANSFORMS, T(A)

Scattering Object, A   Diffraction Pattern, T(A)

1. Row of scatterers, spacing **a**.

1. Parallel fringes, spacing $1/$**a**.

2. Lattice of scatterers, spacings **a** and **b**.

2. Lattice of spots, spacings $1/$**a** and $1/$**b**.

3. Horizontal slits of spacing **a**. (Derived from #2 as **b** goes to 0.)

3. Single vertical row of spots, spacings $1/$**a**. (Derived from #2 as $1/$**b** goes to infinity.)

4. Step function with value +1 between limits, 0 outside.

4. Central peak flanked by dying ripples.

FIGURE 5.5 Depiction of several scattering objects, A, at left, and their diffraction patterns or Fourier transforms, T(A), and right. Case 1 is produced by the experimental setup in Figure 5.2, and case 2 from Figure 5.3. Cases 1 and 3 are the inverse of one another. See text for analysis.

Case 3 in Figure 5.5 is interesting in that it appears to be the inverse of case 1. The transform of a vertical row of scatters is a set of parallel fringes; and the transform of a row of slits is a vertical row of spots. This is because real space and reciprocal space are entirely symmetrical. The transform of the transform of the object is the original object again. Some science fiction writer of the quality of Philip K. Dick might well write a story in which reciprocal space is the *actual* space, and we are only the diffraction patterns of objects therein. How do we know this isn't so?

Case 4 illustrates the transform of a step function, which is useful in placing boundaries on something. A step function is +1 out to some established limit in one, two, or three dimensions, and 0 thereafter. The transform of a step function is a central Gaussian-like peak, flanked by ripples of decreasing magnitude. A circular geologist's sieve with a diameter 5 cm as in Figure 5.4 can logically be described as a screen of infinite extent, multiplied by a step function disk of diameter 5 cm. This explains the extra small spots along the lattice lines in Figure 5.4, as will be clear after introducing one more (the last) new concept: that of the Convolution Product.

A Convolution Product is a three-dimensional spatial analogue of simple multiplication. When you multiply 3×4, you take one 3 for every element of 4, or $3 + 3 + 3 + 3 = 12$. You also could take one 4 for every element of 3, or $4 + 4 + 4 = 12$. Multiplication is commutative: $3 \times 4 = 4 \times 3$. Suppose that you have two different three-dimensional functions, A and B. Each one of them has a particular value $A(x,y,z)$ and $B(x,y,z)$ at every point in space. Now set down a full image of A centered at point (x,y,z) of B, multiplied by the value of $B(x,y,z)$ at that point. Repeat this addition process for every point (x,y,z) of B. This fantastically complex product is known as the Convolution Product, A*B. Convolution also is commutative, A*B = B*A. Exactly the same result would be obtained by setting down a complete image of B centered at every point of A, and multiplied by the value of $A(x,y,z)$ at that point.

To avoid cerebral overload, we will consider only those Convolution Products for which one of the functions, A or B, is a collection or lattice of points. Figure 5.6a shows the convolution of one house with four points, three of weight +1 and the other of weight +2. The result is a residential subdivision with three small houses and one twice as big. (Looking at some of the boringly repetitive modern suburbs in Southern California, one is tempted to speculate that the developers simply chose to convolute one standard house with a topographic grid of the planned estate.)

Figure 5.6b extends convolution to two dimensions, convoluting a single tree with a collection of points, mostly of magnitude +1, but with one zero, one +2 and one enigmatic –1. The result is a forest,

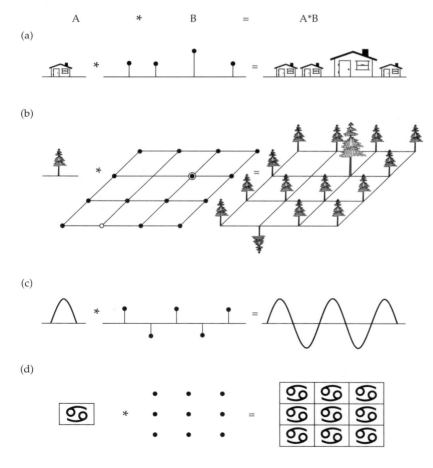

CONVOLUTION PRODUCT, A*B

(a)

(b)

(c)

(d)

FIGURE 5.6 The Convolution Product, A*B, as a spatial analogue of the process of multiplication. In each case function A is a continuous function whereas for simplicity function B is always taken here as a collection of peaks or points. (a) One dimension, involving points of weight 1 and 2 in B. (b) Function B as a two dimensional grid with points of weight +2 (●), +1 (●), 0 ( ) and –1 (○). (c) Generation of a sine wave by convolution a half wave with a row of alternating +1 and –1 points, (d) Generating a crystal by convoluting a lone unit cell with a crystal lattice.

with standard trees at all the +1 locations and a double-size tree at +2. But what is one to think about a "negative tree" at the –1 locus? What is a "negative tree"? One is reminded of the classical definition of a *nebbish*: "A nebbish is someone who, when he comes into a room, it is like somebody just left." Nebbish trees do not exist in forests, but do exist in convolutions. Figure 5.6c shows that a sine wave can be generated by convoluting a half-wave with a set of properly spaced +1 and –1 peaks. And Figure 5.6d illustrates one of the most important concepts in crystallography: *a crystal can be regarded as the convolution of a collection of atoms or molecules (the contents of one unit cell) with a crystal lattice.*

Now we can address the punch line of this long treatment: the Convolution Theorem. This theorem relates convolutions and transforms, and has two versions:

I. The transform of the *convolution* of two functions is just the *product* of the transforms of the individual functions, or: T(A*B) = T(A) × T(B).
II. The transform of the *product* of two functions is just the *convolution* of the transforms of the individual functions, or:
T(A × B) = T(A)*T(B).

The symmetry of these two versions is inevitable, considering that "real space" (our world) and "reciprocal space" (diffraction pattern) are themselves entirely symmmetrical.

Version I is illustrated in Figure 5.7a. In the upper row, a pair of molecules is convoluted with a lattice of spacing **a** to yield a crystal of unit cell **a**. Below the molecules is their molecular Fourier transform T(A), represented by a continuous curve with peaks and valleys. Below the lattice of spacings **a** in the center is the reciprocal lattice, T(B), with spacing 1/**a**. The product of these two transforms, T(A) × T(B), is the transform of the two molecules "sampled" at the reciprocal lattice points. A continuous transform has been converted, by the repetitions within the crystal, into a multi-point sampling of the entire continuous function. This is what x-ray crystallographers have to work with. In principle one might say, "Forget the crystal. Just pass the x-ray beam through a single molecule, and collect the entire continuous molecular transform. Why settle for less?" Unfortunately, this is impossible at present (although people are working on it). One needs the millions upon millions of molecules, all cooperatively scattering in unison within a crystal, to obtain diffraction data strong enough even to detect.

Version II of the Convolution Theorem, Figure 5.7b, is directly applicable to the determination of the structure of DNA. The transform of a continuous helix, A, at left is a pattern of spots along the arms of a cross, T(A). (This will be justified shortly.) Now let us suppose that

CONVOLUTION THEOREM

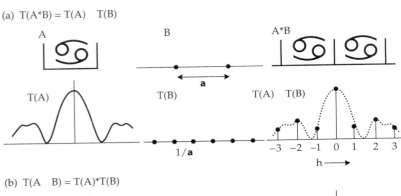

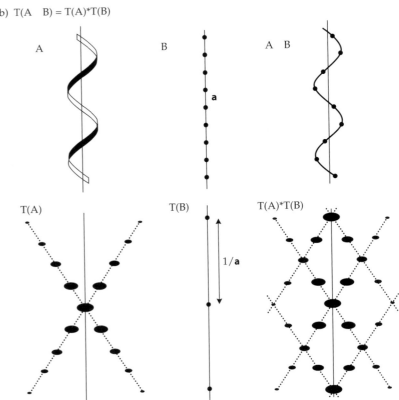

FIGURE 5.7 The two forms of the Convolution Theorem: (a) The transform of the convolution of two functions is the product of their individual transforms, or: T(A*B) = T(A) × T(B). (b) The transform of the product of two functions is the convolution of their individual transforms, or: T(A × B) = T(A)*T(B). The former is particularly relevant to crystals; the latter to fibers.

the helix actually is not continuous, but is built from subunits, B, that repeat at intervals of **a** up the helix axis. Multiplying the density of the continuous helix A by that of the repeat interval B generates the discontinuous helix A × B at right. The transform of the repeat of spacing **a** is T(B), a repeating function of spacing 1/**a**. Version II of the convolution theorem tells us that the *transform* of the discontinuous helix A × B is the *convolution* of T(A) with T(B). This is produced by taking the cross pattern of T(A), and repeating it at intervals of 1/**a** up and down the axis of the diffraction pattern. It was the realization of these relationships that made a structure analysis of B-DNA possible.

But why should the transform of a helix be a pattern of spots along the arms of a cross? Figure 5.8a shows the reason. To a good approximation, the near sides of the helical turns can be thought of as a set of fringes of spacing T, running from lower left to upper right. By case 3 of Figure 5.5, we should expect these to produce a set of reflections along a line perpendicular to these fringes, with an inter-spot spacing of 1/T. Moreover, the far sides of the helical turns at the back of the molecule amount to another set of fringes of similar spacing, but now oriented from upper left to lower right. They generate another row of spots of spacing 1/T, in a direction perpendicular to the rear fringes. The result of these two rows of spots is the observed cross pattern.

Of course we are looking only at a projection of the helix in Figure 5.8, and hence at a cross-section of the diffraction pattern. In the full pattern, each pair of spots to left and right of the vertical axis becomes a ring or doughnut of intensity around that axis, as I have tried to suggest in Figure 4.2 of Chapter 4. But because the fiber is cylindrically symmetrical, a one-dimensional slice through its diffraction pattern contains all the available information.

A few geometrical relationships should be noted. If P is the pitch of the helix, and β is the angle by which the helical turns are inclined, then T is given by the relationship: T = P cos β. The spacing of spots in the diffraction pattern along the arms of the cross will be 1/T, and the fact that the pitch of the helix is P means that all spots also must lie on a set of horizontal layer lines up and down the photograph, with spacing 1/P. The relationship between these quantities is: (1/P) = (1/T) cos δ. But because the internal half-angle δ of the cross pattern is essentially identical to angle β in the helix, the two results, T = P cos β and (1/P) = (1/T) cos δ, are equivalent.

By the convolution theorem, the diffraction pattern or Fourier transform of a discontinuous helix, Figure 5.8b, is obtained by repeating the continuous-helix transform at intervals up and down the vertical axis. If h is the rise per residue along the helix axis (what we have been calling **a**), then the distance from the origin of the photograph to the origin of the shifted image (the

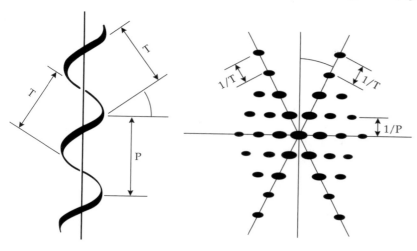

(a) Continous helix

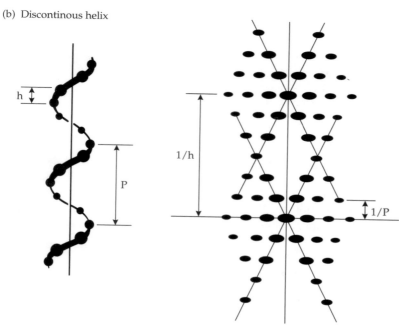

(b) Discontinous helix

FIGURE 5.8 Important parameters in diffraction by (a) continuous and (b) discontinuous helices. See text for details. The smaller subsidiary peaks to right and left along each layer line are present because the cylindrical symmetry of the fiber gives rise to Bessel functions, but they are irrelevant for our purposes and have been omitted for simplicity in Figure 5.7.

(a)
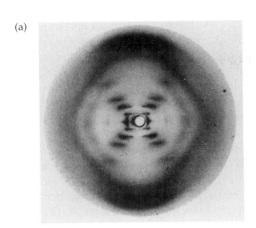

(b)
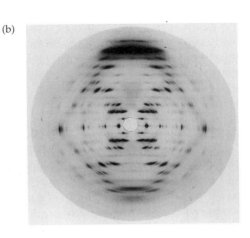

FIGURE 5.9 (a) Diffraction pattern from moist fiber of calf thymus DNA, taken by Rosalind Franklin and Raymond Gosling. It was published by them in 1953 in paper G and in *Acta Cryst.* 6, 673–677, and was later used by Klug in his 1968 *Nature* analysis of Franklin's contributions (Key Papers list). Many of the features expected from Figure 5.8b are visible, but the image is blurred. (From Aaron Klug, 2004.) (b) A superb photograph of the same material, calf thymus DNA, taken nearly 30 years later. The quality of the image now is so high that virtually every feature of Figure 5.8b can be discerned: the 3.4 Å axial reflection on the tenth layer line, all of the layer lines up to that level, and peaks in both the arms of the central cross pattern and those of the shifted crosses. The fiber has been tilted away from the perpendicular as in Figure 4.2b of the previous chapter, in order to obtain a stronger image of the axial 3.4 Å reflection. (From Struther Arnott, 1981.)

first dense reflection up and down the vertical axis) is $1/h$. In the x-ray photograph, the layer line spacing $1/P$ reveals the pitch of the helix itself, and the distance to this first dense axial reflection $1/h$ reveals the rise per residue along the helix. If the first axial image of the origin sits on the 6th layer line, as in this example, this tells us that there are *six repeating units per turn of helix*. Figure 5.7b showed the situation with a fourfold helix, having four repeating units per turn.

Figure 5.9a is the actual x-ray photograph of B-DNA published by Rosalind Franklin and Raymond Gosling in 1953 (G). Disorder in the fiber prevents the pattern from being as clear and detailed as the diagram in Figure 5.8b, but the main features are there. The central cross pattern of spots is obvious. Spots are prominent on the first, second and third layer line above and below the center, but become blurred thereafter. The heavy dark smears on the vertical axis above and below the center are images of the origin in the displaced crosses. The upper one of these has streaks running down to left and right which in fact are the smeared-out images of the spots forming the arms of the shifted cross. Similarly, the lower image has streaks running up to left and right. The distance from one layer line to the next near the center of the pattern tells us that the pitch of the helix is P = 34 Å, and the distance up to the dark axial smear tells us that the rise per residue is h = 3.4 Å. Hence the axial smear sits on the tenth layer line (even though these higher layer lines are not easily distinguishable), and the helix has 10 repeating units (base pairs) per turn. This was the best photo available in 1953 when the structure of B-DNA was being established. But experimental techniques improve with time, and Figure 5.9b shows what could be accomplished 28 years later by people trained in the fiber diffraction group that Wilkins developed at King's. Almost all of the ideal features of Figure 5.8b are visible, and one can easily count layers to establish that B-DNA is a tenfold helix. Figure 5.9b would have been revolutionary in 1953, and it is amazing that so much was discerned at the time from Figure 5.9a.

The Long Road to the DNA Helix

In the 1860's, Miescher in Switzerland isolated nucleic acid from pus and salmon sperm. He noted its high phosphorus content and its acidity. Little attention was paid to nucleic acids until Kossel in the early 1900's determined that nucleic acids were built from phosphate groups, sugar rings, and four kinds of bases: the single-ring pyrimidines: thymine (uracil in RNA) and cytosine, and the double-ring purines: adenine and guanine. In the 1920's it was established that two kinds of nucleic acid existed, RNA with ribose sugars and DNA with deoxyribose. RNA was thought to be characteristic of plants and DNA of animals, which was wrong. Equally wrong, and harmful, was the so-called "tetranucleotide hypothesis," which regarded a nucleic acid as a sterile repetition of the same four-base sequence such as: -(A-T-C-G)-(A-T-C-G)-(A-T-C-G)-(A-T-C-G)-. Even if the tetranucleotide hypothesis were disproven, as later was the case, a polymer with only four symbols was thought to be an unlikely candidate for the storehouse of genetic information, when the 20-symbol proteins were available. (There is a serious fallacy here. Any concept that can be expressed in words can be conveyed by Morse code, which has even fewer symbols than does DNA. Question: Just *how many* symbols are involved in a Morse code message? The answer to this trick question will be found at (2) in the reference list at the end of the chapter.)

In 1944, Oswald Avery at the Rockefeller Institute in New York (now Rockefeller University) injected DNA from donor bacteria into mutant bacteria of the same kind, and found that some mutants were transformed back into the donor type (3). Hence, Avery concluded, it must be the DNA that carries the genetic information. But few people paid much attention to so radical an hypothesis. Eight years later, A. D. Hershey and Martha Chase at Cold Spring Harbor made a greater scientific impact by showing that, when bacteriophage infect *E. coli* bacteria, only the DNA goes inside, while the protein viral coat remains behind, outside the bacterium (4). They accomplished this by radiolabeling both the phosphorus of DNA and the sulfur of proteins, and verifying that the phosphorus entered the bacterium while the sulfur did not. Why did Hershey and Chase have such a greater influence on the field than Avery? Perhaps because the idea of bacteriophage as genetic organisms only evolved slowly in the late 1940's, and the monotonous tetranucleotide hypothesis died hard.

Erwin Chargaff of Columbia University was one of the most important pioneers in the biochemistry of nucleic acids. He began his study in 1945, only a year after Avery's experiment. By 1950 (paper A) he was speculating on the as yet unknown function of nucleic acids in the cell. "....the chemistry of nucleic acids represents one of the remaining major unsolved problems in biochemistry." He came close to attributing a genetic role to them, saying:

> It is impossible to write the history of the cell without considering its geography; and we cannot do this without attention to what may be called the chronology of the cell, i. e. the sequence in which the cellular constituents are laid down and in which they develop from each other. If this is done, nucleic acids will be found pretty much at the beginning. An attempt to say more leads directly into empty speculations in which almost no field abounds more than the chemistry of the cell. Since an ounze [sic] of proof still weighs more than a pound of prediction, the important genetical functions, ascribed—probably quite rightly—to the nucleic acids by many workers, will not be discussed here. Terms such as 'template' or 'matrix' or 'reduplication' will not be found in this lecture.

This excerpt, and paper A in general, illustrate three recurring aspects of Chargaff's style: (a) a meticulous accuracy and thoroughness in his research, (b) a fluid and literate writing style that makes many of his essays well worth reading today, and (c) an almost obstinate unwillingness to speculate about the possible long-range significance of his work. It was the latter which prevented him from possibly solving, or sharing in the solution, of the structure of the DNA helix. Chargaff never forgave Watson and Crick for doing so.

What Chargaff did achieve was a careful analysis of the base composition of DNA from many different sources. He established clearly in Tables II – IV of paper A that A, G, C and T did *not* all occur in the same quantities in a given DNA sample, and hence that the tetranucleotide hypothesis was wrong. He used the differing ratios of the purines A and G to characterize the DNA from different organisms. But although his tables make it quite clear that A and T tend to occur with the same frequencies, as do G and C, Chargaff virtually ignores this fact. He does say on page A.206:

> The results serve to disprove the tetranucleotide hypothesis. It is, however, noteworthy—whether this is more than accidental, cannot yet be said—that in all

desoxypentose nucleic acids examined thus far the molar ratios of total purines to total pyrimidines, and also of adenine to thymine and of guanine to cytosine, were not far from 1.

Chargaff seems to get more pleasure from stating with a flourish that a given fact may be, or indeed may not be, significant, than he does in thinking about what the significance might entail. In another paper published a year later (B), Table 4 fairly shouts that the ratios of A to T, and of G to C, are effectively 1.00 from ox, to man, to birds, to fish, to plants, to bacteria. But the sole conclusion that he draws from this is the statement that:

> There exist a number of regularities. Whether these are merely accidental cannot yet be decided. In almost all DNA preparations studied until now the ratio of total purines to total pyrimidines never was far from 1. Similarly the ratios of adenine to thymine and of guanine to cytosine were near 1.

He never asks why this should be so, or speculates as to what significance these facts might have. Watson and Crick, when they did begin to study the structure of DNA, were kind enough to call these equalities "Chargaff's ratios." But it would be wrong to claim that Chargaff ever published any ideas at that time specifically about base "pairing."

Triple Helices

The story of how Jim Watson and Francis Crick at the Medical Research Council Laboratory of Molecular Biology in Cambridge (known as the MRC Lab) launched an investigation into the structure of DNA, and how Rosalind Franklin and Maurice Wilkins at King's College, London, obtained the beautiful fiber diffraction photos that made the Watson/Crick analysis possible, has been told so many times by so many people that it is hardly worth reiterating here. If you do not know the story, you may want to look at any of the first eight books in the Highly Recommended Reading list of Appendix 2. Everyone knows that Watson and Crick published their DNA bombshell in the 25 April 1953 issue of *Nature*. But what is less well known is that two years earlier they had proposed an absurd inside-out triple helix with sugar-phosphate backbones on the inside, held together by electrostatic attractions between ionized phosphates and sodium atoms, and with bases on the outside of a cylindrical molecule. The story is told on pp. 357-364 of Robert Olby's book "The Path to the Double Helix" (see Appendix 2). The laughter from the King's College people when they were invited to Cambridge to view the model so infuriated Bragg, the Director of the Laboratory, that he ordered Watson and Crick to stop working on DNA and return to other projects. Needless to say, no publications resulted.

It also is almost forgotten that a young doctoral candidate in Wilkins' laboratory, Bruce Fraser, independently proposed a model for DNA in 1951 which, although it too was a triple helix, had the bases interacting via hydrogen bonds on the interior, and the sugar/phosphate backbones on the outside. The paper, C, was never published. Two years later Watson and Crick published their double-helix results alongside papers by Franklin and by Wilkins (E–G). Wilkins suggested that Fraser's work be included as a fourth paper. But Crick demurred because Fraser's triplex structure obviously was incorrect, and so it was only mentioned in passing by Watson and Crick as "(in the press)." Fraser by then had finished his doctorate and left for Australia to become a leader in the study of natural and synthetic polypeptide fibers. The figures that were to have accompanied Fraser's paper, regrettably, have been lost. But the text, from a single copy in Wilkins' personal files, conveys some idea of the structure. The first paragraph obviously was written in haste after Pauling and Corey (not "Grey") published their own DNA model in early 1953 (D). Frazer's helix had three chains, with sugar/phosphate backbones on the surface and bases hydrogen-bonded to one another on the interior. But without figures, the text of his manuscript provides few clues as to how he envisioned the bases interacting.

It is curious enough that, before the double helix was conceived, there should have two attempts to build DNA as a triple helix: by Watson and Crick, and by Fraser. But in fact, there were *three*: the structure that Pauling and Corey proposed in February 1953 was also a triple helix. Their paper, entitled "A Proposed Structure for the Nucleic Acids," (D) began with typical Pauling bravado:

> **The structure is not a vague one, but is precisely predicted; atomic coordinates for the principal atoms are given in table 1. This is the first precisely described structure for the nucleic acids that has been suggested by any investigator.**

Published, yes; suggested, no. The Pauling/Corey helix, seen in stereo in Figure 5.10 and viewed down the helix axis in Figure 6 of paper D, was built around a three-stranded sugar/phosphate core, with bases extending radially outward like branches of a Christmas tree. The "trunk" of the tree was held together by hydrogen bonds between O atoms and –OH groups on the phosphates. Each of the three strands defined a right-handed helix (one strand is shown darker than the others in Figure 5.10), but the way these strands were intertwined created an overall left-handed pattern of ridges and grooves. This arrangement was deemed possibly of biological significance:

> **It is interesting to note that the purine and pyrimidine groups, on the periphery of the molecule, occupy positions such that their hydrogen-bond forming groups are directed radially. This would permit the nucleic acid molecule to interact vigorously with other molecules.......As Astbury has pointed out, the 3.4 Å x-ray reflection, indicating a similar distance along the axis of the molecule, is approximately the length per residue in a nearly extended polypeptide chain, and accordingly the nucleic acids are, with respect to this dimension, well suited to the ordering of amino-acid residues in a protein.**

In other words, they considered the extended bases on the outside of their helix as a possible template for binding an extended polypeptide chain. This DNA could exhibit almost ribosome-like behavior! Completely absent is any idea of base pairing as a cornerstone of DNA replication and heredity.

The arrival of a manuscript of the Caltech paper, brought to Cambridge by Linus Pauling's son Peter, sent shock waves through the laboratory comparable only to those caused in 1951 by the news of Pauling's α-helix. Cambridge had lost the race for the protein helix to Caltech. Was it to lose the race for the DNA helix as well? Bragg was persuaded to unleash Watson and Crick once again (as Aaron Klug has put it), and by April of 1953 the problem was solved. DNA was a double helix with bases paired at the center, and with two sugar/phosphate backbones wrapped around the outside.

But before we turn to the correct structure, it is worthwhile asking why it was that *three separate sets of investigators* independently tried to make a triple helix out of DNA. The answer tells us something of significance about the correct double helix. The distances of the strong axial or meridional reflections above and below the center of the x-ray photo of B-DNA (Figure 5.9a) indicated that the repeat distance along the helix axis was 3.4 Å. Equatorial reflections to left and right of the center of the photo suggested that cylindrical DNA helices were packed roughly 19 Å apart. Hence the repeating slab of a DNA helix was a disk 19 Å in diameter and 3.4 Å thick. But how many nucleotides were contained in one such disk? One? Two? Three? The volume of one such disk is 964 Å³. From the average molecular weight of one nucleotide of ca. 330, and the measured density, 1.62 g/cm³, one can calculate that the volume per nucleotide is ca. 338 Å³. And:

$$964 \text{ Å}^3 / 338 \text{ Å}^3 = 2.85 \approx 3$$

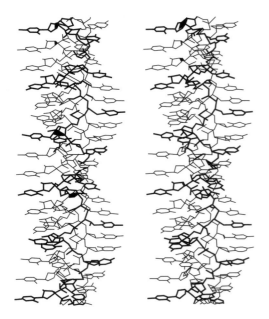

FIGURE 5.10   Stereo pair drawing of the Pauling and Corey 1953 triple helix for DNA (D). Bases project out like branches of a tree around a hydrogen-bonded sugar/phosphate trunk. The striking left-handed rotation of the structure is somewhat misleading. Separate strands, one of which is given in dark line, rotate in a right-handed direction with eight bases per turn. All bases have been drawn for simplicity here as pyrimidines, which is unimportant because base pairing was not an element in the Pauling/Corey helix. Drawn by David S. Goodsell from published coordinates.

One slab of DNA helix has room enough for three nucleotides, suggesting a triple helix. This calculation was sufficient to mislead Watson, Crick, Fraser, Pauling and Corey.

What is the fallacy? Just because the slab has enough room for three nucleotides does not necessarily mean that three are present. In reality, one 3.4 Å thick disk along the helix contains only two nucleotides, the remaining space being filled by water molecules in the deep major and minor grooves. Conventional wisdom tells us that DNA is built from three components: sugars, phosphates, and bases. But later single-crystal x-ray analyses of synthetic DNA molecules have suggested that DNA in fact possesses *four* functional components: sugars, phosphates, bases, and bound water molecules (5, 6). It is these solvent-filled major and minor grooves that provide the channels for recognition and control of particular sequences of DNA without unpairing the bases.

Pauling and Corey were hampered by lack of access to DNA fiber diffraction pictures of the quality that came from King's College. In their words:

> It has recently been reported by Wilkins, Gosling and Seeds that highly oriented fibers of sodium thymonucleate have been prepared, which give sharper x-ray photographs than those of Astbury and Bell. Our own preparations have given photographs somewhat inferior to those of Astbury and Bell. In the present work we have made use of data from our own photographs and from reproductions of the photographs of Astbury and Bell...

They did not publish their x-ray photographs of DNA; in fact, their paper contains no experimental evidence at all, aside from one table concerned with lateral packing of helices.

This apparent detachment from experimental data may explain a curious blunder in their paper. On page 93 they state that "...the identity distance or approximate identity distance is 27.2 Å, corresponding to eight layers." On page 94 they describe their structure as a triple helix, each strand of which contains eight bases 3.4 Å thick per turn of helix, and the erroneous helix repeat distance occurs yet again in their Table 1. In the language of Figure 5.8, h = 3.4 Å, n = 8, and P = 27.2 Å. But where did this putative helix repeat distance of 27.2 Å come from? In Arnott's B-DNA photograph, Figure 5.9b, it is easy to count ten horizontal layer lines, not eight, between the center of the pattern and the large 3.4 Å meridional reflection at the top of the photo. Of course no one had seen such a beautiful photo in 1953, and it is hard to count layer lines in Franklin's own photo, Figure 5.9a. However, careful measurement of spacings between the first three horizontal layer lines above and below the center yields a helix repeat distance of P = 34 Å. How did Pauling and Corey come up with 27.2 Å instead?

An answer is suggested by Aaron Klug's 2004 paper in the Key Papers list at the head of this chapter. Rosalind Franklin was the person who first discovered that DNA could adopt both a low-humidity A form and a high-humidity B form. Previous investigators had not realized that there were two kinds of helix, and sometimes took disordered x-ray photographs of mixtures of the two forms. Franklin initially concentrated mainly on A-DNA because it gave a more ordered diffraction pattern. In a November 1951 seminar, and in her research report for the year 1951, she stated that the absence of meridional reflections in the "crystalline form" (A-DNA) suggested a helical structure, with nucleotides occurring in equivalent positions only at intervals of 27 Å, the length of one turn of the helix. It may well be that this value of P = 27 Å found its way to Pasadena, that Pauling and Corey missed the fact that she was describing the A form rather than the B, and that dividing 27 Å by the known thickness of one base, 3.4 Å, yielded for them the fatal value of 8.

Pauling and Corey were wrong on three counts. Their model for DNA was a bases-out, eightfold triple helix, rather than a bases-in, tenfold double helix. Their

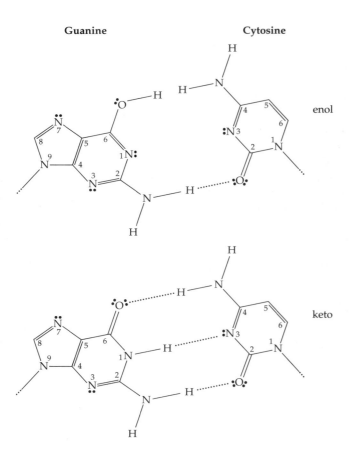

FIGURE 5.11 Base pairing possibilities using alternative tautomers of guanine and thymine. In each case, G-C and A-T, the incorrect enol tautomer is shown first and the correct keto tautomer is directly below it. Enol tautomers were once favored by standard reference books, but are inconsistent with base pairing via hydrogen bonds (dotted lines).

mistake can conceivably be laid at the door of the House Un-American Activities Committee in Washington. Pauling had been invited by the Royal Society to present his protein research in London in May 1952. Had he done so, he certainly would have visited King's College and Cambridge, and would have seen the superb new B-DNA photographs of Franklin and Wilkins. This just might have been enough to put him on the track to the correct DNA structure. But as Eisenberg's reprint in Chapter 4 relates, the U. S. State Department refused to renew his recently-expired passport, apparently under the influence of the House Un-American Activities Committee which had branded Pauling a supporter of the "Communist peace effort." Pauling never saw Franklin's Figure 5.9a, which inspired Watson and Crick.

Jim Watson recalls that when he first read the draft manuscript that had been brought to Cambridge by Peter Pauling, he immediately went to the visiting American, Jerry Donohue, and asked him, "Wouldn't the phosphate groups be *ionized* at pH 7?" Donohue replied that, of course they would. Then the hydrogen bonds that Pauling and Corey invoked to hold together the sugar/phosphate core could not exist, and their structure was wrong. As Watson recalled in his *Double Helix*,

...by mid-March at the latest Linus' paper* would be spread around the world. Then it would be only a matter of days before the error* would be discovered. Thus we had anywhere up to six weeks before Linus again was in full-time pursuit of DNA.

But Watson and Crick won the race. The Pauling/Corey paper had appeared in print on 15 February 1953; the Watson/Crick *Nature* note announcing the true DNA structure came out only ten weeks later, on 25 April.

Double Helices

The high-quality fiber diffraction photos from King's College laid down some reasonably strict blueprints for B-DNA: (a) a helix with ten repeating units, each 3.4 Å thick, per 34 Å turn of helix, and (b) a roughly cylindrical helix with a diameter of around 19 Å. The other vital piece of information was that the helix was a duplex rather than a triplex, and Chargaff's ratios (A=T and G=C) probably were a powerful influence in this regard. But the idea of pairing of bases was not obvious; indeed, with the commonly accepted tautomers of the bases, pairing could not exist. Most standard references on these organic bases maintained that G and T were in what is called the "enol" form as shown in Figure 5.11, with —OH groups adjacent to :N at ring positions 6 and 1 in guanine, and 4 and 3 in thymine. Hence H-atom donor groups, —OH and H_2N—, would oppose one another at the interface between bases, interfering with base pairing.

Jerry Donohue once again came to the rescue. He informed Watson and Crick that enol tautomers of guanine and thymine were now disfavored with respect to keto tautomers, in which a proton was shifted so that adjacent —OH and N: were converted into =O and N—H. Watson has described how, when the keto forms were

---

* "Linus' nonsense" and "the blunder," respectively, in the original typed manuscript as circulated among the principals for commentary.

used, base pairing followed at once, with G/C and A/T pairs having nearly identical size and shape. The double helix with a completely flexible "message" of base sequence was born. There is an interesting sidelight to this issue of tautomers. In response to a recent letter by the author asking about the unpublished 1951 Watson/Crick triple helix, Francis Crick replied with a comment that is relevant to their subsequent double helix:

> **Dear Dick: It's certainly interesting to see exactly what mistakes people made. In my case my biggest mistake was in assuming that the term 'tautomeric forms' implied that the two possible forms occurred equally often.**

But they don't. Keto was the correct choice, and the DNA duplex structure saw the light of day.

The classic and often-published DNA model photograph shows Watson and Crick standing beside a tall model built with wire backbones and sheet metal base pairs. In one version Crick points at the model with a slide rule, a gesture that he maintains was suggested by the photographer. Figure 5.12 shows something quite different. According to Jim Watson, he had been invited to the June 1953 Cold Spring Harbor Symposium to present their new helix. A small traveling model of DNA was built for him by Tony Broad, the laboratory electron engineer. (*"Electron engineer,"* according to Tony, not *"electronics engineer."* His talent was to make beams of electrons go where they were wanted. It was he who designed and built for the MRC Lab the rotating-anode x-ray generators that later were marketed commercially by Siemens.)

Broad's DNA model was beautifully crafted, and housed in a removable clear plastic cylindrical case 11" tall and 5" in diameter. He later made other copies of this model, possibly three or four in total according to both Watson and Crick, to distribute among those who had worked on the structure. The model was not used to illustrate the published version of their Cold Spring Harbor talk; a simple schematic drawing sufficed. But it was the central illustration in a more comprehensive paper in the *Proceedings of the Royal Society* which was submitted two months after the Cold Spring Harbor conference and appeared early in 1954. (See Key Papers for both references.) It also appeared in an article on "The Structure of the Hereditary Material" which Crick published in the October 1954 issue of *Scientific American,* and in a serial publication of Watson's "Double Helix," which came out in the January and February 1968 issues of *Atlantic Monthly.* The scientific importance of the Broad model has faded with time, but it remains an entrancing little work of art.

Publication, Publicity and Credit

It is clear that Watson and Crick could not have come up with their B-DNA structure without having had access to the beautifully detailed fiber diffraction photos taken by Rosalind Franklin and Maurice Wilkins. The story of this rivalry has been told so often that it need not be repeated here. Books 1–8 in Appendix 2 cover this subject thoroughly. More to our point, a careful reading of the three papers published jointly in the 25 April 1953 issue of Nature (E–G) provides insight into the priorities and thought processes of the various participants. Paper E by Watson and Crick is as unlike Chargaff's conservative style as one could imagine. They begin by invoking "novel features which are of considerable biological interest," and then describe their model with care. They employ the keto rather than the enol tautomers in order to make base pairing possible, and point out specifically that the sequence of one chain must be complementary to that of the other. They end with a sentence which can only have been the result of much honing and polishing: "It has not escaped our notice that the specific pairing we have postulated immediately suggests a possible copying mechanism for the genetic material." Compare this

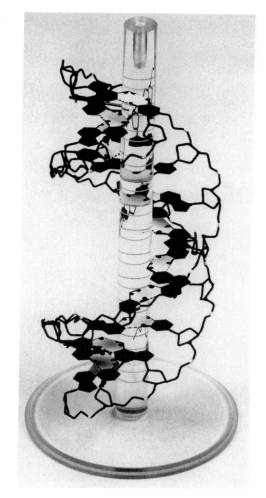

FIGURE 5.12   Earliest display model of the Watson/Crick B-DNA double helix, built for Watson to carry to a Cold Spring Harbor symposium in June 1953, only two months after publication of the *Nature* paper. The viewpoint chosen for this photo is not the conventional one, perpendicular to the helix axis. Rather, it is at an oblique angle, looking down the axes of the grooves as they wind around the left side of the helix. The contrast between the broad major groove and the narrow minor groove above and below it is striking. But the model is somewhat misleading, since base pairs in B-DNA actually sit *on* the helix axis rather than being displaced from it as seen here to accommodate the central support rod. Major and minor grooves in B-DNA in reality are equal in depth although of different width.

with Chargaff's hypercautious remarks in his earlier papers! Donohue is specifically thanked for his advice, and Wilkins and Franklin for their unpublished x-ray data.

Wilkins' paper F goes into the experimental proof that the scattering fiber is helical, a subject that paper E never touches upon, even though Crick shared, with Cochran and Vand, credit for establishing the mathematics of diffraction by a helix (see Key Papers for Chapter 4). Wilkins proposes that bases are "....arranged like a pile of pennies in the central regions of the helical system," but does not take the added step to base pairing. He concludes, "....in general there appears to be reasonable agreement between the experimental data and the kind of model described by Watson and Crick."

Franklin (G) is the only one to mention that DNA exists in two forms, the high-humidity B-DNA, and an A-DNA which can be produced by drying. Because the drier A form gives a more ordered, crystalline diffraction pattern, her crystallographic background led her to favor the A form over the B. She evinces an almost Chargaff-like caution in describing the B form: "While the X-ray evidence cannot, at present, be taken as direct proof that the structure is helical, other considerations discussed below make the existence of a helical structure highly probable." She repeats the calculation mentioned earlier, that a helical disk 3.4 Å thick and 19 Å in diameter has room enough for three nucleotides, and concludes: "It is therefore difficult to decide, on the basis of density measurements alone, whether one repeating until contains ten nucleotides on each of two or on each of three co-axial molecules." But her analysis of the more ordered A form had shown it to possess two chains, not three; and the interconvertibility of A and B forms was a powerful argument that B-DNA also is a duplex. The rest of her analysis is quite thorough. In essence, Franklin arrived at every feature of B-DNA except base pairing.

If Franklin and Wilkins had been left in peace and quiet to work at their own speed, would either of them eventually have thought of the concept of base pairing (probably from Chargaff's ratios), and arrived at the structure of B-DNA on their own? This has been a lively and controversial subject. The books by Sayre and by Maddox in Appendix 2 cover the issue exhaustively. Aaron Klug, who was a scientific colleague of Franklin at Birkbeck College after she left King's, became her scientific executor after her premature death, and preserved all of her papers. Klug has addressed this issue at least three times: in *Nature* papers in 1968 and 1974, and a comprehensive 2004 paper in the *Journal of Molecular Biology* that summarizes the keynote address which he gave at the April 2003 Cambridge Symposium celebrating 50 years of the DNA double helix.

In the 1974 article (reprinted as paper H), Klug reports finding a draft manuscript for *Nature* dated 17 March 1953, which Franklin was preparing before she had heard of the new Cambridge structure. She proposed a double helix, 20 Å in diameter, with bases on the inside and linked by hydrogen bonds. As Klug says in his 2004 paper:

> **In Franklin's draft, it is deduced that the phosphate groups of the backbone lie, as she had long thought, on the outside of the two co-axial helical strands whose geometrical configuration is specified, with the bases arranged on the inside. The two strands are separated by 13 Å (three-eights of the helix pitch in the axial direction). But the draft shows she had not yet grasped that the two chains in B also ran antiparallel as in the A form. Her notebooks show that for fitting the bases into the centre of a double helix, she had already formed the notion of the interchangeability of the two purine bases with each other, and also of the two pyrimidines. She also knew the correct tautomeric forms of at least three of the four bases, and was aware of Chargaff's base ratios. The step from interchangeability to the specific base-pairing postulated by Crick and Watson is a large one, but there is little doubt that Franklin was poised to make it....Crick and I have discussed this several times. We agree she would have solved the structure, but the results would have come out gradually, not as a thunderbolt in a short paper in Nature.**

There is a sad irony here. Her premature death in 1958 at age 38 robbed her of the glory of the 1962 Nobel Prize shared by Crick, Watson and Wilkins. Nobels are never awarded posthumously. But they also are never awarded to more than three individuals. Had Franklin lived, then who among the four would have been honored by Stockholm?

Other papers on DNA structure followed quickly from all of the principals; some of these are listed under Key Papers, but are not reprinted because they add little that is new. The last three reprinted papers, I–K, illuminate the bitter conflict over priorities and propriety that followed the announcement of the double helix. In the spring of 1968 Watson published an autobiographical memoir, "The Double Helix. A Personal Account of the Discovery of the Structure of DNA." (See book 4 in Appendix 2.) Chargaff reviewed the book in March for *Science* (I), and virtually exploded in sarcasm and righteous indignation:

> It is a great pity that the double helix was not discovered ten years earlier: some of the episodes could have been brought to the screen splendidly by the Marx brothers.

> Though I have no profound knowledge of this field, most scientific autobiographies that I have seen give me the impression of having been written for the remainder tables of the bookstores...

> I know of no other document in which the degradation of present-day science to a spectator sport is so clearly brought out. On almost every page, you can see the protagonists racing through the palaestra, as if they were chased by the Hound of Heaven—a Hound of Heaven with a Swedish accent.

> The workers at King's College, and especially Miss Franklin, were naturally reluctant to slake the Cavendish couple's thirst for other people's knowledge, before they themselves had had time to consider the meaning of their findings. The evidence found its way, however, to Cambridge.

This last remark leads to a veiled insinuation that Max Perutz erred in showing Watson and Crick a Medical Research Council report on research at King's which contained their best DNA diffraction photos. This prompted Perutz, Wilkins and Watson all to respond in a June issue of the same magazine (J). Max explains that the report was in no way considered classified or proprietary:

> I was inexperienced and casual in administrative matters and, since the report was not confidential, I saw no reason for withholding it.

He also points out that Maurice Wilkins and Francis Crick maintained a friendly relationship, and that Wilkins on more than one occasion passed information about their diffraction patterns on to Crick. Wilkins presents a capsule summary of the timetable of x-ray photography of the A and B forms of DNA. Watson gives his own account of events, beginning with:

> I am very sorry that, by not pointing out that the Randall report was non-confidential I portrayed Max Perutz in a way which allowed your reviewer to badly misconstrue his actions. The report was never marked 'confidential,' and I should have made the point clear in my text.

Chargaff, perhaps encouraged by the fact that Watson's book achieved great popularity and did *not* immediately wind up on "the remainder tables of the bookstores," published several collections of reminiscences and essays of his own in the 1970's and 80's which still make elegant and entertaining reading (Appendix 2, books 9 through 11). One essay is reprinted from Heraclitean Fire (1978), entitled "Gullible's Troubles" (K). It describes a visit that Chargaff made to Cambridge in May 1953, and is a masterpiece of bitter sarcasm. Chargaff again invokes the Marx

Brothers, and the ensuing description of Crick ("the looks of a fading racing tout") and Watson ("quite undeveloped at twenty-three") tells you more about Chargaff than about those whom he was describing. He says of their efforts:

> **So far as I could make out, they wanted, unencumbered by any knowledge of the chemistry involved, to fit DNA into a helix. The main reason seemed to be Pauling's alpha-helix model of a protein. I do not remember whether I was actually shown their scale model of a polynucleotide chain, but I do not believe so, since they still were unfamiliar with the chemical structures of the nucleotides. They were, however, extremely worried about the correct 'pitch' of their helix.**

In another setting, Chargaff referred to Watson and Crick as "two pitchmen in search of a helix"(7). The remark about their not knowing the chemical structures of the bases arose because, according to Crick, during the visit he once accidentally confused guanine with adenine in conversation.

Chargaff characterizes the duo as exhibiting "enormous ambition and aggressiveness, coupled with an almost complete ignorance of, and a contempt for, chemistry, that most real of exact sciences—a contempt that was later to have a nefarious influence on the development of 'molecular biology'." On more than one other occasion, Chargaff proclaimed scornfully that, "Molecular biology is essentially the practice of biochemistry without a license!" (e.g. *Voices in the Labyrinth*, p. 94)

Perhaps the most serious accusation in the essay was that which began with:

> **I told them all I knew. If they had heard before about the pairing rules, they concealed it. But as they did not seem to know much about anything, I was not unduly surprised.....I believe that the double-stranded model of DNA came about as a consequence of our conversation....**

This is not fair. As far as I can determine, Chargaff did indeed establish that A and T tended to occur in equal quantities in DNA, as did G and C, but he never suggested in print that this was because A and T, or G and C, were *associated or paired in some manner*. He may well have thought this privately, but if so, his innate conservatism seemingly kept him from writing the idea down and publishing it. To speak of "the pairing rules" as something which Watson and Crick learned about from others (by inference, Chargaff himself) is simply unfair.

And so the drama of DNA structure ended, and the still-continuing drama of the biological implications of that structure continues. We shall let Chargaff have the last word, simply because he chooses his words so elegantly. The essay following "Gullible's Troubles" in the same book is entitled "Matches for Herostratos." Herostratos, as Chargaff carefully explains for the ill-informed reader, was a Greek who achieved fame and immortality by committing one horrendous action: he burned the Artemison temple to the ground in 356 B. C. Chargaff wryly compares Herostratos to Watson and Crick without specifically naming them, and then comments:

> **If Herostratos has earned immortality for having burned down the temple of Artemis in Ephesos, maybe the man from whom he got the matches ought not to be entirely forgotten. I am that man.**

References

1. R. E. Dickerson. 1964. *The Proteins* (Hans Neurath, ed.), Academic Press, New York, 2nd edition, Vol. 2, pp. 603–778. "X-Ray Analysis and Protein Structure."
2. On first thought, the Morse code might be regarded as having two symbols: dot and dash. But *three symbols* actually are needed to convey a message: dot, dash, and space.
3. O. T. Avery, C. M. MacLeod and M. McCarty. 1944. *J. Exp. Med.* 79, 137–158. "Studies on the Chemical Transformation of Pneumococcal Types."

4. A. D. Hershey and M. Chase. 1952. *J. Gen. Physiol.* 36, 39–56. "Independent Functions of Viral Proteins and Nucleic Acid in Growth of Bacteriophage."

5. R. E. Dickerson. 1999. In *Oxford Handbook of Nucleic Acid Structure* (S. Neidle, ed.), Oxford U. Press, pp. 145–197. "Helix Structure and Molecular Recognition by B-DNA."

6. R. E. Dickerson. 2001. In *Crystallography of Biological Macromolecules: International Tables for Macromolecular Crystallography*, Vol. 4 (M. G. Rossmann and E. Arnold, eds.), Ch. 23.3, pp. 588–622. "Nucleic Acids."

7. E. Chargaff. 1974. *Nature* 248, 776–779. "Building the Tower of Babble."

Study Questions

1. How does the tilting of the geologist's sieve in Figure 5.4c illustrate the reciprocal or inverse relationship between "real space" and "diffraction pattern space"?

2. Because of this reciprocal relationship, a 2 Å resolution data set from a simple mineral, sodium chloride, contains only 23 x-ray reflections, whereas the same resolution data set from the protein myoglobin contains ten thousand spots. Why does this make good sense from the standpoint of information content?

3. In the diffraction pattern of a fiber of B-DNA, what does the spacing between layer lines tell you about the helix? What tells you the number of repeating units (base pairs) per turn of helix?

4. Why is it reasonable that the diffraction pattern of B-DNA should involve a cross pattern whose arms are rows of equally spaced spots? How is the angle of the arms related to the structure of the helix?

5. Why was the issue of tautomers of bases so crucial in arriving at the structure of DNA?

6. In the story of DNA, and in other branches of science as well, there seem to be two philolosophical camps: data analyzers and theorists or model builders. Linus Pauling once remarked: "If the experimental data appear to contradict a really good theory, then it is time to go back and collect better experimental data." Which camp does that remark put him in? Where would you put Erwin Chargaff? Rosalind Franklin? Jim Watson? Maurice Wilkins?*

7. Why were three different groups led astray into thinking initially that B-DNA had three chains, and a fourth group at least entertained the idea? Who were these respective groups? (C, D, G)

8. How did the pK_a of phosphate groups make Pauling's DNA model impossible? (D)

9. Why did Pauling and Corey propose a DNA model with only eight base steps per turn, whereas everyone else used ten?

10. Why did Rosalind Franklin initially favor working on A-DNA rather than B?

11. What one idea, which could be derived from Chargaff's work, was apparently missing from Franklin's thinking about B-DNA structure? (H)

12. If Franklin had been left alone to work in peace at her own speed, how long do you think it would have been before she arrived at the correct helical structure for B-DNA?

13. What ethical impropriety did Chargaff appear to ascribe to Perutz in the DNA controversy? Were his allegations justified? What really happened? (I, J)

14. Why do you think Erwin Chargaff is so bitter and sarcastic in his published essays?

15. How did the recent history of protein chain folding make the Cambridge group in 1953 so hypersensitive about Caltech and DNA?

---

*A lecturer was making a point about innovation in science: "They said that Alexander Graham Bell was crazy. They said Albert Einstein was crazy. They said Niels Bohr was crazy. They said Linus Pauling was crazy. They said Karl Halbkopf was crazy!" A student called out: "Who is Karl Halbkopf, anyway?," to which the lecturer replied, "Oh, Halbkopf—he *was* crazy!"

EXPERIENTIA

Vol. VI - Fasc. 6   Pag. 201–240   15. VI. 1950

Chemical Specificity of Nucleic Acids and Mechanism of their Enzymatic Degradation[1]

By Erwin Chargaff[2], New York, N.Y.

I. *Introduction*

The last few years have witnessed an enormous revival in interest for the chemical and biological properties of nucleic acids, which are components essential for the life of all cells. This is not particularly surprising, as the chemistry of nucleic acids represents one of the remaining major unsolved problems in biochemistry. It is not easy to say what provided the impulse for this rather sudden rebirth. Was it the fundamental work of E. Hammarsten[3] on the highly polymerized desoxyribonucleic acid of calf thymus? Or did it come from the biological side, for instance the experiments of Brachet[4] and Caspersson[5]? Or was it the very important research of Avery[6] and his collaborators on the transformation of pneumococcal types that started the avalanche?

It is, of course, completely senseless to formulate a hierarchy of cellular constituents and to single out certain compounds as more important than others. The economy of the living cell probably knows no conspicuous waste; proteins and nucleic acids, lipids and polysaccharides, all have the same importance. But one observation may be offered. It is impossible to write the history of the cell without considering its geography; and we cannot do this without attention to what may be called the chronology of the cell, i. e. the sequence in which the cellular constituents are laid down and in which they develop from each other. If this is done, nucleic acids will be found pretty much at the beginning. An attempt to say more leads directly into empty speculations in which almost no field

abounds more than the chemistry of the cell. Since an ounze of proof still weighs more than a pound of prediction, the important genetical functions, ascribed —probably quite rightly—to the nucleic acids by many workers, will not be discussed here. Terms such as "template" or "matrix" or "reduplication" will not be found in this lecture.

II. *Identity and Diversity in High Molecular Cell Constituents*

The determination of the constitution of a complicated compound, composed of many molecules of a number of organic substances, evidently requires the exact knowledge of the nature and proportion of all constituents. This is true for nucleic acids as much as for proteins or polysaccharides. It is, furthermore, clear that the value of such constitutional determinations will depend upon the development of suitable methods of hydrolysis. Otherwise, substances representing an association of many chemical individuals can be described in a qualitative fashion only; precise decisions as to structure remain impossible. When our laboratory, more than four years ago, embarked upon the study of nucleic acids, we became aware of this difficulty immediately.

The state of the nucleic acid problem at that time found its classical expression in Levene's monograph[1]. (A number of shorter reviews, indicative of the development of our conceptions concerning the chemistry of nucleic acids, should also be mentioned[2].) The old tetranucleotide hypothesis—it should never have been called a theory—was still dominant; and this was characteristic of the enormous sway that the organic chemistry of small molecules held over biochemistry. I should like to illustrate what I mean by one example. If in the investigation of a disaccharide consisting of two different hexoses we isolate 0·8 mole of one sugar and 0·7 mole of the other, this will be sufficient for the

[1] This article is based on a series of lectures given before the Chemical Societies of Zürich and Basle (June 29th and 30th, 1949), the Société de chimie biologique at Paris, and the Universities of Uppsala, Stockholm, and Milan.

[2] Department of Biochemistry, College of Physicians and Surgeons, Columbia University, New York. The author wishes to thank the *John Simon Guggenheim Memorial Foundation* for making possible his stay in Europe. The experimental work has been supported by a research grant from the *United States Public Health Service*.

[3] E. Hammarsten, Biochem. Z. *144*, 383 (1924).

[4] J. Brachet in *Nucleic Acid*, Symposia Soc. Exp. Biol. No. 1 (Cambridge University Press, 1947), p. 207. Cp. J. Brachet, in *Nucleic Acids and Nucleoproteins*, Cold Spring Harbor Symp. Quant. Biol. *12*, 18. (Cold Spring Harbor, N.Y., 1947).

[5] T. Caspersson, in *Nucleic Acid*, Symp. Soc. Exp. Biol., No. 1 (Cambridge University Press, 1947), p. 127.

[6] O. T. Avery, C. M. MacLeod, and M. McCarty, J. Exp. Med. *79*, 137 (1944).

[1] P. A. Levene and L. W. Bass, Nucleic Acids (Chemical Catalog Co., New York, 1931).

[2] H. Bredereck, Fortschritte der Chemie organischer Naturstoffe *1*, 121 (1938). – F. G. Fischer, Naturwissensch. *30*, 377 (1942). – R. S. Tipson, Adv. Carbohydrate Chem. *1*, 193 (1945). – J. M. Gulland, G. R. Barker, and D. O. Jordan, Ann. Rev. Biochem. *14*, 175 (1945). – E. Chargaff and E. Vischer, Ann. Rev. Biochem. *17*, 201 (1948). – F. Schlenk, Adv. Enzymol. *9*, 455 (1949).

Table I

Resistance of pyrimidines to treatment with strong acid. A mixture of pyrimidines of known concentration was dissolved in the acids indicated below and heated at 175° in a bomb tube. The concentration shifts of the individual pyrimidines were determined through a comparison of the recoveries of separated pyrimidines before and after the heating of the mixture.

Experiment No.	Acid	Heating time min.	Concentration shift, per cent of starting concentration		
			Uracil	Cytosine	Thymine
1	HCl (10%)	90	+ 62	− 63	+ 3
2	10 N HCOOH +	60	+ 3	− 5	0
3	N HCl (1:1)	120	+ 24	− 19	0
4	HCOOH (98 to 100%)	60	0	− 1	− 2
5		120	0	+ 2	+ 1

the hydrolysis of the pyrimidine nucleotides by means of concentrated formic acid. For the liberation of the purines N sulfuric acid (100°, 1 hour) is employed; for the liberation of the pyrimidines, the purines are first precipitated as the hydrochlorides by treatment with dry HCl gas in methanol and the remaining pyrimidine nucleotides cleaved under pressure with concentrated formic acid (175°, 2 hours). This procedure proved particularly suitable for the investigation of the desoxypentose nucleic acids. For the study of the composition of pentose nucleic acids a different procedure, making use of the separation of the ribonucleotides, was developed more recently, which will be mentioned later.

VII. *Composition of Desoxypentose Nucleic Acids*

It should be stated at the beginning of this discussion that the studies conducted thus far have yielded no indication of the occurrence in the nucleic acids examined in our laboratory of unusual nitrogenous constituents. In all desoxypentose nucleic acids investigated by us the purines were adenine and guanine, the pyrimidines cytosine and thymine. The occurrence in minute amounts of other bases, e.g. 5-methylcytosine, can, however, not yet be excluded. In the pentose nucleic acids uracil occurred instead of thymine.

A survey of the composition of desoxyribose nucleic acid extracted from several organs of the ox is provided

Table II [1]

Composition of desoxyribonucleic acid of ox (in moles of nitrogenous constituent per mole of P).

Constituent	Thymus			Spleen		Liver
	Prep. 1	Prep. 2	Prep. 3	Prep. 1	Prep. 2	
Adenine . .	0·26	0·28	0·30	0·25	0·26	0·26
Guanine . .	0·21	0·24	0·22	0·20	0·21	0·20
Cytosine . .	0·16	0·18	0·17	0·15	0·17	
Thymine . .	0·25	0·24	0·25	0·24	0·24	
Recovery . .	0·88	0·94	0·94	0·84	0·88	

[1] From E. CHARGAFF, E. VISCHER, R. DONIGER, C. GREEN, and F. MISANI, J. Biol. Chem. *177*, 405 (1949); and unpublished results.

in Table II. The molar proportions reported in each case represent averages of several hydrolysis experiments. The composition of desoxypentose nucleic acids from human tissues is similarly illustrated in Table III. The preparations from human liver were obtained from a pathological specimen in which it was possible, thanks to the kind cooperation of M. FABER, to separate portions of unaffected hepatic tissue from carcinomatous tissue consisting of metastases from the sigmoid colon, previous to the isolation of the nucleic acids[1].

Table III [2]

Composition of desoxypentose nucleic acid of man (in moles of nitrogenous constituent per mole of P).

Constituent	Sperm		Thymus	Liver	
	Prep. 1	Prep. 2		Normal	Carcinoma
Adenine . . .	0·29	0·27	0·28	0·27	0·27
Guanine . . .	0·18	0·17	0·19	0·19	0·18
Cytosine . . .	0·18	0·18	0·16		0·15
Thymine . . .	0·31	0·30	0·28		0·27
Recovery . . .	0·96	0·92	0·91		0·87

In order to show examples far removed from mammalian organs, the composition of two desoxyribonucleic acids of microbial origin, namely from yeast[3] and from avian tubercle bacilli[4], is summarized in Table IV.

Table IV [5]

Composition of two microbial desoxyribonucleic acids.

Constituent	Yeast		Avian tubercle bacilli
	Prep. 1	Prep. 2	
Adenine	0·24	0·30	0·12
Guanine	0·14	0·18	0·28
Cytosine	0·13	0·15	0·26
Thymine	0·25	0·29	0·11
Recovery	0·76	0·92	0·77

The very far-reaching differences in the composition of desoxypentose nucleic acids of different species are best illustrated by a comparison of the ratios of adenine to guanine and of thymine to cytosine as given in Table V. It will be seen that in all cases where enough material for statistical analysis was available highly significant differences were found. The analytical figures on which Table V is based were derived by comparing the ratios found for individual nucleic acid hydrolysates of one species regardless of the organ from which the preparation was isolated. This procedure assumes that there is no organ specificity with

[1] Unpublished experiments.

[2] From E. CHARGAFF, S. ZAMENHOF, and C. GREEN, Nature (in press); and unpublished results.

[3] E. CHARGAFF and S. ZAMENHOF, J. Biol. Chem. *173*, 327 (1948).

[4] E. CHARGAFF and H. F. SAIDEL, J. Biol. Chem. *177*, 417 (1949).

[5] From E. VISCHER, S. ZAMENHOF, and E. CHARGAFF, J. Biol. Chem. *177*, 429 (1949); and unpublished results.

Table V

Molar proportions of purines and pyrimidines in desoxypentose nucleic acids from different species.

Species	Number of different organs	Number of different preparations	Adenine/Guanine			Thymine/Cytosine		
			Number of hydrolyses[3]	Mean ratio	Standard error	Number of hydrolyses[3]	Mean ratio	Standard error
Ox[1]	3	7	20	1·29	0·013	6	1·43	0·03
Man[2]	2	3	6	1·56	0·008	5	1·75	0·03
Yeast	1	2	3	1·72	0·02	2	1·9	
Avian tubercles bacillus	1	1	2	0·4		1	0·4	

[1] Preparations from thymus, spleen, and liver served for the purine determinations, the first two organs for the estimation of pyrimidines.

[2] Preparations from spermatozoa and thymus were analysed.

[3] In each hydrolysis between 12 and 24 determinations of individual purines and pyrimidines were performed.

respect to the composition of desoxypentose nucleic acids of the same species. That this appears indeed to be the case may be gathered from Tables II and III and even better from Table VI where the average purine and pyrimidine ratios in individual tissues of the same species are compared. That the isolation of nucleic acids did not entail an appreciable fractionation is shown by the finding that when whole defatted human spermatozoa, after being washed with cold 10% trichloroacetic acid, were analyzed, the same ratios of adenine to guanine and of thymine to cytosine were found as are reported in Tables V and VI. It should also be mentioned that all preparations, with the exception of those from human liver, were derived from pooled starting material representing a number, and in the case of human spermatozoa a very large number, of individuals.

Table VI

Molar proportions of purines and pyrimidines in desoxypentose nucleic acids from different organs of one species.

Species	Organ	Adenine/ Guanine	Thymine/ Cytosine
Ox	Thymus	1·3	1·4
	Spleen	1·2	1·5
	Liver	1·3	
Man	Thymus	1·5	1·8
	Sperm	1·6	1·7
	Liver (normal)	1·5	1·8
	Liver (carcinoma)	1·5	1·8

The desoxypentose nucleic acids extracted from different species thus appear to be different substances or mixtures of closely related substances of a composition constant for different organs of the same species and characteristic of the species.

The results serve to disprove the tetranucleotide hypothesis. It is, however, noteworthy—whether this is more than accidental, cannot yet be said—that in all desoxypentose nucleic acids examined thus far the molar ratios of total purines to total pyrimidines, and also of adenine to thymine and of guanine to cytosine, were not far from 1.

VIII. *Composition of Pentose Nucleic Acids*

Here a sharp distinction must be drawn between the prototype of all pentose nucleic acid investigations—the ribonucleic acid of yeast—and the pentose nucleic acids of animal cells. Nothing is known as yet about bacterial pentose nucleic acids. In view of the incompleteness of our information on the homogeneity of pentose nucleic acids, which I have stressed before, I feel that the analytical results on these preparations do not command the same degree of confidence as do those obtained for the desoxypentose nucleic acids.

Table VIII[1]

Composition of pentose nucleic acids from animal tissues.

Constituent	Calf liver	Ox liver	Sheep liver	Pig liver	Pig pancreas
Guanylic acid . . .	16·3	14·7	16·7	16·2	22·5
Adenylic acid . . .	10	10	10	10	10
Cytidylic acid . . .	11·1	10·9	13·4	16·1	9·8
Uridylic acid . . .	5·3	6·6	5·6	7·7	4·6
Purines : pyrimidines	1·6	1·4	1·4	1·1	2·5

Three procedures, to which reference is made in Tables VII and VIII, were employed in our laboratory for the analysis of pentose nucleic acids. In *Procedure 1*, the pentose nucleic acid was hydrolysed to the nucleotide stage with alkali, at p_H 13·5 and 30°, and the nucleotides, following adjustment to about p_H 5, separated by chromatography with aqueous ammonium isobutyrate-isobutyric acid as the solvent. Under these conditions, guanylic acid shares its position on the chromatogram with uridylic acid; but it is possible to determine the concentrations of the two components in the eluates by simultaneous equations based on the ultraviolet absorption of the pure nucleotides[2]. The very good recoveries of nucleotides obtained in terms of both nucleic acid phosphorus and nitrogen show the cleavage by mild alkali treatment of pentose nucleic acids to be practically quantitative.—In

---

[1] Unpublished results.

[2] E. VISCHER, B. MAGASANIK, and E. CHARGAFF, Federation Proc. *8*, 263 (1949). – E. CHARGAFF, B. MAGASANIK, R. DONIGER, and E. VISCHER, J. Amer. Chem. Soc. *71*, 1513 (1949).

characterized by greatly increased ratios of adenine to guanine, thymine to cytosine, purines to pyrimidines, and by greater resistance to enzymatic attack. Preliminary results of studies of this type on wheat germ DNA are presented in table 3. The analytical findings on the intact DNA are in good agreement with analyses recently reported for the same nucleic acid from

TABLE 1. SALMON SPERM DNA; PROPORTIONS (IN MOLES OF NITROGENOUS CONSTITUENT PER MOLE OF P IN HYDROLYSATE)

CONSTITUENT	MEAN PROPORTION	STANDARD ERROR
Adenine	0.280	0.005
Guanine	0.196	0.004
Cytosine	0.192	0.006
Thymine	0.274	0.005

TABLE 2. SALMON SPERM DNA; MOLAR RELATIONSHIPS

	MOLAR RATIO
Adenine to guanine	1.43
Thymine to cytosine	1.43
Adenine to thymine	1.02
Guanine to cytosine	1.02
Purines to pyrimidines	1.02
P accounted for as percentage of P in hydrolysate	95.8 ($\pm$1.6)
Average no. of gm.-atoms N per mole constituent	3.7
Atomic N:P ratio in DNA preparations	3.6, 3.7

two other laboratories (30, 31). The figures given here for the '19% core' and the '8% core' refer to the dialysis residues recovered when 81 and 92 per cent of the DNA respectively had been converted by the enzyme to dialyzable fragments. The trend of degradation appears similar to that observed with calf thymus DNA.

The molar ratios found in the DNA specimens studied in our laboratory are compared in table 4. The tendency toward certain regularities will be observed. The figures for hen DNA (chicken erythrocytes) and for the DNA from the K-12 strain of *B. coli* must be considered as preliminary. Both components were studied in collaboration with B. Gandelman. The DNA of *Hemophilus influenzae, type c*, was studied by S. Zamenhof. A few points are noteworthy. In the case of wheat germ DNA, methylcytosine and cytosine apparently must be considered together, if the regular ratios, observed in most other instances, are to be obtained. Another remarkable fact is that the regularities are maintained even in the nucleic acids of the 'GC type' despite the complete inversion in individual ratios.

As was already mentioned before, it is almost impossible to decide at present whether these regularities are entirely fortuitous or whether they reflect the existence in all DNA preparations of certain common structural principles, irrespective of far-reaching differences in their individual composition and the absence of an easily recognizable periodicity. It may be assumed that the nucleic acids as we know them today, as for that matter also the proteins, are the result of an age-long selection process in the course of which many less suitable or less stable components must have been eliminated. One could speak of the survival of the fittest nucleic acids. Such macromolecules will exhibit diversity and uniformity at the same time, since they are called upon to perform, in diverse species, the same tasks. It is, therefore, perhaps not astonishing that the nucleic acids will share certain features that may be directly connected with their stability or their ability to form conjugated nucleoproteins. One property which is quite striking is the uniform absorption spectrum in the ultraviolet of all highly polymerized DNA specimens, both with respect to the position and the intensity of the

TABLE 3. WHEAT GERM DNA; INTACT PREPARATION AND ENZYMATICALLY PRODUCED CORES (IN MOLES OF NITROGENOUS CONSTITUENT PER MOLE OF P IN HYDROLYSATE)

CONSTITUENT	INTACT DNA	19% CORE	8% CORE
Adenine	0.27	0.33	0.35
Guanine	0.22	0.20	0.20
Cytosine	0.16	0.12	0.10
5-Methylcytosine	0.06	0.04	0.04
Thymine	0.27	0.26	0.23
Total purines	0.49	0.53	0.55
Total pyrimidines	0.49	0.42	0.37
Recovery	0.98	0.95	0.92

absorption maximum. The center of absorption fluctuates only between 257 and 261 mμ and the (ϵ)P is around 6600. Another surprising feature is the balance between amino groups and enolic hydroxyls in all DNA preparations examined by us. (See last column of table 4.) Even in the core preparations, this ratio changed only very little.

But let us return for a moment to the other outstanding characteristic of nucleic acids, viz. their diversity. If we accept the evidence of the existence of species-specific DNA, then there arise many new questions, both of a biological and chemical nature. DNA presumably is an important part of the chromosomes and may be

periodicity in a nucleic acid chain. Any simplified assumption with respect to periodicity has been disproved by the studies on the course of action of desoxyribonuclease on the DNA of calf thymus (29) and of wheat germ. The composition of both the dialyzable degradation products and the dialysis residues, the 'cores,' exhibited continuous and characteristic changes with respect to the distribution of purines and pyrimidines. One must conclude that the sequence is highly aperiodic and that it is not inconceivable that the same cellular DNA could give rise to many different nucleoproteins, depending upon the shape and configuration of the particular protein.

TABLE 4. MOLAR RATIOS IN DNA PREPARATIONS OF DIFFERENT ORIGIN

SOURCE	ADENINE TO GUANINE	THYMINE TO CYTOSINE	ADENINE TO THYMINE	GUANINE TO CYTOSINE	PURINES TO PYRIMIDINES	AMINO GROUPS TO ENOLIC HYDROXYLS
Ox	1.29	1.43	1.04	1.00	1.1	1.4
Man	1.56	1.75	1.00	1.00	1.0	1.3
Hen	1.45	1.29	1.06	0.91	0.99	1.5
Salmon	1.43	1.43	1.02	1.02	1.02	1.4
Wheat	1.22	1.18[1]	1.00	0.97[1]	0.99	1.4
Yeast	1.67	1.92	1.03	1.20	1.0	1.3
Hemophilus influenzae, type C	1.74	1.54	1.07	0.91	1.0	1.5
B. coli K-12	1.05	0.95	1.09	0.99	1.0	1.6
Avian tubercle bacillus	0.4	0.4	1.09	1.08	1.1	1.7
Serratia marcescens	0.7	0.7	0.95	0.86	0.9	1.6
Hydrogen organism Bacillus Schatz	0.7	0.6	1.12	0.89	1.0	1.7

[1] In these computations the sum of cytosine and methylcytosine was used. If cytosine alone is considered, the thymine to cytosine ratio is 1.62 and that of guanine to cytosine 1.33.

surmised to be involved in their biological functions. Does this mean that a cell contains as many different DNA individuals as it contains genes? Or can one and the same species-specific DNA form so many three-dimensional structures, in connection with the proteins to which it is attached, that the genic requirements are fulfilled? (For a more detailed discussion of some of these points, cf. 32.) This question, as so many others in this field, cannot yet be answered. No way has as yet been found to fractionate a family of very similar macromolecules which may differ in no more than the sequence of a few of their component nucleotides. But one could perhaps say that the more regular the arrangement of nucleotides is in a given DNA, the less the chance of its forming many different specific structures. And this brings us again to the very important question of nucleotide sequence and

It must, moreover, be understood that the recognition of periodicity, i.e. the presence of recurring units, will be particularly difficult in a macromolecule of the type of DNA. If, for instance, in a chain composed of 3000 nucleotides a particular sequence of 100 consecutive nucleotides were repeated 30 times, this periodicity could not be recognized, unless we had a method producing cleavage only at the points where these repeating units are joined. In other words, the perception of periodicity would require the proper distance for a bird's-eye view which will not be easy to attain.

Another approach to the problem of sequence analysis in DNA may be seen in the study of its controlled chemical degradation. That the purines can be detached from a nucleic acid with much greater ease than can the pyrimidines, has long been known; but the resulting end product,

Available online at www.sciencedirect.com

SCIENCE @ DIRECT®

Journal of Structural Biology 145 (2004) 181–183

ELSEVIER

Journal of
**Structural
Biology**

www.elsevier.com/locate/yjsbi

Editorial

The third man and the fourth paper

The 50th anniversary of the momentous work leading to elucidation of the structure of DNA has been celebrated with an outpouring of publications, documentaries, and symposia. Included in and perhaps swept along by this stream has been the autobiography of one of the principals, Maurice Wilkins. The enigmatic title of his memoir, of which Graham Greene would surely have approved, is *The Third Man of DNA*. This volume is reviewed in our current issue by David Parry, an Editorial Board member who by coincidence shares a Kiwi connection with Wilkins.

The breakthrough work was published in three short but resoundingly significant papers in *Nature* in April 1953. There was, however, a fourth paper by Bruce Fraser, a graduate student at King's College, that was cited by Francis Crick and Jim Watson as "in press." It described a model-building exercise carried out in 1951 that was almost correct but for the number of chains, which was assumed to be three.

The circumstances behind the failure to publish this paper in 1951 are described in excerpts from an article by Wilkins in an edited volume and the book by Brenda Maddox that are cited below; with permission. Certainly, Wilkins, who was well aware of the paper's significance, urged that it be published in 1953. Fraser, by then in Australia, prepared a manuscript but, under pressure from Crick (according to Maddox), Wilkins decided not to submit it to *Nature*. Nevertheless, it is appropriate, we believe, that this fourth paper should also form part of the historical record. It was recovered in 2002, regrettably without figures, by Wilkins from his personal archives and is reproduced in this issue. Although Fraser's model of DNA did not correctly describe the B-form of duplex DNA, it came close to describing the triplex polynucleotide forms that have since been characterized, first for RNA in 1957 (Felsenfeld et al., 1957) and then as a "high energy" state of DNA (H-DNA). DNA triplexes have attracted considerable interest in view of their perceived potential as agents for gene regulation and as "antigenes" (Gowers and Fox, 1999; Neidle, 1997).

Among their many other distinctions, this quartet—the three 1953 *Nature* papers and the long-dormant Fraser paper—serve a timely reminder of the value of model-building as a means—perhaps the only means—of systematically integrating information from diverse experimental sources and of serving as a platform for designing further experiments to test and refine a working hypothesis, i.e., the current model.

As noted above, Bruce Fraser and his wife, Mary, an MRC research biochemist at King's College who was preparing DNA for Wilkins, emigrated to Australia at a critical point in the drama. There his focus transferred to fibrous proteins and, in particular, his penchant for three-stranded molecules was fully requited in the case of collagen. Bruce Fraser initiated a program of structural studies in the Division of Protein Chemistry at CSIRO Parkville, Victoria, which became the premier institution for the study of naturally occurring and synthetic fibrous polypeptides by diverse biophysical approaches, but with a strong emphasis on X-ray fiber diffraction. In 1973, he published the definitive book on this subject *Conformation in Fibrous Proteins and Related Synthetic Polypetides*. This tome was co-authored by his long-time collaborator Tom MacRae and published by Academic Press. Bruce retired from CSIRO in 1987 but retains an active interest in the structures and properties of fibrous proteins (e.g., Fraser and Parry, 2003; Fraser et al., 2003a,b).

Editorial / Journal of Structural Biology 145 (2004) 181–183

References

Felsenfeld, G., Davies, D.R., Rich, A., 1957. J. Am. Chem. Soc. 79, 2023–2024.
Fraser, R.D.B., Parry, D.A.D., 2003. Macrofibril assembly in trichocyte (hard alpha-) keratins. J. Struct. Biol. 142, 319–325.
Fraser, R.D.B., Rogers, G.E., Parry, D.A.D., 2003a. Nucleation and growth of macrofibrils in trichocyte (hard-alpha) keratins. J. Struct. Biol. 143, 85–93.
Fraser, R.D.B., Steinert, P.M., Parry, D.A.D., 2003b. Structural changes in trichocyte keratin intermediate filaments during keratinization. J. Struct. Biol. 142, 266–271.
Gowers, D.M., Fox, K.R., 1999. Towards mixed sequence recognition by triple helix formation. Nucleic Acids Res. 27, 1569–1581.
Neidle, S., 1997. Recent developments in triple-helix regulation of gene expression. Anticancer Drug Des. 12, 433–442.

Alasdair C. Steven

Wolfgang Baumeister

The following excerpts from books by Maurice Wilkins and Brenda Maddox describe some of the circumstances behind the failure to publish the fourth paper.

Extract from "Origins of DNA research at Kings College London," by M.H.F. Wilkins in *DNA Structure: Genesis of a Discovery,* edited by Seweryn Chomet, King's College, London. Published by Newman-Hemisphere, 1995.

Pages 22–23

At King's, very late in 1951 (but a year and a half before the Double Helix), Bruce Fraser, a physics research student in Bill Price's spectroscopy group, built a DNA model which showed very well the general way we were thinking in our laboratory. Bill Price knew a great deal about chemical bonds and energies, and this knowledge helped Fraser to build his model. It was helical with phosphates on the outside and the flat bases stacked inside. The bases on one chain were hydrogen bonded to those on the other chains. The model had most of the features of the Double Helix except, of course, that it had three chains instead of two and, because of that, could not have the very special and unique system of base pairs which was such a staggering feature of the Watson and Crick model.

The reason why the Fraser model had three chains was that we at King's thought that density and water content data on DNA meant that there were three chains. We barely considered the possibility that there were only two chains. We were not alone in making that mistake: both Pauling and Astbury fell into the same physical chemistry trap. The Fraser model looked very promising, but because it had three chains, it could not be made to fit the X-ray data. That was very disappointing.

Extracts from *Rosalind Franklin—The Dark Lady of DNA,* by Brenda Maddox. Published by HarperCollins, 2002.

Pages 161–162

Not everybody at King's was opposed to model-building. In early November, Bruce Fraser, working for his doctorate with Dr William Price in the spectroscopy group (allied to the Biophysics Unit), decided to put together a model to summarise the latest DNA thinking. Fraser, who sported a black handlebar moustache testifying to his war service as a pilot for the RAF, was neutral in the Franklin–Wilkins feud. Such contact as he and his wife Mary had with Rosalind was entirely friendly. Rosalind discussed the model with him just as she had read Sven Furberg's thesis on which it was based. When Fraser asked her to guess how many chains there might be

Editorial / Journal of Structural Biology 145 (2004) 181–183

in the molecule, she ventured three. Wilkins said the same, for the same reasons: the measurements of density and water content of DNA suggested more than one chain and probably three. Two would not be enough to fill the space. (As Linus Pauling later came up with a three-chain answer, the reasoning was sound, if erroneous.)

Fraser's model of DNA, completed very quickly, was a simple structure that had what would turn out to be all main features correct except for the number of chains. It had a helical shape, phosphates on the outside, and bases stacked like a pile of pennies, separated by the 3.4 Å distance worked out by Astbury. Rosalind saw it, and her view was what she felt about all models: 'That's very nice—how are you going to prove it is the solution?'

A big step in the right direction, Fraser's November 1951 model was another glaring example of King's' institutional hesitancy. Its details were never published, just as Stokes's calculations on helical diffraction were never published. Wilkins himself did not think it worth pursuing because, although Fraser's model was full of potential, there was no structural hypothesis behind it; it did not explain anything. In a few months Bruce and Mary Fraser, with their new baby, prepared to emigrate to Australia—out of the drama until its very end.

Pages 209–210

In the ensuing panic to get King's papers ready, Wilkins cabled—a dramatic gesture in those austere times—Bruce Fraser in Australia and asked him to write up his model as quickly as possible in a note suitable for publication in *Nature*. Fraser complied. In view of the urgency, he sat up all night typing the very paper Wilkins had suppressed two years earlier and sketching the diagrams. There were no photocopying machines in 1953, and had Fraser not done it by hand, the manuscript would have lain several days in a queue before being photographed for duplication. Then, even more expensively, Fraser had his work cabled off to London next morning.

But when the Fraser paper arrived, Crick vetoed it: what was the point of publishing wrong data? Instead, he and Watson appended an acknowledgement to Fraser at the end of their own paper, describing the Fraser model as 'rather ill-defined' and the Fraser paper as 'in the press.' It was not, and was never published.

The manuscript sent to Maurice Wilkins in 1953

As Brenda Maddox explained, Fraser did not keep a copy of the manucript. However, in 2002, Fraser and Brenda Maddox independently approached Maurice Wilkins to see if he could unearth the original, which he kindly did, and a copy is reproduced here—unfortunately the accompanying figures were missing.

Available online at www.sciencedirect.com

Journal of Structural Biology 145 (2004) 184–186

ELSEVIER

Journal of
**Structural
Biology**

www.elsevier.com/locate/yjsbi

THE STRUCTURE OF DEOXYRIBOSE NUCLEIC ACID

In a recent letter to this journal[1] Pauling and Grey report the formulation of a structure for the nucleic acids involving three intertwined helical polynucleotide chains. A considerable body of evidence *on* this topic has accumulated during the past few years in this department from studies of birefringence and extensibility[2], ultra-violet[3] and infra-red dichroism[4,] x-ray diffraction, and physico-chemical properties. The task of integrating these observations to give a detailed picture of the structure of the nucleic acids is a formidable one, and is by no means complete. However, in view of the letter by Pauling and Grey it seems worth describing a type of structure that we have considered, which, although it involves three intertwined helical polynucleotide chains, differs considerably from that formulated by them.

The structure was arrived at by considering the contributions of various attractive forces between nucleotides to the stability of the deoxyribonucleic acid molecule. This material is normally prepared and studied in the form of the sodium salt, and the chief forces were presumed to be

(a) electrostatic attractions between the negatively charged phosphate groups and the sodium ions,

(b) Van der Waals attractions between the planar purine and pyrimidine residues,

(e) hydrogen bonds formed between the C=O, NH_2, NH and OH groups of the purine and pyrimidine residues.

Models of polynucleotide chains were built assuming that the β-deoxyribofuranosides were joined by 3′,5′ phosphate diester linkages, as in Figure 1.

- 2 -

The orientation of the sugar residue with respect to the base residue corresponded to that shown by Furberg[5] to exist in ribose nucleosides.

R.D.B. Fraser / Journal of Structural Biology 145 (2004) 184–186

No satisfactory configurations of a single polynucleotide chain could be found, and various structures involving three chains were considered. Of these, the type which appeared most stable, and explained much of the chemical and physico-chemical behaviour consisted of three intertwined helical chains in which the phosphate groups were arranged on the surface of s pseudo-cylindrical molecule, with the base residues directed towards the axis. The molecule is maintained in this configuration by the Van der Waals attraction between the base residues and the ring of hydrogen bonds around the axle of the cylinder. A diagrammatic section perpendicular to the molecular axis is given in Figure 2, in which the relation between nucleotides of the three chains A, B and C is shown. Cytosine and thymine residual are shown for convenience, but similar systems of hydrogen bonds are possible for the purine residues. The next layer of base residues is obtained by a translation of 3.4A and an anticlockwise rotation about the axis sufficient to avoid steric interference between the sugar residues, leading to a helical configuration of the polynucleotide chain. Each layer of base residues is, therefore, in contact with that immediately above and below. In an alternative arrangement each base residue is tilted slightly and the hydrogen bonding scheme forms a spiral, i.e. in Figure 2 base residue A is hydrogen bonded to B and the residue below C, B to C, and C to the residue above A.

The order and proportions of the purine and pyrimidine residues in each chain will lead to a departure from the cylindrical molecule envisaged above, and regularities may result in long-range periodicities. The cohesive forces between individual chains in the sodium salt of deoxyribonucleic acid will result from the electrostatic attraction between the negatively charged surface of the molecules and the sodium ions.

- 3 -

I wish to thank Professor J.T. Randall F.R.S. for his encouragement in this work and Drs. M.H.F. Wilkins, R. Franklin and W.C. Price for many helpful discussions.

Present Address:
Wool Textile Research
Laboratory, C.S.I.R.O.,
Parkville, Melbourne,
Australia

R.D.B. Fraser,
Wheatstone Physics Laboratory
King's College,
Strand, W.C.2.
March I7 1953.

REFERENCES

1   Pauling, L., and Corey, R. B. Nature, <u>171</u>, 346 (1953).

2   Wilkins, M.H.F. Gosling, R.G., and Seeds, W.E., Nature,
     <u>167</u>, 759 (1951).

3   Wilkins M.H.F., and Seeds, W.E, Farad, Soc. Discussion, 9
     417 (1950).

4   Fraser, M.J., and Fraser, R.D.B., Nature, <u>167</u>, 761 (1951).

5   Furberg, S., Acta Chem. Scand., <u>4</u>, 751 (1950).

84 *CHEMISTRY: PAULING AND COREY* Proc. N. A. S.

[9] de Stevens, G., and Nord, F. F., *J. Am. Chem. Soc.*, **75**, in press (1953).

[10] Brauns, F. E., *Ibid.*, **61**, 2120 (1939).

[11] (a) Schubert, W. J., and Nord, F. F., *Ibid.*, **72**, 977 (1950); (b) Kudzin, S. F., and Nord, F. F., *Ibid.*, **73**, 4619 (1951).

[12] Nord, F. F., and de Stevens, G., *Naturwiss.*, **39**, 479 (1952).

[13] Mäule, C., *Beiträge wiss. Bot.*, **4**, 166 (1900).

[14] Kudzin, S. F., and Nord, F. F., *J. Am. Chem. Soc.*, **73**, 690 (1951).

[15] Klason, P., *Svensk Kem. Tidsk.*, **9**, 135 (1897).

[16] Freudenberg, K., *Sitzungsber. Heidelberger Akademie Wissensch.* (1949), No. 5.

[17] Hägglund, E., *Chemistry of Wood*, p. 344, Academic Press, New York (1951).

[18] Vitucci, J. C., and Nord, F. F., *Arch. Biochem.*, **14**, 243 (1947).

[19] Nord, F. F., and Vitucci, J. C., *Advances in Enzymol.*, **8**, 253 (1948).

[20] Vitucci, J. C., and Nord, F. F., *Arch. Biochem.*, **15**, 465 (1947).

[21] Birkinshaw, J. H., and Findlay, W. P. K., *Biochem. J.*, **34**, 82 (1940).

[22] Byerrum, R. U., and Flokstra, J. H., *Federation Proceedings*, **11**, 193 (1952).

[23] Nord, F. F., and Schubert, W. J., *Holzforschung*, **5**, 8 (1951).

[24] Glading, R. E., *Paper Trade J.*, **111** (No. 23), 32 (1940).

A PROPOSED STRUCTURE FOR THE NUCLEIC ACIDS

By Linus Pauling and Robert B. Corey

Gates and Crellin Laboratories of Chemistry,* California Institute of Technology

Communicated December 31, 1952

The nucleic acids, as constituents of living organisms, are comparable in importance to the proteins. There is evidence that they are involved in the processes of cell division and growth, that they participate in the transmission of hereditary characters, and that they are important constituents of viruses. An understanding of the molecular structure of the nucleic acids should be of value in the effort to understand the fundamental phenomena of life.

We have now formulated a promising structure for the nucleic acids, by making use of the general principles of molecular structure and the available information about the nucleic acids themselves. The structure is not a vague one, but is precisely predicted; atomic coordinates for the principal atoms are given in table 1. This is the first precisely described structure for the nucleic acids that has been suggested by any investigator. The structure accounts for some of the features of the x-ray photographs; but detailed intensity calculations have not yet been made, and the structure cannot be considered to have been proved to be correct.

The Formulation of the Structure.—Only recently has reasonably complete information been gathered about the chemical nature of the nucleic acids. The nucleic acids are giant molecules, composed of complex units. Each unit consists of a phosphate ion, HPO_4^{--}, a sugar (ribose in the ribonucleic

Vol. 39, 1953 *CHEMISTRY: PAULING AND COREY* 85

acids, deoxyribose in the deoxyribonucleic acids), and a purine or pyrimidine side chain (adenine, guanine, thymine, cytosine, uracil, 5-methylcytosine). The purine or pyrimidine group is attached to carbon atom $1'$ of the sugar, through the ring nitrogen atom 3 in the case of the pyrimidine nucleotides,[1] and the ring nitrogen atom 9 in the case of the purine nucleotides.[2] Good evidence has recently been obtained as to the nature of the linkage between the sugar and the phosphate, through the investigations of Todd and his collaborators;[3] it seems likely that the phosphate ester links involve carbon atoms $3'$ and $5'$ of the ribose or deoxyribose. New chemical evidence that the natural ribonucleosides have the β-D-ribofuranose configuration has also been reported by Todd and his collaborators,[4] and spectroscopic evidence indicating that the deoxyribonucleosides have the same configuration as the ribonucleosides has been obtained.[5] The β-D-ribofuranose configuration has been verified for cytidine by the determination of the structure of

TABLE 1
ATOMIC COORDINATES FOR NUCLEIC ACID

ATOM	ρ	ϕ	z	ATOM	ρ	ϕ	z
P	2.65 Å	0.0°	0.00 Å	O_1'	4.4 Å	45.4°	2.65 Å
O_I	2.00	28.3°	-0.67	O_2'	6.1	81.0°	2.1
O_{II}	2.00	$-28.3°$	0.67	N_3	6.7	52.8°	2.8
$O_{III} = O_5'$	3.72	13.5°	0.93	C_4	7.85	59.3°	2.8
$O_{IV} = O_3'$	3.72	$-13.5°$	-0.93	C_5	9.1	55.2°	2.8
C_5'	3.4	35.3°	0.7	C_6	9.35	46.9°	2.8
C_4'	3.2	51.6°	1.9	N_6	10.7	44.9°	2.8
C_3'	3.8	74.6°	1.55	N_1	8.45	39.9°	2.8
C_2'	5.3	70.3°	1.75	C_2	7.05	41.5°	2.8
C_1'	5.3	58.2°	2.8	O_2	6.35	32.4°	2.8

Identity distance along z axis = 27.2 Å.

Twenty-four atoms of each kind, with cylindrical coordinates (right-handed axes).

$\rho, \phi + n \cdot 105.0°, n \cdot 3.40 + z; \ \rho, \phi + n \cdot 105.0° + 120°, n \cdot 3.40 + z; \ \rho, \phi + n \cdot 105.0° + 240°, n \cdot 3.40 + z; \ n = 0, 1, 2, 3, 4, 5, 6, 7.$

the crystal by x-ray diffraction; cytidine is the only nucleoside for which a complete x-ray structure determination has been reported.[6]

X-ray photographs have been made of sodium thymonucleate and other preparations of the nucleic acids by Astbury and Bell.[7, 8] It has recently been reported by Wilkins, Gosling, and Seeds[9] that highly oriented fibers of sodium thymonucleate have been prepared, which give sharper x-ray photographs than those of Astbury and Bell. Our own preparations have given photographs somewhat inferior to those of Astbury and Bell. In the present work we have made use of data from our own photographs and from reproductions of the photographs of Astbury and Bell, especially those published by Astbury.[10] Astbury has pointed out that some information about the nature of the nucleic acid structure can be obtained from the x-ray photographs, but it has not been found possible to derive the structure from x-ray data alone.

A configuration of polypeptide chains in many proteins is the α helix.[11] In this structure the amino-acid residues are equivalent (except for differences in the side chains); there is only one type of relation between a residue and neighboring residues, one operation which converts a residue into a following residue. Through the continued application of this operation, a rotation-translation, the α helix is built up. It seems not unlikely that a single general operation is also involved in the construction of nucleic acids, polynucleotides, from their asymmetric fundamental units, the nucleotide residues. The general operation involved would be a rotation-reflection, and its application would lead to a helical structure. We assume, accordingly, that the structure to be formulated is a helix. The giant molecule would thus be cylindrical, with approximately circular cross section.

Some evidence in support of this assumption is provided by the electron micrographs of preparations of sodium thymonucleate described by Williams.[12] The preparation seen in the shadowed electron micrograph is clearly fibrous in nature. The small fibrils or molecules seem to be circular in cross-section, and their diameter is apparently constant; there is no evidence that the molecules are ribbon-like. The diameter as estimated from the length of the shadow is 15 or 20 Å. Similar electron micrographs, leading to the estimated molecular diameter 15 ± 5 Å, have been obtained by Kahler and Lloyd.[13] Also, estimates of the diameter of the molecules of native thymonucleic acid in the range 18 to 20 Å have been made[14, 15] on the basis of sedimentation velocity in the ultracentrifuge and other physicochemical data. The molecular weights reported are in the range 1 million to 4 million.

The x-ray photographs of sodium thymonucleate show a series of equatorial reflections compatible with a hexagonal lattice. The principal equatorial reflection, corresponding to the form 10·0, has spacing 16.2 Å or larger, the larger values corresponding to a higher degree of hydration of the substance. The minimum value,[7] 16.2 Å, corresponds to the molecular diameter 18.7 Å. From the average residue weight of sodium thymonucleate, about 330, and the density, about 1.62 g. cm.$^{-3}$, we calculate that the volume per residue is 338 Å.^{3} The cross-sectional area per residue is 303 Å^2; hence the length per residue along the fiber axis is about 1.12 Å.

The x-ray photographs show a very strong meridional reflection, with spacing about 3.40 Å. This reflection corresponds to a distance along the fiber axis equal to three times the distance per residue. Accordingly, the reflection is to be attributed to a unit consisting of three residues.

If the molecule of nucleic acid were a single helix, the reflection at 3.4 Å would have to be attributed to a regularity in the purine-pyrimidine sequence, or to some other structural feature causing the three nucleotides in the structural unit to be different from one another. It seems unlikely that there is a structural unit composed of three non-equivalent nucleotides.

The alternative explanation of the x-ray data is that the cylindrical molecule is formed of three chains, which are coiled about one another. The structure that we propose is a three-chain structure, each chain being a helix with fundamental translation equal to 3.4 Å, and the three chains being related to one another (except for differences in the nitrogen bases) by the operations of a threefold axis.

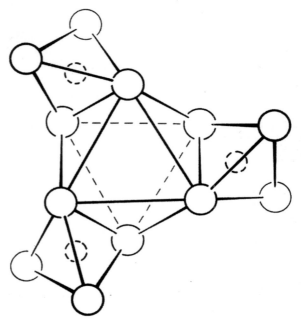

FIGURE 1

A group of three phosphate tetrahedra near the axis of the nucleic acid molecule. Oxygen atoms are indicated by full circles and phosphorus atoms by dashed circles.

The first question to be answered is that as to the nature of the core of the three-chain helical molecule—the part of the molecule closest to the axis. It is important for stability of the molecule that atoms be well packed together, and the problem of packing atoms together is a more difficult one to solve in the neighborhood of the axis than at a distance away from the axis, where there is a larger distance between an atom and the equivalent atom in the next unit.   (An example of a helical structure which seems to satisfy all of the structural requirements except that of close packing of atoms in the region near the helical axis is the 5.2-residue helix of polypeptide chains.   This structure seems not to be represented in proteins, whereas

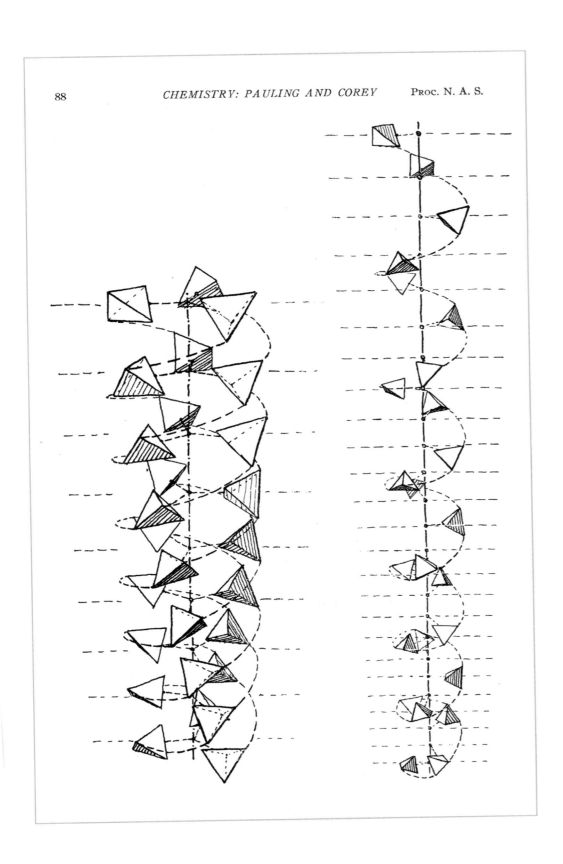

Vol. 39, 1953   *CHEMISTRY: PAULING AND COREY*   89

the similar α helix, in which the atoms are packed in a satisfactorily close manner about the axis, is an important protein structure.)   There are three possibilities as to the composition of the core: it may consist of the purine-pyrimidine groups, the sugar residues, or the phosphate groups.   It is found by trial that, because of their varied nature, the purine-pyrimidine groups cannot be packed along the axis of the helix in such a way that suitable bonds can be formed between the sugar residues and the phosphate groups; this choice is accordingly eliminated.   It is also unlikely that the sugar groups constitute the core of the molecule; the shape of the ribofuranose group and the deoxyribofuranose group is such that close packing of these groups along a helical axis is difficult, and no satisfactory way of packing them has been found.   An example that shows the difficulty of achieving close packing is provided by the polysaccharide starch, which forms helixes with a hole along the axis, into which iodine molecules can fit.   We conclude that the core of the molecule is probably formed of the phosphate groups.

A close-packed core of phosphoric acid residues, HPO_4^{--}, can easily be constructed.   At each level along the fiber axis there are three phosphate groups.   These are packed together in the way shown in figure 1.   Six oxygen atoms, two from each tetrahedral phosphate group, form an octahedron, the trigonal axis of which is the axis of the three-chain helical molecule.   A similar complex of three phosphate tetrahedra can be superimposed on this one, with translation by 3.4 Å along the fiber axis, and only a small change in azimuth.   The neighborhood of the axis of the molecule is then filled with oxygen atoms, arranged in groups of three, which change their azimuthal orientation by about 60° from layer to layer, in such a way as to produce approximate closest packing of these atoms.

The height (between two opposite edges) of a phosphate tetrahedron is about 1.7 Å.   If the same distance were preserved between the next oxygen layers, the basal-plane distance along the fiber axis would be 3.4 Å.   This value is the spacing observed for the principal meridional reflection.

It is to be expected that the outer oxygen atoms of the complex of three phosphate groups would be attached to the ribofuranose or deoxyribofuranose residues, and that the hydrogen atom of the HPO_4^{--} residues

---

FIGURE 2

Figure 2 (*left*).   A 24-residue 7-turn helix representing a single polynucleotide chain in the proposed structure for nucleic acid.   The phosphate groups are represented by tetrahedra, and the ribofuranose groups by dashed arcs connecting them.

FIGURE 3

Figure 3 (*right*).   One unit of the 3-chain nucleic acid structure.   Eight nucleotide residues of each of the three chains are included within this unit.   Each chain executes $3\frac{1}{3}$ turns in this unit.

would be attached to one of the two inner oxygen atoms, and presumably would be involved in hydrogen-bond formation with another of the inner oxygen atoms, of an adjoining phosphate group. The length of the O–H ···O bond should be close to that observed in potassium dihydrogen phosphate, 2.55 Å. The angle P—O—H should be approximately the tetrahedral angle. It is found that the spacing 3.4 Å is not compatible with this bond angle, if the hydrogen bonds are formed between one phosphate group

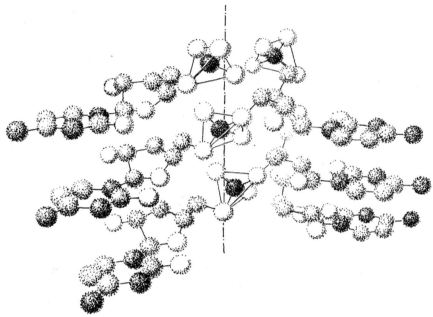

FIGURE 4

Perspective drawing of a portion of the nucleic acid structure, showing the phosphate tetrahedra near the axis of the molecule, the β-D-ribofuranose rings connecting the tetrahedra into chains, and the attached purine and pyrimidine rings (represented as purine rings in this drawing). The molecule is inverted with respect to the coordinates given in table 1.

and a group in the layer above or below it. Accordingly we assume that hydrogen bonds are formed between the oxygen atoms of the phosphate groups in the same basal plane, along outer edges of the octahedron in figure 1.

The maximum distance between the oxygen atoms 3' and 5' of a ribofuranose or deoxyribofuranose residue permitted by the accepted structural parameters (C—C = 1.54 Å, C—O = 2.43 Å, bond angles tetrahedral, with the minimum distortion required by the five-membered ring, one atom of the five-membered ring 0.5 Å from the plane of the other four, as reported by Furberg[6] for cytidine) is 4.95 Å.   It is found that it is very difficult to assign atomic positions in such a way that the residues can form a bridge between an outer oxygen atom of one phosphate group and an outer oxygen atom of a phosphate group in the layer above, without bringing some atoms into closer contact than is normal.   The atomic parameters given in Table

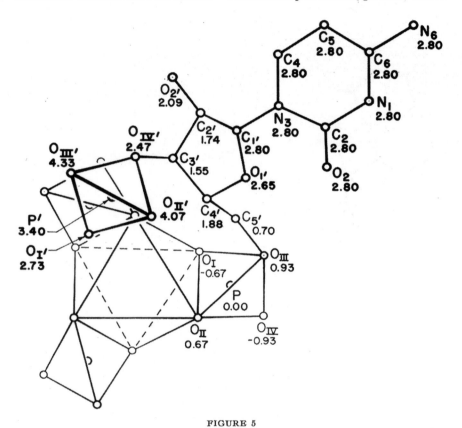

FIGURE 5

A plan of the nucleic acid structure, showing four of the phosphate groups, one ribo-furanose group, and one pyrimidine group.

1 represent the best solution of this problem that we have found; these parameters, however, probably are capable of further refinement.   The structure is an extraordinarily tight one, with little opportunity for change in position of the atoms.

The phosphate groups are unsymmetrical: the P—O distance is 1.45 Å for the two inner oxygen atoms, and 1.60 Å for the two outer oxygen atoms,

92   *CHEMISTRY: PAULING AND COREY*   Proc. N. A. S.

which are involved in ester linkages. This distortion of the phosphate
group from the regular tetrahedral configuration is not supported by direct
experimental evidence; unfortunately no precise structure determinations
have been made of any phosphate di-esters. The distortion, which corresponds to a larger amount of double bond character for the inner oxygen
atoms than for the oxygen atoms involved in the ester linkages, is a reason-

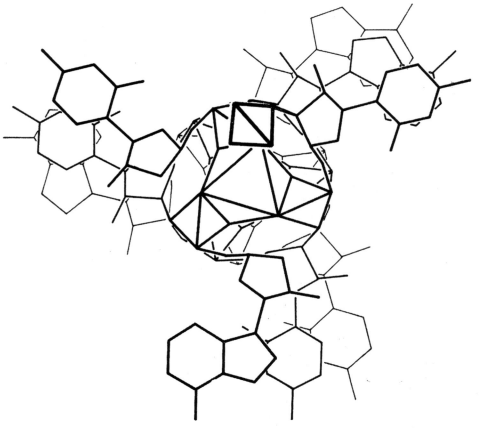

FIGURE 6

Plan of the nucleic acid structure, showing several nucleotide residues.

able one, and the assumed distances are those indicated by the observed
values for somewhat similar substances, especially the ring compound
S_3O_9, in which each sulfur atom is surrounded by a tetrahedron of four
oxygen atoms, two of which are shared with adjacent tetrahedra, and two
unshared. The O—O distances within the phosphate tetrahedron are 2.32
Å (between the two inner oxygen atoms), 2.46 Å, 2.55 Å, and 2.60 Å. The

Vol. 39, 1953   *CHEMISTRY: PAULING AND COREY*   93

hydrogen-bond distance is 2.50 Å, and each phosphate tetrahedron has two O—O contacts at 2.50 Å, with tetrahedra in the layer above. The group of three phosphate tetrahedra in each layer is obtained from that in the layer below by translation upward by 3.40 Å, and rotation in the direction corresponding to a left-handed screw by the azimuthal angle 15°. Thus there are strings of phosphate tetrahedra that are nearly superimposed, and execute a slow twist to the left. These strings are not connected together into a single polynucleotide chain, however. The sugar residues connect each phosphate group with the phosphate group in the layer above that is obtained from it by the translation by 3.40 Å and rotation through the azimuthal angle 105°, in the direction corresponding to a right-handed screw, as shown in figure 2. This gives rise to a helical chain, with pitch 11.65 Å, and with 3.43 residues per turn of the helix. The chain has an identity distance or approximate identity distance of 81.5 Å, corresponding to 24 nucleotide residues in seven turns, as shown in figure 3. The three chains of the molecule interpenetrate in such a way that the pitch of the triple helix is 3.88 Å, and the identity distance or approximate identity distance is 27.2 Å, corresponding to eight layers (see also Figs. 4, 5, and 6).

The structure requires that the sugar residues have the β-furanose configuration; steric hindrance would prevent the introduction of purine or pyrimidine groups in the positions corresponding to the α configuration. The planes of the purine and pyrimidine residues may be perpendicular or nearly perpendicular to the axis of the molecule. This causes these groups to be superimposed in layers that execute a slow left-handed turn about the molecule, the distance between the planes of successive groups being 3.4 Å. The orientation of the groups is accordingly that required by the observed strong negative birefringence of the nucleic acid fibers. The assignment of the sense of the helical molecules corresponding to the right-handed screw is required by the nature of the structure (the packing of the atoms near the axis, and the absolute configuration of the sugar, as given by the recent experimental determination[16] that absolute configurations are correctly given by the Fischer convention).

The structure bears some resemblance to the structures that have been suggested earlier, and described in a general way, without atomic coordinates. Astbury and Bell[7] suggested that the nucleic acid molecule consists of a column of nucleotide residues, with the purine and pyrimidine groups arranged directly above one another, in planes 3.4 Å apart. Astbury[10] considered the possibility that the nucleotides are arranged in a spiral around the long axis of the molecule, and rejected it, on the grounds that it does not lead to a sufficiently close packing of the groups, as is required by the high density of the substance. He pointed out that it is unlikely that adjacent molecules could interleave their purine and pyrimidine residues in such a way as to lead to the high density. Our structure solves this problem by

94   *CHEMISTRY: PAULING AND COREY*   Proc. N. A. S.

the device of intertwining three helical polynucleotide chains, in such a
way that there are three nearly vertical purine-pyrimidine columns, consisting of purine and pyrimidine residues from the three chains in alternation.
Furberg[17] suggested two single helical configurations, each resembling in a
general way one of our helical polynucleotide chains, but his structures involve orientations of phosphate tetrahedra and the ribofuranose rings that
are quite different from ours, and it is doubtful that three chains with either
of the configurations indicated in his drawing could be intertwined.

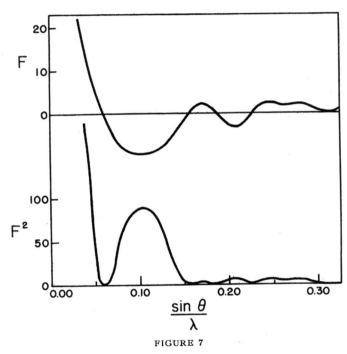

FIGURE 7

The calculated x-ray form factor F and its square F^2 for equatorial
reflections of nucleic acid.

The proposed structure accounts moderately well for the principal features of the x-ray patterns of sodium thymonucleate and other nucleic acid
derivatives.   The spacing 3.40 Å between successive layers of three nucleotides along the molecular axis is required to within about 0.10 Å by the
structural parameters of the nucleotides.   The prediction that the helixes
have 24 nucleotide residues per turn, corresponding to identity distance
8 × 3.4 Å in the direction of the fiber axis, is in good agreement with the
fact that the x-ray diagrams can be reasonably well indexed by placing the
3.4 Å meridional reflection on the eighth layer line.   The formula of Cochran,

Crick, and Vand[18] for the form factor for helical structures requires that the orders of Bessel functions for the successive layer lines from 0 to 8 be 0, 3, 6, 9, 12, 9, 6, 3, and 0. The layer-line intensities agree satisfactorily with this prediction, in the region from layer line 4 to layer line 8. There is an unexplained blackening near the meridian for layer lines 2 to 4, which, however, differs in nature for sodium thymonucleate and clupein thymonucleate, and which probably is to be attributed to material between the polynucleotide chains.

The distribution of intensity along the equator can be accounted for satisfactorily. In figure 7 there are shown the calculated form factor in the

TABLE 2

CALCULATED AND OBSERVED EQUATORIAL X-RAY REFLECTIONS FOR SODIUM THYMO-
NUCLEATE.   HEXAGONAL UNIT WITH $a_0 = 22.1$ Å

hkl	$d_{calc.}$	F_1	F_2	pF_2^2	$I_{obs.}$[a]	$d_{obs.}$[a]
10.0	19.1 Å	55	47	6600	m	18.1 Å
11.0	11.0	9.6	21	1350	m	11.2
20.0	9.5	4.7	−1.0	3		
21.0	7.22	−3.4	−8.9	480	w	7.16
30.0	6.37	−7.7	3.1	29		
22.0	5.52	−9.2	1.3	5		
31.0	5.30	−9.4	−14.6	1280	m	5.30
40.0	4.78	−9.3	−14.4	620		
32.0	4.38	−9.1	−14.1	1200 ⎫	m	19[b]
41.0	4.17	−8.8	1.0	6 ⎭		
50.0	3.83	−6.1	−10.8	350		
33.0	3.68	−5.1	4.2	53		
42.0	3.61	−4.3	−8.9	480 ⎫	vw	3.57
51.0	3.43	−2.6	−7.1	300 ⎭		

The symbol p in column 5 is the frequency factor for the form.

[a] The observed intensity values and interplanar distances are those reported by Astbury and Bell.

[b] The reflection covers the angular range corresponding to interplanar distances 4.0 to 4.4 Å, and may arise in part from overlapping from the adjacent layer lines.

equatorial direction, and the square of the form factor. It is seen that the form factor vanishes at a spacing of about 8 Å, and has a maximum in the region near 5 Å. Calculated intensities, given in table 2, are obtained by making a correction for interstitial material, at the coordinates 1/3 2/3 and 2/3 1/3, the amount of this material being taken as corresponding in scattering power to 1.5 oxygen atoms per nucleotide residue. There is reasonably satisfactory agreement with the experimental values; on the other hand, similar agreement might be given by any cylindrical molecule with approximately the same diameter. A comparison of observed and calculated radial distribution functions would provide a more reliable test of the structure; this comparison has not yet been carried out.

It is interesting to note that the purine and pyrimidine groups, on the periphery of the molecule, occupy positions such that their hydrogen-bond forming groups are directed radially.   This would permit the nucleic acid molecule to interact vigorously with other molecules.   Moreover, there is enough room in the region of each nitrogen base to permit the arbitrary choice of any one of the alternative groups; steric hindrance would not interfere with the arbitrary ordering of the residues.   The proposed structure accordingly permits the maximum number of nucleic acids to be constructed, providing the possibility of high specificity.   As Astbury has pointed out, the 3.4-Å x-ray reflection, indicating a similar distance along the axis of the molecule, is approximately the length per residue in a nearly extended polypeptide chain, and accordingly the nucleic acids are, with respect to this dimension, well suited to the ordering of amino-acid residues in a protein.   The positions of the amino-acid residues might well be at the centers of the parallelograms of which the corners are occupied by four nitrogen bases.   The 256 different kinds of parallelograms (neglecting the possibility of two different orientations of each nitrogen base) would permit considerable power of selection for each position.

(Added in proof.)   Support of the assumed phosphorus-oxygen distances in the phosphate di-ester group is provided by the results of the determination of the structure of ammonium tetrametaphosphate.[19, 20]   In this crystal there are P_4O_{12} complexes, consisting of four tetrahedra each of which shares two oxygen atoms with other tetrahedra.   The phosphorus-oxygen distance is 1.46 Å for the oxygen atoms that are not shared, and 1.62 Å for those that are shared.   These values are to be compared with the values that we have assumed, 1.45 Å for the inner oxygen atoms (which are not shared), and 1.60 Å for the outer ones, which have bonds to carbon atoms.

This investigation was aided by grants from The National Foundation for Infantile Paralysis and The Rockefeller Foundation.

* Contribution No. 1766.

[1] Levene, P. A., and Tipson, R. S., *J. Biol. Chem.*, **94**, 809 (1932); **97**, 491 (1932); **101**, 529 (1933).

[2] Gulland, J. M., and Story, L. F., *J. Chem. Soc.*, **1938**, 259.

[3] Brown, D. M., and Todd, A. R., *Ibid.*, **1952**, 52.

[4] Clark, V. M., Todd, A. R., and Zussman, J., *Ibid.*, **1951**, 2952.

[5] Manson, L. A., and Lampen, J. P., *J. Biol. Chem.*, **191**, 87 (1951).

[6] Furberg, S., *Acta Cryst.*, **3**, 325 (1950).

[7] Astbury, W. T., and Bell, F. O., *Nature*, **141**, 747 (1938); *Cold Spring Harbor Symp. Quant. Biol.*, **6**, 109 (1938).

[8] Astbury, W. T., and Bell, F. O., *Tabulae Biologicae*, **17**, 90 (1939).

[9] Wilkins, N. H. F., Gosling, R. J., and Seeds, W. E., *Nature*, **167**, 759 (1951).

[10] Astbury, W. T., in *Nucleic Acids, Symposia of the Society for Experimental Biology, No. 1*, Cambridge University Press (1947).

GENETICS: BROWN AND CAVE   97

[11] Pauling, L., and Corey, R. B., these PROCEEDINGS, **37**, 235 (1951).
[12] Williams, R. C., *Biochimica et Biophysica Acta*, **9**, 237 (1952).
[13] Kahler, H., and Lloyd, B. J., Jr., *Biochim. et Biophys. Acta*, in press.
[14] Cecil, R., and Ogston, A. G., *J. Chem. Soc.,* 1948, 1382.
[15] Kahler, H., *J. Phys. Colloid Chem.*, **52**, 207 (1948).
[16] Bijvoet, J. M., Peerdeman, A. F., and van Bommel, A. J., *Ibid.*, **168**, 271 (1951).
[17] Furberg, S., *Acta Chemica Scand.*, **6**, 634 (1952).
[18] Cochran, W., Crick, F. H. C., and Vand, V., *Acta Cryst.*, **5**, 581 (1952).
[19] Romers, C., Ketelaar, J. A. A., and MacGillavry, C. H., *Nature* **164**, 960 (1949).
[20] Romers, C., dissertation, Amsterdam. 1948.

---

*INDUCED DOMINANT LETHALITY IN LILIUM**

BY SPENCER W. BROWN AND MARION S. CAVE

DEPARTMENT OF GENETICS AND DEPARTMENT OF BOTANY, UNIVERSITY OF CALIFORNIA

Communicated by R. E. Clausen, November 14, 1952

In 1927, H. J. Muller[1] noticed a marked decrease in the fertility of female *Drosophila melanogaster* mated to irradiated males, and attributed this decrease to dominant lethals induced by the x-ray treatment.

As pointed out by Stancati,[2] in 1932, the primary choice under such circumstances is between the induction of dominant lethals and the inactivation of sperm.   Either effect of x-rays would give the observed decrease in fertility.   Although Muller and Settles[3] had proved in 1927 that sperm could function normally even though carrying chromosome sets which would be lethal in combination with the normal chromosome complement, Stancati's work with *Habrobracon* was the first to demonstrate conclusively that dominant lethality was the result, not the lack, of fertilization by irradiated sperm.   After treatment, the frequency of the biparental females (arising from fertilized eggs) decreased while the frequency of the matroclinous, uniparental males (arising from unfertilized eggs) was neither diminished nor increased.   The missing biparental females were those which would have appeared if dominant lethals had not been induced in the sperm.

Later Demerec and Kaufmann[4] examined cytologically Drosophila eggs presumably fertilized by sperm treated with 5000 r.   Failure of fertilization was apparent in only one case out of the 99 studied.   Therefore the high degree of sterility at this dose was assignable to dominant lethality rather than sperm inactivation.   On the other hand, any studies of dominant lethality at yet higher doses would again require that the sperm be shown to act in fertilizing.

Work with *Habrobracon* also gave information on the dosage necessary to achieve sperm inactivation.   At the relatively low doses generally used, the

equipment, and to Dr. G. E. R. Deacon and the captain and officers of R.R.S. *Discovery II* for their part in making the observations.

[1] Young, F. B., Gerrard, H., and Jevons, W., *Phil. Mag.*, **40**, 149 (1920).
[2] Longuet-Higgins, M. S., *Mon. Not. Roy. Astro. Soc., Geophys. Supp.*, **5**, 285 (1949).
[3] Von Arx, W. S., Woods Hole Papers in Phys. Oceanog. Meteor., **11** (3) (1950).
[4] Ekman, V. W., *Arkiv. Mat. Astron. Fysik.* (Stockholm), **2** (11) (1905).

MOLECULAR STRUCTURE OF NUCLEIC ACIDS

A Structure for Deoxyribose Nucleic Acid

WE wish to suggest a structure for the salt of deoxyribose nucleic acid (D.N.A.). This structure has novel features which are of considerable biological interest.

A structure for nucleic acid has already been proposed by Pauling and Corey[1]. They kindly made their manuscript available to us in advance of publication. Their model consists of three intertwined chains, with the phosphates near the fibre axis, and the bases on the outside. In our opinion, this structure is unsatisfactory for two reasons : (1) We believe that the material which gives the X-ray diagrams is the salt, not the free acid. Without the acidic hydrogen atoms it is not clear what forces would hold the structure together, especially as the negatively charged phosphates near the axis will repel each other. (2) Some of the van der Waals distances appear to be too small.

Another three-chain structure has also been suggested by Fraser (in the press). In his model the phosphates are on the outside and the bases on the inside, linked together by hydrogen bonds. This structure as described is rather ill-defined, and for this reason we shall not comment on it.

We wish to put forward a radically different structure for the salt of deoxyribose nucleic acid. This structure has two helical chains each coiled round the same axis (see diagram). We have made the usual chemical assumptions, namely, that each chain consists of phosphate diester groups joining β-D-deoxyribofuranose residues with 3′,5′ linkages. The two chains (but not their bases) are related by a dyad perpendicular to the fibre axis. Both chains follow right-handed helices, but owing to the dyad the sequences of the atoms in the two chains run in opposite directions. Each chain loosely resembles Furberg's[2] model No. 1 ; that is, the bases are on the inside of the helix and the phosphates on the outside. The configuration of the sugar and the atoms near it is close to Furberg's 'standard configuration', the sugar being roughly perpendicular to the attached base. There

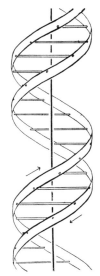

This figure is purely diagrammatic. The two ribbons symbolize the two phosphate—sugar chains, and the horizontal rods the pairs of bases holding the chains together. The vertical line marks the fibre axis

is a residue on each chain every 3·4 A. in the z-direction. We have assumed an angle of 36° between adjacent residues in the same chain, so that the structure repeats after 10 residues on each chain, that is, after 34 A. The distance of a phosphorus atom from the fibre axis is 10 A. As the phosphates are on the outside, cations have easy access to them.

The structure is an open one, and its water content is rather high. At lower water contents we would expect the bases to tilt so that the structure could become more compact.

The novel feature of the structure is the manner in which the two chains are held together by the purine and pyrimidine bases. The planes of the bases are perpendicular to the fibre axis. They are joined together in pairs, a single base from one chain being hydrogen-bonded to a single base from the other chain, so that the two lie side by side with identical z-co-ordinates. One of the pair must be a purine and the other a pyrimidine for bonding to occur. The hydrogen bonds are made as follows : purine position 1 to pyrimidine position 1 ; purine position 6 to pyrimidine position 6.

If it is assumed that the bases only occur in the structure in the most plausible tautomeric forms (that is, with the keto rather than the enol configurations) it is found that only specific pairs of bases can bond together. These pairs are : adenine (purine) with thymine (pyrimidine), and guanine (purine) with cytosine (pyrimidine).

In other words, if an adenine forms one member of a pair, on either chain, then on these assumptions the other member must be thymine ; similarly for guanine and cytosine. The sequence of bases on a single chain does not appear to be restricted in any way. However, if only specific pairs of bases can be formed, it follows that if the sequence of bases on one chain is given, then the sequence on the other chain is automatically determined.

It has been found experimentally[3,4] that the ratio of the amounts of adenine to thymine, and the ratio of guanine to cytosine, are always very close to unity for deoxyribose nucleic acid.

It is probably impossible to build this structure with a ribose sugar in place of the deoxyribose, as the extra oxygen atom would make too close a van der Waals contact.

The previously published X-ray data[5,6] on deoxyribose nucleic acid are insufficient for a rigorous test of our structure. So far as we can tell, it is roughly compatible with the experimental data, but it must be regarded as unproved until it has been checked against more exact results. Some of these are given in the following communications. We were not aware of the details of the results presented there when we devised our structure, which rests mainly though not entirely on published experimental data and stereochemical arguments.

It has not escaped our notice that the specific pairing we have postulated immediately suggests a possible copying mechanism for the genetic material.

Full details of the structure, including the conditions assumed in building it, together with a set of co-ordinates for the atoms, will be published elsewhere.

We are much indebted to Dr. Jerry Donohue for constant advice and criticism, especially on interatomic distances. We have also been stimulated by a knowledge of the general nature of the unpublished experimental results and ideas of Dr. M. H. F. Wilkins, Dr. R. E. Franklin and their co-workers at

738   N A T U R E   April 25, 1953   VOL. 171

King's College, London. One of us (J. D. W.) has been aided by a fellowship from the National Foundation for Infantile Paralysis.

J. D. WATSON
F. H. C. CRICK

Medical Research Council Unit for the
Study of the Molecular Structure of
Biological Systems,
Cavendish Laboratory, Cambridge.
April 2.

[1] Pauling, L., and Corey, R. B., *Nature*, **171**, 346 (1953); *Proc. U.S. Nat. Acad. Sci.*, **39**, 84 (1953).

[2] Furberg, S., *Acta Chem. Scand.*, **6**, 634 (1952).

[3] Chargaff, E., for references see Zamenhof, S., Brawerman, G., and Chargaff, E., *Biochim. et Biophys. Acta*, **9**, 402 (1952).

[4] Wyatt, G. R., *J. Gen. Physiol.*, **36**, 201 (1952).

[5] Astbury, W. T., Symp. Soc. Exp. Biol. 1, Nucleic Acid, 66 (Camb. Univ. Press, 1947).

[6] Wilkins, M. H. F., and Randall, J. T., *Biochim. et Biophys. Acta*, **10**, 192 (1953).

Molecular Structure of Deoxypentose Nucleic Acids

WHILE the biological properties of deoxypentose nucleic acid suggest a molecular structure containing great complexity, X-ray diffraction studies described here (cf. Astbury[1]) show the basic molecular configuration has great simplicity. The purpose of this communication is to describe, in a preliminary way, some of the experimental evidence for the polynucleotide chain configuration being helical, and existing in this form when in the natural state. A fuller account of the work will be published shortly.

The structure of deoxypentose nucleic acid is the same in all species (although the nitrogen base ratios alter considerably) in nucleoprotein, extracted or in cells, and in purified nucleate. The same linear group of polynucleotide chains may pack together parallel in different ways to give crystalline[1-3], semi-crystalline or paracrystalline material. In all cases the X-ray diffraction photograph consists of two regions, one determined largely by the regular spacing of nucleotides along the chain, and the other by the longer spacings of the chain configuration. The sequence of different nitrogen bases along the chain is not made visible.

Oriented paracrystalline deoxypentose nucleic acid ('structure B' in the following communication by Franklin and Gosling) gives a fibre diagram as shown in Fig. 1 (cf. ref. 4). Astbury suggested that the strong 3·4-A. reflexion corresponded to the internucleotide repeat along the fibre axis. The ∼ 34 A. layer lines, however, are not due to a repeat of a polynucleotide composition, but to the chain configuration repeat, which causes strong diffraction as the nucleotide chains have higher density than the interstitial water. The absence of reflexions on or near the meridian immediately suggests a helical structure with axis parallel to fibre length.

Diffraction by Helices

It may be shown[5] (also Stokes, unpublished) that the intensity distribution in the diffraction pattern of a series of points equally spaced along a helix is given by the squares of Bessel functions. A uniform continuous helix gives a series of layer lines of spacing corresponding to the helix pitch, the intensity distribution along the nth layer line being proportional to the square of J_n, the nth order Bessel function. A straight line may be drawn approximately through

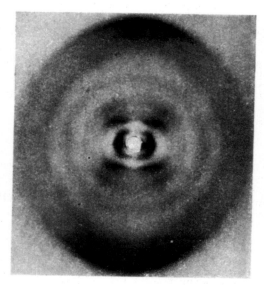

Fig. 1. Fibre diagram of deoxypentose nucleic acid from *B. coli.* Fibre axis vertical

the innermost maxima of each Bessel function and the origin. The angle this line makes with the equator is roughly equal to the angle between an element of the helix and the helix axis. If a unit repeats n times along the helix there will be a meridional reflexion $(J_0{}^2)$ on the nth layer line. The helical configuration produces side-bands on this fundamental frequency, the effect[5] being to reproduce the intensity distribution about the origin around the new origin, on the nth layer line, corresponding to C in Fig. 2.

We will now briefly analyse in physical terms some of the effects of the shape and size of the repeat unit or nucleotide on the diffraction pattern. First, if the nucleotide consists of a unit having circular symmetry about an axis parallel to the helix axis, the whole diffraction pattern is modified by the form factor of the nucleotide. Second, if the nucleotide consists of a series of points on a radius at right-angles to the helix axis, the phases of radiation scattered by the helices of different diameter passing through each point are the same. Summation of the corresponding Bessel functions gives reinforcement for the inner-

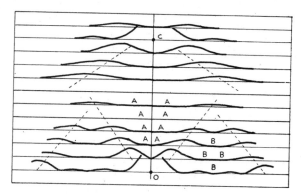

Fig. 2. Diffraction pattern of system of helices corresponding to structure of deoxypentose nucleic acid. The squares of Bessel functions are plotted about 0 on the equator and on the first, second, third and fifth layer lines for half of the nucleotide mass at 20 A. diameter and remainder distributed along a radius, the mass at a given radius being proportional to the radius. About C on the tenth layer line similar functions are plotted for an outer diameter of 12 A.

most maxima and, in general, owing to phase difference, cancellation of all other maxima. Such a system of helices (corresponding to a spiral staircase with the core removed) diffracts mainly over a limited angular range, behaving, in fact, like a periodic arrangement of flat plates inclined at a fixed angle to the axis. Third, if the nucleotide is extended as an arc of a circle in a plane at right-angles to the helix axis, and with centre at the axis, the intensity of the system of Bessel function layer-line streaks emanating from the origin is modified owing to the phase differences of radiation from the helices drawn through each point on the nucleotide. The form factor is that of the series of points in which the helices intersect a plane drawn through the helix axis. This part of the diffraction pattern is then repeated as a whole with origin at C (Fig. 2). Hence this aspect of nucleotide shape affects the central and peripheral regions of each layer line differently.

Interpretation of the X-Ray Photograph

It must first be decided whether the structure consists of essentially one helix giving an intensity distribution along the layer lines corresponding to $J_1, J_2, J_3 \ldots$, or two similar co-axial helices of twice the above size and relatively displaced along the axis a distance equal to half the pitch giving $J_2, J_4, J_6 \ldots$, or three helices, etc. Examination of the width of the layer-line streaks suggests the intensities correspond more closely to J_1^2, J_2^2, J_3^2 than to $J_2^2, J_4^2, J_6^2 \ldots$ Hence the dominant helix has a pitch of ~ 34 A., and, from the angle of the helix, its diameter is found to be ~ 20 A. The strong equatorial reflexion at ~ 17 A. suggests that the helices have a maximum diameter of ~ 20 A. and are hexagonally packed with little interpenetration. Apart from the width of the Bessel function streaks, the possibility of the helices having twice the above dimensions is also made unlikely by the absence of an equatorial reflexion at ~ 34 A. To obtain a reasonable number of nucleotides per unit volume in the fibre, two or three intertwined coaxial helices are required, there being ten nucleotides on one turn of each helix.

The absence of reflexions on or near the meridian (an empty region AAA on Fig. 2) is a direct consequence of the helical structure. On the photograph there is also a relatively empty region on and near the equator, corresponding to region BBB on Fig. 2. As discussed above, this absence of secondary Bessel function maxima can be produced by a radial distribution of the nucleotide shape. To make the layer-line streaks sufficiently narrow, it is necessary to place a large fraction of the nucleotide mass at ~ 20 A. diameter. In Fig. 2 the squares of Bessel functions are plotted for half the mass at 20 A. diameter, and the rest distributed along a radius, the mass at a given radius being proportional to the radius.

On the zero layer line there appears to be a marked J_{10}^2, and on the first, second and third layer lines, $J_9^2 + J_{11}^2$, $J_8^2 + J_{12}^2$, etc., respectively. This means that, in projection on a plane at right-angles to the fibre axis, the outer part of the nucleotide is relatively concentrated, giving rise to high-density regions spaced $c.$ 6 A. apart around the circumference of a circle of 20 A. diameter. On the fifth layer line two J_5 functions overlap and produce a strong reflexion. On the sixth, seventh and eighth layer lines the maxima correspond to a helix of diameter ~ 12 A. Apparently it is only the central region of the helix structure which is well divided by the 3·4-A. spacing, the outer

parts of the nucleotide overlapping to form a continuous helix. This suggests the presence of nitrogen bases arranged like a pile of pennies[1] in the central regions of the helical system.

There is a marked absence of reflexions on layer lines beyond the tenth. Disorientation in the specimen will cause more extension along the layer lines of the Bessel function streaks on the eleventh, twelfth and thirteenth layer lines than on the ninth, eighth and seventh. For this reason the reflexions on the higher-order layer lines will be less readily visible. The form factor of the nucleotide is also probably causing diminution of intensity in this region. Tilting of the nitrogen bases could have such an effect.

Reflexions on the equator are rather inadequate for determination of the radial distribution of density in the helical system. There are, however, indications that a high-density shell, as suggested above, occurs at diameter ~ 20 A.

The material is apparently not completely paracrystalline, as sharp spots appear in the central region of the second layer line, indicating a partial degree of order of the helical units relative to one another in the direction of the helix axis. Photographs similar to Fig. 1 have been obtained from sodium nucleate from calf and pig thymus, wheat germ, herring sperm, human tissue and T_2 bacteriophage. The most marked correspondence with Fig. 2 is shown by the exceptional photograph obtained by our colleagues, R. E. Franklin and R. G. Gosling, from calf thymus deoxypentose nucleate (see following communication).

It must be stressed that some of the above discussion is not without ambiguity, but in general there appears to be reasonable agreement between the experimental data and the kind of model described by Watson and Crick (see also preceding communication).

It is interesting to note that if there are ten phosphate groups arranged on each helix of diameter 20 A. and pitch 34 A., the phosphate ester backbone chain is in an almost fully extended state. Hence, when sodium nucleate fibres are stretched[3], the helix is evidently extended in length like a spiral spring in tension.

Structure *in vivo*

The biological significance of a two-chain nucleic acid unit has been noted (see preceding communication). The evidence that the helical structure discussed above does, in fact, exist in intact biological systems is briefly as follows:

Sperm heads. It may be shown that the intensity of the X-ray spectra from crystalline sperm heads is determined by the helical form-function in Fig. 2. Centrifuged trout semen give the same pattern as the dried and rehydrated or washed sperm heads used previously[6]. The sperm head fibre diagram is also given by extracted or synthetic[1] nucleoprotamine or extracted calf thymus nucleohistone.

Bacteriophage. Centrifuged wet pellets of T_2 phage photographed with X-rays while sealed in a cell with mica windows give a diffraction pattern containing the main features of paracrystalline sodium nucleate as distinct from that of crystalline nucleoprotein. This confirms current ideas of phage structure.

Transforming principle (in collaboration with H. Ephrussi-Taylor). Active deoxypentose nucleate allowed to dry at ~ 60 per cent humidity has the same crystalline structure as certain samples[3] of sodium thymonucleate.

740   N A T U R E   April 25, 1953   VOL. 171

We wish to thank Prof. J. T. Randall for encouragement ; Profs. E. Chargaff, R. Signer, J. A. V. Butler and Drs. J. D. Watson, J. D. Smith, L. Hamilton, J. C. White and G. R. Wyatt for supplying material without which this work would have been impossible ; also Drs. J. D. Watson and Mr. F. H. C. Crick for stimulation, and our colleagues R. E. Franklin, R. G. Gosling, G. L. Brown and W. E. Seeds for discussion. One of us (H. R. W.) wishes to acknowledge the award of a University of Wales Fellowship.

M. H. F. WILKINS

Medical Research Council Biophysics
        Research Unit,

A. R. STOKES
H. R. WILSON

Wheatstone Physics Laboratory,
    King's College, London.
        April 2.

[1] Astbury, W. T., Symp. Soc. Exp. Biol., 1, Nucleic Acid (Cambridge Univ. Press, 1947).
[2] Riley, D. P., and Oster, G., Biochim. et Biophys. Acta, 7, 526 (1951).
[3] Wilkins, M. H. F., Gosling, R. G., and Seeds, W. E., Nature, 167, 759 (1951).
[4] Astbury, W. T., and Bell, F. O., Cold Spring Harb. Symp. Quant. Biol., 6, 109 (1938).
[5] Cochran, W., Crick, F. H. C., and Vand, V., Acta Cryst., 5, 581 (1952).
[6] Wilkins, M. H. F., and Randall, J. T., Biochim. et Biophys. Acta, 10, 192 (1953).

Molecular Configuration in Sodium Thymonucleate

SODIUM thymonucleate fibres give two distinct types of X-ray diagram. The first corresponds to a crystalline form, structure A, obtained at about 75 per cent relative humidity ; a study of this is described in detail elsewhere[1]. At higher humidities a different structure, structure B, showing a lower degree of order, appears and persists over a wide range of ambient humidity. The change from A to B is reversible. The water content of structure B fibres which undergo this reversible change may vary from 40–50 per cent to several hundred per cent of the dry weight. Moreover, some fibres never show structure A, and in these structure B can be obtained with an even lower water content.

The X-ray diagram of structure B (see photograph) shows in striking manner the features characteristic of helical structures, first worked out in this laboratory by Stokes (unpublished) and by Crick, Cochran and Vand[2]. Stokes and Wilkins were the first to propose such structures for nucleic acid as a result of direct studies of nucleic acid fibres, although a helical structure had been previously suggested by Furberg (thesis, London, 1949) on the basis of X-ray studies of nucleosides and nucleotides.

While the X-ray evidence cannot, at present, be taken as direct proof that the structure is helical, other considerations discussed below make the existence of a helical structure highly probable.

Structure B is derived from the crystalline structure A when the sodium thymonucleate fibres take up quantities of water in excess of about 40 per cent of their weight. The change is accompanied by an increase of about 30 per cent in the length of the fibre, and by a substantial re-arrangement of the molecule. It therefore seems reasonable to suppose that in structure B the structural units of sodium thymonucleate (molecules on groups of molecules) are relatively free from the influence of neighbouring

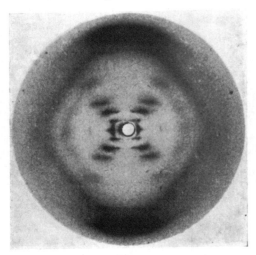

Sodium deoxyribose nucleate from calf thymus. Structure B

molecules, each unit being shielded by a sheath of water. Each unit is then free to take up its least-energy configuration independently of its neighbours and, in view of the nature of the long-chain molecules involved, it is highly likely that the general form will be helical[3]. If we adopt the hypothesis of a helical structure, it is immediately possible, from the X-ray diagram of structure B, to make certain deductions as to the nature and dimensions of the helix.

The innermost maxima on the first, second, third and fifth layer lines lie approximately on straight lines radiating from the origin. For a smooth single-strand helix the structure factor on the nth layer line is given by :

$$F_n = J_n(2\pi rR) \exp i\, n(\psi + \tfrac{1}{2}\pi),$$

where $J_n(u)$ is the nth-order Bessel function of u, r is the radius of the helix, and R and ψ are the radial and azimuthal co-ordinates in reciprocal space[2]; this expression leads to an approximately linear array of intensity maxima of the type observed, corresponding to the first maxima in the functions J_1, J_2, J_3, etc.

If, instead of a smooth helix, we consider a series of residues equally spaced along the helix, the transform in the general case treated by Crick, Cochran and Vand is more complicated. But if there is a whole number, m, of residues per turn, the form of the transform is as for a smooth helix with the addition, only, of the same pattern repeated with its origin at heights mc^*, $2mc^*$ . . . etc. (c is the fibre-axis period).

In the present case the fibre-axis period is 34 A. and the very strong reflexion at 3·4 A. lies on the tenth layer line. Moreover, lines of maxima radiating from the 3·4-A. reflexion as from the origin are visible on the fifth and lower layer lines, having a J_5 maximum coincident with that of the origin series on the fifth layer line. (The strong outer streaks which apparently radiate from the 3·4-A. maximum are not, however, so easily explained.) This suggests strongly that there are exactly 10 residues per turn of the helix. If this is so, then from a measurement of R_n the position of the first maximum on the nth layer line (for $n \leqslant 5$), the radius of the helix, can be obtained. In the present instance, measurements of R_1, R_2, R_3 and R_5 all lead to values of r of about 10 A.

No. 4356   **April 25, 1953**   NATURE   741

Since this linear array of maxima is one of the strongest features of the X-ray diagram, we must conclude that a crystallographically important part of the molecule lies on a helix of this diameter. This can only be the phosphate groups or phosphorus atoms.

If ten phosphorus atoms lie on one turn of a helix of radius 10 A., the distance between neighbouring phosphorus atoms in a molecule is 7·1 A. This corresponds to the P . . . P distance in a fully extended molecule, and therefore provides a further indication that the phosphates lie on the outside of the structural unit.

Thus, our conclusions differ from those of Pauling and Corey[4], who proposed for the nucleic acids a helical structure in which the phosphate groups form a dense core.

We must now consider briefly the equatorial reflexions. For a single helix the series of equatorial maxima should correspond to the maxima in $J_0(2\pi rR)$. The maxima on our photograph do not, however, fit this function for the value of r deduced above. There is a very strong reflexion at about 24 A. and then only a faint sharp reflexion at 9·0 A. and two diffuse bands around 5·5 A. and 4·0 A. This lack of agreement is, however, to be expected, for we know that the helix so far considered can only be the most important member of a series of coaxial helices of different radii ; the non-phosphate parts of the molecule will lie on inner co-axial helices, and it can be shown that, whereas these will not appreciably influence the innermost maxima on the layer lines, they may have the effect of destroying or shifting both the equatorial maxima and the outer maxima on other layer lines.

Thus, if the structure is helical, we find that the phosphate groups or phosphorus atoms lie on a helix of diameter about 20 A., and the sugar and base groups must accordingly be turned inwards towards the helical axis.

Considerations of density show, however, that a cylindrical repeat unit of height 34 A. and diameter 20 A. must contain many more than ten nucleotides.

Since structure B often exists in fibres with low water content, it seems that the density of the helical unit cannot differ greatly from that of dry sodium thymonucleate, 1·63 gm./cm.³ [1,5], the water in fibres of high water-content being situated outside the structural unit. On this basis we find that a cylinder of radius 10 A. and height 34 A. would contain thirty-two nucleotides. However, there might possibly be some slight inter-penetration of the cylindrical units in the dry state making their effective radius rather less. It is therefore difficult to decide, on the basis of density measurements alone, whether one repeating unit contains ten nucleotides on each of two or on each of three co-axial molecules. (If the effective radius were 8 A. the cylinder would contain twenty nucleotides.) Two other arguments, however, make it highly probable that there are only two co-axial molecules.

First, a study of the Patterson function of structure A, using superposition methods, has indicated[6] that there are only two chains passing through a primitive unit cell in this structure. Since the $A \rightleftharpoons B$ transformation is readily reversible, it seems very unlikely that the molecules would be grouped in threes in structure B. Secondly, from measurements on the X-ray diagram of structure B it can readily be shown that, whether the number of chains per unit is two or three, the chains are not equally spaced along the fibre axis. For example, three equally spaced chains would mean that the nth layer line depended on J_{3n}, and would lead to a helix of diameter about 60 A. This is many times larger than the primitive unit cell in structure A, and absurdly large in relation to the dimensions of nucleotides. Three unequally spaced chains, on the other hand, would be crystallographically non-equivalent, and this, again, seems unlikely. It therefore seems probable that there are only two co-axial molecules and that these are unequally spaced along the fibre axis.

Thus, while we do not attempt to offer a complete interpretation of the fibre-diagram of structure B, we may state the following conclusions. The structure is probably helical. The phosphate groups lie on the outside of the structural unit, on a helix of diameter about 20 A. The structural unit probably consists of two co-axial molecules which are not equally spaced along the fibre axis, their mutual displacement being such as to account for the variation of observed intensities of the innermost maxima on the layer lines ; if one molecule is displaced from the other by about three-eighths of the fibre-axis period, this would account for the absence of the fourth layer line maxima and the weakness of the sixth. Thus our general ideas are not inconsistent with the model proposed by Watson and Crick in the preceding communication.

The conclusion that the phosphate groups lie on the outside of the structural unit has been reached previously by quite other reasoning[1]. Two principal lines of argument were invoked. The first derives from the work of Gulland and his collaborators[7], who showed that even in aqueous solution the —CO and —NH₂ groups of the bases are inaccessible and cannot be titrated, whereas the phosphate groups are fully accessible. The second is based on our own observations[1] on the way in which the structural units in structures A and B are progressively separated by an excess of water, the process being a continuous one which leads to the formation first of a gel and ultimately to a solution. The hygroscopic part of the molecule may be presumed to lie in the phosphate groups ($(C_2H_5O)_2PO_2Na$ and $(C_3H_7O)_2PO_2Na$ are highly hygroscopic[8]), and the simplest explanation of the above process is that these groups lie on the outside of the structural units. Moreover, the ready availability of the phosphate groups for interaction with proteins can most easily be explained in this way.

We are grateful to Prof. J. T. Randall for his interest and to Drs. F. H. C. Crick, A. R. Stokes and M. H. F. Wilkins for discussion. One of us (R. E. F.) acknowledges the award of a Turner and Newall Fellowship.

ROSALIND E. FRANKLIN*
R. G. GOSLING

Wheatstone Physics Laboratory,
  King's College, London.
    April 2.

* Now at Birkbeck College Research Laboratories, 21 Torrington Square, London, W.C.1.

[1] Franklin, R. E., and Gosling, R. G. (in the press).
[2] Cochran, W., Crick, F. H. C., and Vand, V., *Acta Cryst.*, **5**, 501 (1952).
[3] Pauling, L., Corey, R. B., and Bransom, H. R., *Proc. U.S. Nat. Acad. Sci.*, **37**, 205 (1951).
[4] Pauling, L., and Corey, R. B., *Proc. U.S. Nat. Acad. Sci.*, **39**, 84 (1953).
[5] Astbury, W. T., Cold Spring Harbor Symp. on Quant. Biol., **12**, 56 (1947).
[6] Franklin, R. E., and Gosling, R. G. (to be published).
[7] Gulland, J. M., and Jordan, D. O., Cold Spring Harbor Symp. on Quant. Biol., **12**, 5 (1947).
[8] Drushel, W. A., and Felty, A. R., *Chem. Zent.*, **89**, 1016 (1918).

Nature Vol. 248 April 26 1974   *Molecular Biology*   **787**

could determine the time of action of the gene. Could we gain a detailed understanding of development from this approach? One view is that this might allow us to deduce the logical structure of the genetic programme by seeing whether there are partitioned subsets of genes which correspond in some interesting way to the phenotype. Questions of this sort are not strictly molecular being concerned more with the software of organisms than with their hardware. In the long run, however, we must find the molecular implementations. This would involve identifying the products of the genes and finding out which cells have them and where and when they work to produce their effects. But to do this requires high resolution protein spectroscopy and this technology only exists in a crude and cumbersome form today. Not many people would be willing to embark on such a programme because of its intrinsic magnitude, and there are also many workers who would object on grounds of principle to this approach. There is in fact a fairly large antigenetic school of molecular biologists who believe that the work should proceed from the biochemistry and not the other way around. The only difficulty is that, except in special cases, nobody knows what biochemistry ought to be done.

Cell surface

This brings me finally to the question of the cell surface. One of those nebulous communally circulating ideas is that the cell surface contains the key to the mechanisms of development. These notions have been generated from many different sources. Some come by extrapolation from cell interactions in the immune system, others from the development or regeneration of neural connections, and yet others from the study of the reaggregation properties of embryonic cells. The central thought is that cells or groups of cells carry a specific chemical code on their surfaces and by this code they know themselves and their neighbours. In the pages of this journal there have appeared various baroque forms of this theory in which gene stitching, reverse transcriptase and other pieces of molecular clockwork have been grafted onto the idea of specific chemical labelling of cells. Nobody will deny that cells do have surface proteins and that some of these will be involved in recognising specific chemical signals such as nerve transmitters or hormones. The real question is whether there really exists a system which, by one mechanism or another, can generate 10^6 chemical names and plant these on the right cells. There seems to be no convincing general argument either for or against this view and only time will tell.

Much has been written about the philosophical consequences of molecular biology. I think it is now quite clear what the enterprise is about. We are looking at a rather special part of the physical universe which contains special mechanisms none of which conflict at all with the laws of physics. That there would be new laws of Nature to be found in biological systems was a misjudged view and that hope or fear has just vanished. Our job is simply to find out how these interesting pieces of machinery work, how they get built and how they came to be the way they are. In one sense, the answers already exist and all we have to do is to find out how to look them up in Nature. That is why molecular biology seems to me to be the art of the inevitable.

Rosalind Franklin and the double helix

A. Klug
MRC Laboratory of Molecular Biology, Cambridge, UK

A draft manuscript shows how near Rosalind Franklin came to finding the correct structure of DNA.

SOME years ago I gave an account[1] of Rosalind Franklin's contribution to the discovery of the structure of DNA, based on published sources supplemented by references to her notebooks and reports. I pointed out how close Franklin had come in the progress of her work to various features of the structure contained in the correct solution. At the time, however, I did not know of the existence of a draft manuscript which confirms how close she had got to the answer. This only came to light later. I therefore take the opportunity of this 21st Anniversary issue to fill out the record and to highlight a dramatic element in the 'race' for DNA.

In my article I told how the analysis of the B form of DNA in terms of helical diffraction theory, which is given in Franklin's paper with Gosling in *Nature*[2] on April 25, 1953, can be found in her notebook for the period January to March 1953, that is before the Watson–Crick structure had become known to her. I went on to say, however, that she apparently did not feel convinced enough of the relevance of this analysis to publish it (because she had not solved the A form). This is erroneous. The typescript I have found is dated March 17, 1953 and is clearly a draft of the *Nature* paper. This suggests a different explanation from the one I gave, namely that she was proposing to publish what she knew on the B form, two other papers on the A form having already been submitted (before March 6) to *Acta Crystallographica*. This draft contains all the essentials of the *Nature* paper, and much of the wording is carried over intact. It required only slight modifications to take into account the Watson–Crick structure, news of which reached King's College on March 18, one day later.[4] In the final (undated) typescript, there is inserted by hand the sentence "Thus our general ideas are not inclnsistent with the model proposed by Watson and Crick in the preceding communication."

In the draft it is deduced that the phosphate groups lie on a helix of diameter 20 Å, that is, on the outside of the molecule, in accordance with the conclusion Franklin had reached earlier on the basis of physicochemical reasoning, including her own work on the water uptake by DNA fibres undergoing the A–B transition. Moreover, on the basis of the intensity distribution in the X-ray pattern, she concludes that there is not one chain in the helix, but probably two, coaxially arranged, and that these are separated by 3/8 of the period, or 13 Å, along the fibre axis direction. The wording is so couched, however, as to show Franklin had not yet understood that the two chains must run in opposite directions, although she had already observed in her notebook that this must be the case in the crystalline A form (which has a two-fold axis of symmetry). I have already argued that it would not have taken long for her to see the true relationship between the two forms—at the time she was thinking of the A form as an unwound version of the helices in the B state (rather, I imagine, like the β-sheet structure is to the α helix in polypeptides).

This would have brought her to the final and crucial step necessary to the complete solution, base pairing. Franklin had thought from an early stage that the bases must lie on the inside of the molecule and be linked by hydrogen bonds, and she had already formed the notion that the two purines were interchangeable with each other and also the two pyrimidines[3]. An entry in her notebook shows that she was considering Chargaff's analytical data, though there is nothing

788   *Molecular Biology*   *Nature Vol. 248 April 26 1974*

to show that she knew the correct tautomeric forms of the bases. The step from base interchangeability to base pairing is a large one, but the idea would have been essential to fitting the variable parts of the structure, the bases, in to the regularly repeating part, the double helix of phosphate-sugar chains at which she had arrived by March 1953.

[1] Klug, A., *Nature*, **219**, 808–810, 843–844; also 880 (1968).
[2] Franklin, R. E., and Gosling, R. G., *Nature*, **171**, 740–741 (1953).
[3] Franklin, R. E., and Gosling, R. G., *Acta Crystallogr.*, **6**, 673–677 (1953).
[4] Olby, R. C., *The Path to the Double Helix* (Macmillan, London, in the press).

Molecular biologists come of age in Aries

Donald A. Windsor

Post Office Box 604, Norwich, New York 13815

It seems that more molecular biologists are born under Aries than any other sign.

ONE major tenet of astrology is that the 'Sun sign' (the constellation of the zodiac in which the Sun was located at the moment of a person's birth) has a powerful influence on personality[1]. I have compared the birthdays of two very different kinds of biologists—taxonomists and molecular biologists. All names listed under all of the taxonomy terms (cyto-, insect, and plant taxonomy; systematic botany, entomology, icthyology, and zoology; systematics; taxonomic botany, and taxonomy) and all of the molecular biology terms (molecular biology, biophysics, genetics, pharmacology)

TABLE 1   Relative frequencies of births under the different Sun signs

Sun sign	Taxonomists No.	%	Index	Molecular Biologists No.	%	Index	General Population Month	%	Index
Aries	28	8.2	98.2	58	12.3	148.0	March	8.4	101.4
Taurus	30	8.8	105.3	32	6.8	81.6	April	8.1	97.1
Gemini	31	9.1	108.8	39	8.3	99.5	May	8.1	97.5
Cancer	38	11.1	133.3	41	8.7	104.6	June	8.2	98.9
Leo	32	9.4	112.3	32	6.8	81.6	July	8.6	103.7
Virgo	31	9.1	108.8	42	8.9	107.1	Aug	8.7	104.8
Libra	25	7.3	87.7	41	8.7	104.6	Sept	8.8	105.4
Scorpio	18	5.3	63.2	41	8.7	104.6	Oct	8.2	98.4
Sagittarius	27	7.9	94.7	40	8.5	102.0	Nov	8.0	95.6
Capricorn	25	7.3	87.7	33	7.0	84.2	Dec	7.9	94.6
Aquarius	26	7.6	91.2	35	7.4	89.3	Jan	8.3	100.0
Pisces	31	9.1	108.8	36	7.7	91.8	Feb	8.6	102.7

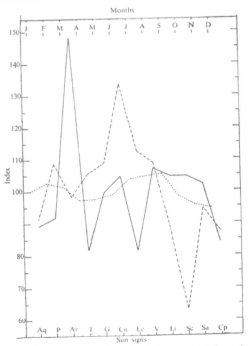

FIG. 1 Relative frequencies of births, expressed as index values, under the different Sun signs for taxonomists (– – –) and molecular biologists (———); also shown are monthly birth data for the United States in 1934 (. . .).

were taken from the Discipline Index[2] of the twelfth edition of *American Men and Women of Science*[3]. The birthdays were obtained from the main reference portion and the appropriate sun signs were assigned. Names appearing more than once were only counted once. A great deal of variation occurs in the dates for these signs because the sun does not always change signs at midnight, so I used the dates given by Goodman[1]. Not all scientists listed had birthdays given. The numbers of persons born under each sign were tallied and the relative frequencies were expressed as an index (each value divided by the average, times 100)[4] (Table 1, Fig. 1). The monthly birth data for the general population of the United States for 1934 are given for comparison; this is the year closest to the births of most of the scientists now recorded, for which good data are available.

More molecular biologists were born under the sign of Aries than any other sign. More taxonomists were born under the sign of Cancer than any other sign and relatively few were born under Scorpio. Since these peaks do not coincide with the peaks for the general population and are sometimes contrary to them, they are even more remarkable.

[1] Goodman, L., *Linda Goodman's Sun Signs* (Bantam Books, Taplinger Publishing Co., New York, 1968).
[2] *Discipline Index. American Men and Women of Science. The Physical and Biological Sciences*, twelfth ed. (edit. by Jaques Cattell Press), 135, 231, 304–306, 469, 511–512 (Bowker, New York, 1973).
[3] *American Men and Women of Science. The Physical and Biological Sciences*, twelfth ed. (edit. by Jaques Cattell Press) (Bowker, New York; A–C **1**:1–1288 (1971), D–G **2**:1289–2392 (1972), H–K **3**:2399–3496 (1972), L–O **4**:3497–4730 (1972), P–Sr **5**:4731–6044 (1972), St–Z **6**:6045–7184 (1973).
[4] Rosenberg, H. M., *Vital Health Statistics*, **21**(9), 1–59 (1966).

Book Reviews

A Quick Climb Up Mount Olympus

The Double Helix. A Personal Account of the Discovery of the Structure of DNA. JAMES D. WATSON. Atheneum, New York, 1968. xvi + 238 pp., illus. $5.95.

Unfortunately, I hear it very often said of a scientist, "He's got charisma." What is meant by "charisma" is not easy to say. It seems to refer to some sort of ambrosial body odor: an emanation that can be recognized most easily by the fact that "charismatic" individuals expect to be paid at least two-ninths more than the rest, unless Schweitzer or Einstein chairs are available. But what does one do if two men share one charisma?

This would certainly seem to be the case with the two who popularized base-pairing in DNA and conceived the celebrated structural model that has become the emblem of a new science, molecular biology. This model furnishes the title of this "personal account," and Watson describes it, without undue modesty, as "perhaps the most famous event in biology since Darwin's book." Whether Gregor Mendel's ghost concurred in this rodomontade is not stated. The book as a whole testifies, however, to a regrettable degree of strand separation which one would not have thought possible between heavenly twins; for what is Castor without Pollux?

This is the beginning of chapter 1 of Watson's book:

I have never seen Francis Crick in a modest mood. Perhaps in other company he is that way, but I have never had reason so to judge him. It has nothing to do with his present fame. Already he is much talked about, usually with reverence, and someday he may be considered in the category of Rutherford or Bohr. But this was not true when, in the fall of 1951, I came to the Cavendish Laboratory of Cambridge University. . . .

As we read on, the impression grows that we are being taken on a sentimental journey; and if the book lacks the champagne sparkle of Sterne's garrulous prose, it bubbles at least like soda water: a beverage that some people are reported to like more than others. The patter is maintained throughout, and habitual readers of gossip columns will like the book immensely: it is a sort of molecular Cholly Knickerbocker. They will be happy to hear all about the marital difficulties of one distinguished scientist (p. 26), the kissing habits of another (p. 66), or the stomach troubles of a third (p. 136). The names are preserved for posterity; only I have omitted them here. Do you wish to accompany the founders of a new science as they run after the "Cambridge popsies"? Or do you want to share with them an important truth? "An important truth was slowly entering my head: a scientist's life might be interesting socially as well as intellectually."

In a foreword to Watson's book Sir Lawrence Bragg praises its "Pepys-like frankness," omitting the not inconsiderable fact that Pepys did not publish his diaries; they were first printed more than a hundred years after his death. Reticence has not been absent from the minds of many as they set out to write accounts of their lives. Thus Edward Gibbon, starting his memoirs:

My own amusement is my motive and will be my reward; and, if these sheets are communicated to some discreet and indulgent friends, they will be secreted from the public eye till the author shall be removed beyond the reach of criticism or ridicule.

But less discreet contemporaries would probably have been delighted had there been a book in which Galilei said nasty things about Kepler. Most things in Watson's book are, of course, not exactly nasty—except perhaps the treatment accorded the late Rosalind Franklin—and some are quite funny, for instance, the description of Sir Lawrence's futile attempts to escape Crick's armor-piercing voice and laughter. It is a great pity that the double helix was not discovered ten years earlier: some of the episodes could have been brought to the screen splendidly by the Marx brothers.

As we read about John and Peter, Francis and Herman, Rosy, Odile, Elizabeth, Linus, and Max and Maurice, we may often get the impression that we are made to look through a keyhole at scenes with which we have no business. This is perhaps unavoidable in an autobiography; but then the intensity of vision must redeem the banality of content. This requirement can hardly be said to be met by Watson's book, which may, however, have a strong coterie appeal, as our sciences are dominated more than ever by multiple cliques. Some of those will undoubtedly be interested in a book in which so many names, and usually first names, appear that are known to them.

This is then a scientific autobiography; and to the extent that it is nothing else, it belongs to a most awkward literary genre. If the difficulties facing a man trying to record his life are great—and few have overcome them successfully—they are compounded in the case of scientists, of whom many lead monotonous and uneventful lives and who, besides, often do not know how to write. Though I have no profound knowledge of this field, most scientific autobiographies that I have seen give me the impression of having been written for the remainder tables of the bookstores, reaching them almost before they are published. There are, of course, exceptions; but even Darwin and his circle come to life much more convincingly in Mrs. Raverat's charming recollections of a Cambridge childhood than in his own autobiography, remarkable a book though it is. When Darwin, hypochondriacally wrapped in his shivering plaid, wrote his memoirs, he was in the last years of his life. This touches on another characteristic facet: scientists write their life's history usually after they have retired from active life, in the solemn moment when they feel that they have not much else to say. This is what makes these books so sad to read: the eagerness has gone; the beaverness remains. In this respect, Watson's book is quite exceptional: when it begins he is 23, and 25 when it ends; and it was written by a man not yet 40.

There may also be profounder reasons for the general triteness of scientific autobiographies. *Timon of Athens* could not have been written, *Les De-*

moiselles d'Avignon not have been painted, had Shakespeare and Picasso not existed. But of how many scientific achievements can this be claimed? One could almost say that, with very few exceptions, it is not the men that make science; it is science that makes the men. What A does today, B or C or D could surely do tomorrow.

Hence the feverish and unscrupulous haste that Watson's book reflects on nearly every page. On page 4: "Then DNA was still a mystery, up for grabs, and no one was sure who would get it and whether he would deserve it. . . . But now the race was over and, as one of the winners, I knew the tale was not simple. . . ." And on page 184: "I explained how I was racing Peter's father [Pauling] for the Nobel Prize." Again on page 199: "I had probably beaten Pauling to the gate." These are just a few of many similar instances. I know of no other document in which the degradation of present-day science to a spectator sport is so clearly brought out. On almost every page, you can see the protagonists racing through the palaestra, as if they were chased by the Hound of Heaven—a Hound of Heaven with a Swedish accent.

There were, of course, good reasons for the hurry, for these long-distance runners were far from lonely. They carried, however, considerably less baggage than others whom they considered, sometimes probably quite wrongly, as their competitors. Quite a bit was known about DNA: the discovery of the base-pairing regularities pointed to a dual structure; the impact of Pauling's α-helix prepared the mind for the interpretation of the x-ray data produced by Wilkins, Franklin, and their collaborators at King's College without which, of course, no structural formulation was possible. The workers at King's College, and especially Miss Franklin, were naturally reluctant to slake the Cavendish couple's thirst for other people's knowledge, before they themselves had had time to consider the meaning of their findings. The evidence found its way, however, to Cambridge. One passage must be quoted. Watson goes to see the (rather poor) film *Ecstasy* (p. 181):

> Even during good films I found it almost impossible to forget the bases. The fact that we had at last produced a stereochemically reasonable configuration for the backbone was always in the back of my head. Moreover, there was no longer any fear that it would be incompatible with the experimental data. By then it had been checked out with Rosy's precise measure-

ments. Rosy, of course, did not directly give us her data. For that matter, no one at King's realized they were in our hands. We came upon them because of Max's membership on a committee appointed by the Medical Research Council to look into the research activities of Randall's lab. Since Randall wished to convince the outside committee that he had a productive research group, he had instructed his people to draw up a comprehensive summary of their accomplishments. In due time this was prepared in mimeograph form and sent routinely to all the committee members. As soon as Max saw the sections by Rosy and Maurice, he brought the report in to Francis and me. Quickly scanning its contents, Francis sensed with relief that following my return from King's I had correctly reported to him the essential features of the B pattern. Thus only minor modifications were necessary in our backbone configuration.

Rosy is Rosalind Franklin, Max stands for Perutz.

As can be gathered from this astonishing paragraph, Watson's book is quite frank. Without indulging in excesses of self-laceration, he is not a "stuffed shirt" and seems to tell what he considers the truth, at any rate, so far as it concerns the others. In many respects, this book is less a scientific autobiography than a document that should be of interest to a sociologist or a psychologist, who could give an assessment that I am not able to supply. Such an analysis would also have to take account of the merciless persiflage concerning "Rosy" (not redeemed by a cloying epilogue) which goes on throughout the book. I knew Miss Franklin personally, as I have known almost all the others appearing in this book; she was a good scientist and made crucial contributions to the understanding of the structure of DNA. A careful reading even of this book will bear this out.

It is perhaps not realized generally to what extent the "heroes" of Watson's book represent a new kind of scientist, and one that could hardly have been thought of before science became a mass occupation, subject to, and forming part of, all the vulgarities of the communications media. These scientists resemble what Ortega y Gasset once called *the vertical invaders*, appearing on the scene through a trap door, as it were. "He [Crick] could claim no clear-cut intellectual achievements, and he was still without his Ph.D." "Already for thirty-five years he [Crick] had not stopped talking and almost nothing of fundamental value had emerged." I believe it is only recently that such terms as the stunt or the scoop have entered the vocabulary of scientists, who also were not in the habit before of referring to each other as smart cookies. But now, the modern version of King Midas has become all too familiar: whatever he touches turns into a publicity release. Under these circumstances, is it a wonder that what is produced may resemble a Horatio Alger story, but will not be a *Sidereus Nuncius*? To the extent, however, that Watson's book may contribute to the much-needed demythologization of modern science, it is to be welcomed.

ERWIN CHARGAFF
Department of Biochemistry,
Columbia University, New York City

Alaska: The Measureless Wealth

Glacier Bay. The Land and the Silence. DAVE BOHN. DAVID BROWER, Ed. Sierra Club, San Francisco, 1967. 165 pp., illus. $25.

In *Glacier Bay*, the Sierra Club once again turns to the task of stimulating public awareness of the natural world and of imparting respect for the land. This magnificently illustrated and sensitively written volume, along with such earlier Sierra Club books as those on the Grand Canyon, the Big Sur coast, and the High Sierra, allow one to *see* and to marvel.

The wondrous scenes these volumes contain are themselves the best of all arguments for resisting needless encroachment on them by the mining companies, the loggers, and the dam builders. Although economic analysis is becoming increasingly useful in shaping policy on the use and conservation of natural resources, economists know no way to make benefit-cost analysis adequately reflect the intangible values of wilderness and other natural environments. A view of, say, the Grand Canyon's inner gorge is indisputably of value, but it is not a marketable masterpiece to be sold at auction. Indeed, to put a price on such a scene is to play into the hands of those who would plug the gorge with concrete and flood it. In the realm of benefit-cost analysis, as in the marketplace, the demand is not for abstractions but for ready coin.

Although some of them are keenly appreciative of natural values, economists seem not to have had much suc-

DNA Helix

I recently came across Dr. E. Chargaff's review (*1*) of J. D. Watson's book *The Double Helix* (*2*). I was disturbed by his quotation of an episode which relates how I handed to Watson and Crick an allegedly confidential report by Professor J. T. Randall with vital information about the x-ray diffraction pattern of DNA.

As this might indicate a breach of faith on my part, I have tried to discover what historical accuracy there is in Watson's version of the story, which reads as follows (*3*):

Even during good films I found it almost impossible to forget the bases. The fact that we had at last produced a stereochemically reasonable configuration for the backbone was always at the back of my head. Moreover, there was no longer any fear that it would be incompatible with the experimental data. By then it had been checked out with Rosy's precise measurements. Rosy, of course, did not directly give her data. For that matter, no one at King's realized they were in our hands. We came upon them because of Max's membership on a committee appointed by the Medical Research Council to look into the research activities of Randall's lab. Since Randall wished to convince the outside committee that he had a productive research group, he had instructed his people to draw up a comprehensive summary of their accomplishments. In due time this was prepared in mimeographed form and sent routinely to all committee members. As soon as Max saw the sections by Rosy and Maurice, he brought the report in to Francis and me. Quickly scanning its contents Francis sensed with relief that following my return from King's I had correctly reported to him the essential features of the "B" pattern. Thus only minor modifications were necessary in our backbone configuration.

Watson showed me his book twice in manuscript; I regret that I failed to notice how this passage would be interpreted by others and did not ask him to alter it. The incident, as told by Watson, does an injustice to the history of one of the greatest discoveries of the century. It pictures Wilkins and Miss Franklin jealously trying to keep their data secret, and Watson and Crick getting hold of them in an underhand way, through a confidential report passed on by me. What historical evidence I have been able to collect does not corroborate this story. In summary, the committee of which I was a member did not exist to "look into the research activities of Randall's lab," but to bring the different Medical Research Council units working in the field of biophysics into touch with each other. The report

was not confidential and contained no data that Watson had not already heard about from Miss Franklin and Wilkins themselves. It did contain one important piece of crystallographic information useful to Crick; however, Crick might have had this more than a year earlier if Watson had taken notes at a seminar given by Miss Franklin.

I discarded the papers of the committee many years ago but the Medical Research Council kindly found them for me in their archives. According to their records there were, in fact, two committees. First, the Biophysics Research Unit Advisory Committee, set up at the beginning of 1947 "to advise regarding the scheme of research in biophysics under the direction of Professor J. T. Randall." Neither Randall nor I were members of that committee; I did not know of its existence until recently. It held its final meeting in October 1947, 5 years before the episode related by Watson. Later that year the Council set up the Biophysics Committee "to advise and assist the Council in promoting research work over the whole field of biophysics in relation to medicine." This new committee consisted mainly of the heads of all the Medical Research Council units related to biophysics, and included Randall and myself. We visited each laboratory in turn; the director would tell the others about the research in his unit and circulate a report. The reports were not confidential. The committee served to exchange information but was not a review body; we were never asked for an opinion of the work we saw. The Medical Research Council dissolved it in 1954, in the words of the official letter because "the Committee has fulfilled the purpose for which it was set up, namely to establish contact between the groups of people working for the Council in this field" (Appendix 1).

On 15 December 1952, we met in Randall's laboratory where he gave us a talk and also circulated the report referred to in Watson's book. As far as I can remember, Crick heard about its existence from Wilkins, with whom he had frequent contact, and either he or Watson asked me if they could see it. I realized later that, as a matter of courtesy, I should have asked Randall for permission to show it to Watson and Crick, but in 1953 I was inexperienced and casual in administrative matters and, since the report was not confidential, I saw no reason for withholding it.

I now come to the technical details of the report. It includes one short section describing Wilkins' work on DNA and nucleoprotein structures and then another on "X-ray studies of calf thymus DNA" by R. E. Franklin and R. G. Gosling. They are reproduced in Appendix 2 below. Note that they contain only two pieces of numerical data. One is the length of the fiber axis repeat of 34 Å in the wet or "B" form of DNA; this is the biologically more important form, solved by Watson and Crick. The other piece consists of the unit-cell dimensions and symmetry of the partially dried "A" form, which was the one discovered and worked on by Wilkins and Miss Franklin, to be solved later by Wilkins and his colleagues. The report contained no copies of the x-ray diffraction photographs of either form.

We can now ask if this section really contained "Rosy's precise measurements needed to check out" Watson and Crick's tentative model and whether it is true that "Rosy did not give us her data . . . and no one at King's realized that they were in our hands." In fact, the report contained no details of the vital "B" pattern apart from the 34 Å repeat, but Watson, according to his own account heard them from Wilkins himself, shortly before he saw the report. This story is told in chapter 23, relating Watson's visit to King's College in late January 1953 where Miss Franklin supposedly tried to hit him and where Wilkins showed him a print of one of her exciting new x-ray photographs of the "B" form of DNA. The next chapter (24) begins as follows: "Bragg was in Max's office when I rushed in the next day to blurt out what I had learned. Francis was not yet in, for it was a Saturday morning and he was home in bed glancing at the *Nature* that had come in the morning mail. Quickly I started to run through the details of the "B" form of DNA, making a rough sketch to show the evidence that DNA was a helix which repeated its pattern every 34 Å along the helical axis." The incident of the report comes in the following chapter (25) and is dated early 1953.

It is interesting that a drawing of the "B" patterns from squid sperm is also contained in a letter from Wilkins to Crick written before Christmas 1952. All this clearly shows that Wilkins disclosed many, even though perhaps not all, of the data obtained at King's to either Watson or Crick.

Turning now to the x-ray pattern of the "A" form, this had been the subject of a seminar given by Miss Franklin at King's in November 1951, an occasion described by Watson in chapter 10. After Miss Franklin's tragic death in 1958, her colleague, Dr. A. Klug, preserved her scientific papers; among these are her notes for that seminar, which he now kindly showed me. These notes include the unit-cell dimensions and symmetry of the "A" form which were circulated in the report a year later.

Watson, according to his own account, had failed to take notes at Miss Franklin's seminar, so that he could not give the unit-cell dimensions and symmetry to Crick afterward. Crick tells me now that the report did bring the monoclinic symmetry of the unit cell home to him for the first time. This really was an important clue as it suggested the existence of twofold symmetry axes running normal to the fiber axis, requiring the two chains of a double helical model to run in opposite directions, but he could clearly have had this clue much earlier.

MAX F. PERUTZ

42 Sedley Taylor Road,
Cambridge, England

References and Notes

1. E. Chargaff, *Science* **159**, 1448 (1968).
2. J. D. Watson, *The Double Helix, A Personal Account of the Discovery of the Structure of DNA* (Atheneum, New York, 1968).
3. ———, *ibid.*, p. 181.
4. I thank the Medical Research Council, Dr. A. Klug, and Dr. R. Olby for supplying me with historical documents, and Sir J. Randall, Professor M. H. F. Wilkins, and Dr. R. G. Gosling for permission to publish their report.

10 April 1969

Appendix 1

27 April 1954

Dear Perutz

The Council have been considering the future of their Biophysics Committee, which was appointed in 1947 and would be due for reconstitution if it were to be kept in being. After consultation with the Chairman and others, they have come to the conclusion that *the Committee has fulfilled the purpose for which it was set up, namely to establish contact between the different groups of people working for the Council in this field.* It has accordingly been decided that the Committee should now be discharged. I am asked by the Council to send you their best thanks for all the help that you have given to their work by serving on this Committee.

Yours sincerely,
Landsborough Thomson
(Secretary to the Biophysics Committee)

1538

Appendix 2

Report by Professor J. T. Randall to the Medical Research Council, dated December 1952

Nucleic Acid Research

The research on nucleic acids, like that on collagen, has both a structural and a biological interest. Some time ago Wilkins found that fibres from sodium desoxyribonucleate gave remarkably good x-ray fibre diagrams. He also examined the optical properties of the fibres in relation to their molecular structure. The detailed examination of the structure has been continued by Miss Franklin and R. G. Gosling, and Wilkins has concentrated on a study of the oriented nucleoprotein of sperm heads. The biological implications of this work are indicated later in this section.

The study of nucleic acids in living cells has been continued by Walker (tissue cultures) and by Chayen (plant root meristem cells); and lately Wilkins and Davies have been measuring the dry weight of material in *Tradescantia* pollen grains during the course of cell division by means of interference microscopy. Thus, while the work of Walker on nucleic acid content of nuclei relates only to part of the cell contents, the interference microscope enables the total content of the cell, other than water, to be measured.

Desoxyribose Nucleic Acid and Nucleoprotein Structure (M. H. F. Wilkins)

A molecular structure approach has been made to the question of the function of nucleic acid in cells.

First, x-ray evidence shows that DNA from all kinds of sources has the same basic molecular configuration which is little (if at all) dependent on the nucleotide ratio. Some grouping of polynucleotide chains takes place to give ~ 20 Å diameter rod-shaped units, and the internal chemical binding which holds each unit together is not affected much by the normal extraction procedure. The basic point is to find the general nature of this structure and the hydrogen bonding etc. in it. Using two dimensional data, the most reasonable interpretation was in terms of a helical structure and the experimental evidence for such helices was much clearer than that obtained for any protein. The crystalline material gives an x-ray picture with considerable elements of simplicity which could be accounted for by the helical ideas, but three dimensional data show apparently that the basic physical explanation of the simplicity of the picture lies in some quite different and, a priori, much less likely structural characteristic. The 20 Å units, while roughly round in cross-section, appear to have highly asymmetric internal structure.

The same general configuration appears to exist in intact sperm heads and synthetic or extracted nucleoprotein, and in bacteriophage (and not in insect virus

where the protein is different). It appears that the protein is probably bound electrostatically on the outside of the nucleic acid units and does not alter their structure. In some sperm the whole head has a crystalline (but somewhat imperfect) structure. In these sperm, the protein has very low molecular weight and it will be especially interesting to find if any high molecular weight protein exists in such sperm heads. If not, all the genetical characteristics may be supposed to lie in the DNA (as in bacteriophage). Biochemical study of the composition of the protein is planned. In other kinds of cell nucleus with different biological function the proteins are quite different. The main idea is to find the structure of the DNA first, then how it is linked to protein in the crystalline sperm heads, and then attempt to elucidate the more complex structure of the other kinds of cell nuclei. It may be that the characteristic x-ray picture of DNA is especially related to a particular function of the nuclear nucleoprotein. In this way molecular structure and cytochemical studies begin to overlap.

X-ray Studies of Calf Thymus DNA (R. E. Franklin and R. G. Gosling)

(*a*) *The Role of Water*: The crystalline form of calf thymus DNA is obtained at about 75 percent RH and contains about 20 percent by weight of water.

Increasing the water content leads to the formation of a different structural modification which is less highly ordered. The water content of this form is ill-defined.

The change from the first to the second structure is accompanied by a change in the fibre-axis repeat period of 28 Å to 34 Å and a corresponding microscopic length-change of the fibre of about 20 percent.

Decreasing the water-content below 20 percent leads to a gradual fading out of the crystalline x-ray pattern and a corresponding increase in the diffuse background scattering. After strong drying only diffuse scattering is observed.

All these changes are readily reversible. The following explanation is suggested:

The phosphate groups, being the most polar part of the structure would be expected to associate with one another and also with the water molecules. Phosphate-phosphate bonds are considered to be responsible for intermolecular linking in the crystalline structure. The water molecules are grouped around these bonds (approximately four water molecules per phosphorus atom). Increased water content weakens these bonds and leads, first, to a less highly ordered structure and, ultimately, to gel formation and solution. Drying leaves the phosphate-phosphate links intact but leads to the formation of holes in the structure with resulting strain and deformation. The three-dimensional skeleton is preserved in distorted form and crystalline order is restored when the humidity is again increased.

(b) *The Cylindrically Symmetrical Patterson Function*: It was apparent that the crystalline form was based on a face-centered monoclinic unit cell with the *c*-axis parallel to the fibre axis. But it was not found possible, by direct inspection, to allot all the lattice parameters accurately and unambiguously. To obtain the unit cell with certainty the cylindrically symmetrical Patterson function was calculated. This function is periodic in the fibre-axis direction only.

Special techniques were developed for the measurement of the positions and intensities of the reflections. This was necessary, firstly because all measurements had to be made on micro-photographs, and secondly because the observed reflections were of a variety of shapes and sizes so that integrated intensities could not be directly measured.

On the Patterson function obtained, the lattice translations could be readily identified. On the basis of a unit cell defined by

$$a = 22.0 \text{ Å}$$
$$b = 39.8 \text{ Å}$$
$$c = 28.1 \text{ Å}$$
$$\beta = 96.5°$$

the 66 independent reflections observed could all be indexed with an error of less than 1 percent.

A very satisfactory confirmation of the correctness of the unit cell and the indexing was provided by a fortunate accident which it has so far not been possible to reproduce. One fibre was obtained which gave a photograph showing strong double orientation. It was found that in this photograph those spots which had been indexed *hkl* were strongest in one pair of quadrants while those indexed *hkl* were strongest in the other pair.

(c) *The Three-Dimensional Patterson Function*: Having established the unit cell with certainty, it is now possible to calculate Patterson sections in the normal way. Work on these is in progress.

In Dr. M. F. Perutz's letter, extracts from a Medical Research Council report are published for the first time. For those interested in the history of the early x-ray studies of DNA at King's college, I give here the main facts which form the background to the report.

Early in 1951 "A" patterns of DNA and very diffuse "B" patterns from DNA and from sperm heads indicated (as I described at a meeting at Cambridge in 1951) that DNA was helical. Shortly afterward, when Rosalind Franklin began experimental work on DNA, she almost immediately obtained (in September 1951) the first clear "B" patterns [described at a seminar in 1951 and published in 1953 (*1*)]. By the beginning of 1952 I had obtained basically similar patterns from DNA from various sources and from sperm heads. The resemblance (*2*) of the "B" patterns of DNA and those of

sperm was very clear at that time. The helical interpretation was very obvious too, and it was proposed in general terms in Franklin's fellowship report (*3*). The "B" patterns of DNA that I obtained at that time were quite adequate for a detailed helical interpretation. This was given later (*4*), with one of the patterns, alongside the Watson and Crick description (*5*) of their model. The best, and most helical-looking "B" pattern, was obtained by Franklin in the first half of 1952 and was published in 1953 (*6*), also with a helical interpretation and alongside the Watson-Crick paper. Confusion arose because, during the summer of 1952, Franklin presented, in our laboratory, "A"-type data (in three dimensions) which showed that the DNA molecule was asymmetrical and therefore nonhelical. Later in the year I wrote for the Medical Research Council report a summary of the DNA x-ray work as a whole in our laboratory. Since our previous emphasis had been entirely on helices, I drew attention in the report to the nonhelical interpretation. In 1953, after the Watson-Crick model had been built and when we had more precise "A" data, I reexamined the question of DNA being nonhelical and found that the data gave no support for the molecule being nonhelical (*7*).

M. H. F. WILKINS

Medical Research Council,
Biophysics Research Unit,
King's College, London

References

1. R. E. Franklin and R. G. Gosling, *Acta Cryst.* 6, 673 (1953).
2. M. H. F. Wilkins and J. T. Randall, *Biochim. Biophys. Acta* 10, 192 (1953).
3. A. Klug, *Nature* 219, 808 (1968).
4. M. H. F. Wilkins, A. R. Stokes, H. R. Wilson, ibid 171, 738 (1953).
5. J. D. Watson and F. H. C. Crick, *ibid.*, p. 737.
6. R. E. Franklin and R. G. Gosling, *ibid.*, p. 740.
7. M. H. F. Wilkins, W. E. Seeds, A. R. Stokes, H. R. Wilson, *ibid.* 172, 759 (1953).

10 April 1969; revised 26 May 1969

I am very sorry that, by not pointing out that the Randall report was nonconfidential, I portrayed Max Perutz in a way which allowed your reviewer [*Science* 159, 1448 (1968)] to badly misconstrue his actions. The report was never marked "confidential," and I should have made the point clear in my text [*The Double Helix* (Athenum, New York, 1968)]. It was my intention to reconstruct the story accurately, and so most people mentioned in the story were given the manuscript, either in first draft or in one of the subsequent

revisions, and asked for their detailed comments.

I must also make the following comments.

1) While I was at Cambridge (1951–53) I was led to believe by general lab gossip that the MRC (Medical Research Council) Biophysics Committee's real function was to oversee the MRC–King's College effort, then its biggest venture into pure science. I regret that Perutz did not ask me to change this point.

2) The Randall report was really very useful, especially to Francis [Crick]. In writing the book I often underdescribed the science involved, since a full description would kill the book for the general reader. So I did not emphasize, on page 181, the difference between "A" and "B" patterns. The relevant fact is not that in November 1951 I *could have* copied down Rosalind's seminar data on the unit cell dimensions and symmetry, but that I *did not*. When Francis was rereading the report, after we realized the significance of the base pairs and were building a model for the "B" structure, he suddenly appreciated the diad axis and its implication for a two-chained structure. Also, the report's explicit mention of the "B" form and its obvious relation to the expansion of DNA fiber length with increase of the surrounding humidity was a relief to Francis, who disliked my habit of never writing anything on paper which I hear at meetings or from friends. The fiasco of November 1951 arose largely from my misinterpretation of Rosy's talk, and with my knowledge of crystallography not really much solider, I might have easily been mistaken again. Thus the report, while not necessary, was very, very helpful. And if Max had not been a member of the committee, I feel that neither Francis nor I would have seen the report; and so, it was a fluke that we saw it.

3) Lastly, Max's implication that the King's lab was generally open with all their data badly oversimplifies a situation which, in my book, I attempted to show was highly complicated in very human ways.

All these points aside, I regret and apologize to Perutz for the unfortunate passage.

JAMES D. WATSON

The Biological Laboratories,
Harvard University,
Cambridge, Massachusetts

19 May 1969

white felt that resulted from the evaporation was excellent for chemical studies, but there must have been severe depolymerization. Actually, this was not of great consequence. What I did not then realize was that we were on the threshold of a new kind of science: a normative biology in which reality only serves to corroborate predictions; and if it fails to do so, it is replaced by another reality. And as to dogmas, they are in no need of experiments. What is currently considered as the structure of deoxyribonucleic acid was established by people who required no recourse to actual DNA preparations, whether polymerized or degraded.

Gullible's Troubles

WHEN I FIRST MET F. H. C. Crick and J. D. Watson in Cambridge, in the last days of May, 1952, they seemed to me an ill-matched pair. This intrinsically unmemorable event has so often been painted—"Caesar Falling into the Rubicon"—repainted, touched up, or varnished in the several auto- and allo-hagiographies[15, 16] that even I, with my good memory for comic incidents and great admiration for the Marx Brothers films, find it difficult to scrape off the entire legendary overlay. I hope that the resulting portraits will be in sharper focus than the famous picture of Parmigianino in the Vienna museum.

This is the way it all came about. The summer of 1952 promised to be an unusually busy time for me: the biochemistry congress in Paris; lectures at the Weizmann Institute and in several European cities; trying unsuccessfully, as twice before, for a professorship in Switzerland. My first talk was scheduled in Glasgow, and on the way there I spent May 24 to 27 in Cambridge, where John Kendrew put me up in Peterhouse. He asked me to speak with two people in the Cavendish Laboratory

100 *More Foolish and More Wise*

who were trying to do something with the nucleic acids. What they were trying to do was not clear to him; he did not sound very promising.

The first impression was indeed far from favorable; and it was not improved by the many farcical elements that enlivened the ensuing conversation, if that is the correct description of what was in parts a staccato harangue. Lest I be accused of *crimen laesarum maiestatum*, I have to point out that mythological or historical couples — Castor and Pollux, Harmodios and Aristogeiton, Romeo and Juliet — must have appeared quite differently before the deed than after. In any event, I seem to have missed the shiver of recognition of a historical moment: a change in the rhythm of the heartbeats of biology. Moreover, the statistical likelihood of two geniuses getting together before my eyes here at Cavendish seemed so small that I did not even consider it. My diagnosis was certainly rapid and possibly wrong.

The impression: one, thirty-five years old; the looks of a fading racing tout, something out of Hogarth ("The Rake's Progress"); Cruikshank, Daumier; an incessant falsetto, with occasional nuggets glittering in the turbid stream of prattle. The other, quite undeveloped at twenty-three, a grin, more sly than sheepish; saying little, nothing of consequence; a "gawky young figure, so reminiscent of one of the apprentice cobblers out of Nestroy's *Lumpazivagabundus*."[17] I recognized a variety act, with the two partners at that time showing excellent teamwork, although in later years helical duplicity diminished considerably. The repertory was, however, unexpected.

So far as I could make out, they wanted, unencumbered by any knowledge of the chemistry involved, to fit DNA into a helix. The main reason seemed to be Pauling's alpha-helix model of a protein. I do not remember whether I was actually shown their scale model of a polynucleotide chain, but I do not believe so, since they still were unfamiliar with the chemical structures of the nucleotides. They were, however, extremely

More Foolish and More Wise 101

worried about the correct "pitch" of their helix. I do not recall how much of the X-ray evidence of King's College (Rosalind Franklin, Wilkins) was mentioned. Because—at that time, at any rate—I set little trust in the biological relevance of X-ray photographs of stretched and pickled high-polymer preparations, I may not have paid sufficient attention.

It was clear to me that I was faced with a novelty: enormous ambition and aggressiveness, coupled with an almost complete ignorance of, and a contempt for, chemistry, that most real of exact sciences—a contempt that was later to have a nefarious influence on the development of "molecular biology." Thinking of the many sweaty years of making preparations of nucleic acids and of the innumerable hours spent on analyzing them, I could not help being baffled. I am sure that, had I had more contact with, for instance, theoretical physicists, my astonishment would have been less great. In any event, there they were, speculating, pondering, angling for information. So it appeared at least to me, a man of notoriously restricted vision.

I told them all I knew. If they had heard before about the pairing rules, they concealed it. But as they did not seem to know much about anything, I was not unduly surprised. I mentioned our early attempts to explain the complementarity relationships by the assumption that, in the nucleic acid chain, adenylic was always next to thymidylic acid and cytidylic next to guanylic acid. This had come to nought when we found that gradual enzymic digestion produced a completely aperiodic pattern; for if the nucleic acid chain had been composed of an arrangement of A-T and G-C dinucleotides, the regularities should have persisted.

I believe that the double-stranded model of DNA came about as a consequence of our conversation; but such things are only susceptible of a later judgment:

> *Quando Iudex est venturus*
> *Cuncta stricte discussurus!*

102   *More Foolish and More Wise*

When, in 1953, Watson and Crick published their first note on the double helix,[18] they did not acknowledge my help and cited only a short paper of ours which had appeared in 1952 shortly before theirs, but not, as would have been natural, my 1950 or 1951 reviews.[12, 14]

Later, when molecular prestidigitation ran wild, I was often asked by more or less well-meaning people why I had not discovered the celebrated model. My answer has always been that I was too dumb, but that, if Rosalind Franklin and I could have collaborated, we might have come up with something of the sort in one or two years. I doubt, however, that we could ever have elevated the double helix into "the mighty symbol that has replaced the cross as the signature of the biological analphabet."[19]

Matches for Herostratos

WHEN the Artemision—one of the world wonders of antiquity—went up in flames in 356 B.C., a man was apprehended who confessed that he had done it in order to make his name immortal. The judges, in condemning him, decreed that his name must remain unknown. But soon after, the historian Theopompos claimed that the name was Herostratos. Whether this really was the name or whether Theopompos merely wanted to annoy, say, his father-in-law, cannot be ascertained. Recently, when I mentioned Herostratos in an article, the editor called up to say that nobody in the editorial office had ever heard of him, thus giving belated satisfaction to the judges of Ephesos.

If Herostratos has earned immortality for having burned down the temple of Artemis in Ephesos, maybe the man from whom he got the matches ought not to be entirely forgotten. I am that man.

More Foolish and More Wise 103

-6-

How to Solve a Protein Structure

*"In this life one should Perutz high goals,
but ultimately settle for what one Kendrew."*

KEY PAPERS

(A)   1953   J. Donohue and J. Briekopf, *Max F. Perutz reprint file*, "The X-Ray Structure Analysis of α-Globlglobin" [unpublished, 10 pages].

1956   D. Harker, *Acta Cryst.* 9, 1–9. "The Determination of the Phases of the Structure Factors of Non-Centrosymmetric Crystals by the Method of Double Isomorphous Replacement."

The Hemoglobin Papers: *Proc. Roy. Soc. London* [title/abstract page only]:

I   1947   J. Boyes-Watson, E. Davidson and M. F. Perutz, A191, 83–132. "An X-ray Study of Horse Methaemoglobin, I."

II   1949   M. F. Perutz, A195, 474–499. "An X-ray study of horse methaemoglobin, II."

III   1954   M. F. Perutz, A225, 264–286. "The Structure of Haemoglobin, III. Direct Determination of the Molecular Transform."

IV   1954   D. W. Green, V. M. Ingram and M. F. Perutz, A225, 287–307. "The Structure of Haemoglobin, IV. Sign Determination by the Isomorphous Replacement Method."

V   1954   E. R. Howells and M. F. Perutz, A225, 308–314. "The Structure of Haemoglobin, V. Imidazole-methaemoglobin: A Further Check of the Signs."

VI   1954   W. L. Bragg and M. F. Perutz, A225, 315–329. "The Structure of Haemoglobin, VI. Fourier Projections on the 010 Plane."

VII   1958   D. M. Blow, A247, 302–336. "The Structure of Haemoglobin, VII. Determination of Phase Angles in the Non-centrosymmetric [100] Zone."

VIII   1961   A. F. Cullis, H. Muirhead, M. F. Perutz, F. R. S. and M. G. Rossmann, A265, 15–38. "The Structure of Haemoglobin, VIII. A Three-Dimensional Fourier Synthesis at 5.5 Å Resolution: Determination of the Phase Angles."

IX   1962   A. F. Cullis, H. Muirhead, M. F. Perutz, F. R. S. and M. G. Rossmann. A265, 161–187. "The Structure of Haemoglobin, IX. A Three-Dimensional Fourier Synthesis at 5.5 Å Resolution: Description of the Structure."

(B)   1958   J. C. Kendrew, G. Bodo, H. M. Dintzis, R. G. Parrish, H. Wyckoff and D. C. Phillips, *Nature* 181, 662–666. "A Three-Dimensional Model of the Myoglobin Molecule Obtained by X-Ray Analysis."

(C)   1959   G. Bodo, H. M. Dintzis, J. C. Kendrew and H. W. Wyckoff, *Proc. Roy. Soc. London* A253, 70–102. "The Crystal Structure of Myoglobin. V. A Low-Resolution Three-Dimensional Fourier Synthesis of Sperm Whale Myoglobin Crystals" [7 pages of 33].

While Crick and Watson worked out the structure of DNA, Max Perutz and John Kendrew were pursuing the primary goal of the laboratory: protein crystal structure analysis. Since his graduate days with J. D. Bernal in the 1930's, Perutz had been studying the oxygen-transport horse hemoglobin molecule, with molecular weight 64,650. Kendrew remained at Cambridge after completing his doctorate, and worked on the smaller single-chain oxygen-storage protein myoglobin, roughly one-quarter the size. Today everyone knows that a molecule of horse hemoglobin is built from two alpha chains of 141 amino acids each, and two beta chains of 146 amino acids. These chains and that of whale myoglobin, which has 153 residues, share similar amino acid sequences and a common chain folding. Hemoglobin and myoglobin are evolutionary cousins obtained by divergence from a common gene ancestor. But in 1947 none of this was known: neither the amino acid sequence nor the length of any of the polypeptide chains, and in fact not even the number of chains that were present in hemoglobin.

Standard methods of crystallographic structure analysis simply did not work for such large molecules. With Sir Lawrence Bragg as its Director, the Cavendish Laboratory was the best conceivable location for studying problems of this magnitude. But the problem of protein crystal structure analysis seemed insoluble. Indeed, Linus Pauling himself had said in his 1939 paper with Niemann: *"....the great complexity of proteins makes it unlikely that a complete structure determination for a protein will ever be made by X-ray methods alone."*

Then in 1953, Max Perutz made an amazing discovery which, although it sounds trivial, proved to be the key to protein crystal structure analysis: If one binds *a single atom* of a metal such as silver, gold or mercury to each protein molecule in the crystal, this is enough to produce measurable changes in intensities of the spots in its diffraction pattern. This led to the technique of Multiple Isomorphous Replacement or MIR phase analysis. What that means, how it works, and what it accomplished are the subject of this chapter.

The Structure Analysis of the Protein Globlglobin

But before we turn to MIR phasing of hemoglobin, it is worth looking at a satirical paper that the ubiquitous Jerry Donohue wrote in early 1953 (A), reflecting everyone's frustration with protein structure analysis, just on the eve of Max's new methodology. It is entitled "The X-Ray Structure Analysis of α-Globlglobin," and apparently was circulated in carbon copy among members of the MRC Lab. In 1958 I made my own typescript of the paper from the copy in Perutz' bound reprint collection, and later retyped it and submitted it for publication (unsuccessfully) in the satirical *Journal of Irreproducible Results.*

The paper is a parody of the x-ray methods that existed just prior to MIR phasing, in particular the ideas of Francis Crick about coiled-coils and the mathematics of diffraction by a helix. (See the Key Papers of Chapter 4.) Most of the people mentioned in the paper are real (Table 6.1).

I have circulated Donohue's manuscript among 50 molecular biologists of that generation, and have not found anyone who can hazard a guess as to the identity of Bearish and Fouls. Four important players who do not seem to be included are Sir Lawrence Bragg, Max Perutz, John Kendrew and Linus Pauling, although Corey shows up in the cast.

Donohue shows flashes of pure inspiration in this paper, which reflects the frustration of not yet knowing how to obtain usable crystals, or how to solve a protein structure once crystals are obtained. The various sources of globlglobin protein on page A.1 are inspiring. Perutz and Kendrew indeed searched through various species for the most favorable sources of hemoglobin and myoglobin. Horse myoglobin was tried before Kendrew realized that the richest source of this oxygen-storage protein would be the muscles of diving mammals. Seals had their day, but when the high-resolution 2 Å analysis of myoglobin was carried out in 1958-9, its material came from a metal garbage can full of frozen whale meat from Peru, stored in a cold room at the Molteno Institute.

Although the original typescript was undated, one can pinpoint pretty closely when it was completed. Donohue must have written it before the significance of Perutz' new MIR method was realized, or else he would have made it a major part of his parody. But the globlglobin paper must also have been written no earlier than the spring of 1953, as its summary remark: "*It has not escaped our notice that* the paper chain structure provides a possible method of impulse transfer in the adrenal cortex" recalls Watson and Crick's note to *Nature* of April 1953: "*It has not escaped our notice that* the specific pairing we have postulated immediately suggests a possibly copying mechanism for the genetic material."

Pages 3 and 4 of the globlglobin paper contain a parody of important work by Cochran, Crick and Vand (*Acta Cryst.* 5, 581–586, 1952) on the mathematics of x-ray diffraction by a helical fiber. Equation 1 on page 3 ("using an obvious notation") is not all that more complex than the real equation, which is:

$$F_{(R,\psi,Z)} = \Sigma\Sigma f_j J_q\left(2\pi R r_j\right)\exp i\left\{q\left(\psi - \phi_j + \pi/2\right) + 2\pi Z z_j\right\}$$

TABLE 6.1	
Manuscript alias	**Real person**
J. Donohue	(real name)
J. Briekopf ("Cheesehead")	A frequent Donohue alias in some of his other parodies
Mr. I. M. A. Crock	F. H. C. Crick
S. Holmes	J. Watson
Sorey	Robert Corey of Caltech
Prof. V. Cobbler	Verner Schomaker
J. D. Lenin	J. D. Bernal, both a brilliant scientist and a radical Marxian socialist
F. Bonamie	Herbert "Freddy" Gutfreund ("bon-ami"). Gutfreund, according to Crick, was a chemist in the Department of Colloid Science, who worked on the molecular weight of insulin, and who was closely associated with Fred Sanger.
M. Joseph Fourier	(real name, inventor of Fourier series)
Dr. A. L. Patterson	(real name, inventor of the Patterson function)
Dr. O. N. Bearish	???
Mr. E. Fouls	???

And Donohue's paper-chain/paper-chain structure on page 4 recalls Crick's coiled-coil for α-keratin. The complicated expression above applies to fiber diffraction, not to single crystals. It actually is irrelevant to any structure of "globlglobin," but was an exciting new idea at the time. This is additional evidence that Donohue wrote his essay *before* Perutz had developed MIR phasing.

The reference to Sorey seems to be a jab at Pauling and Corey at Caltech:

The only previous work on globlglobin is that of Sorey and co-workers, whose brilliant and painstaking researches were able to establish unequivocally beyond any doubt that it was not possible with the data at hand to come to any definite conclusions concerning the size and shape of the molecules, or their position in the unit cell.

Recall that the Donohue manuscript probably was written two or three months after the Pauling/Corey DNA paper appeared in print. Indeed, Pauling and Corey never did manage to solve a protein structure together at Caltech.

Donohue's paper is full of bad (i.e., good) puns and jokes. The table comparing observed and calculated intensities of x-ray reflections (h,k,l) makes a mockery of crystallography. Each discrepancy between measured intensities and those calculated from the globlglobin model is given its own special excuse, except for the final reflection in the list, for which the agreement between observed and calculated intensities is beyond belief.

Donohue's spoof has an interesting subsequent history. Of the surviving participants, only Jack Dunitz and Francis Crick remember having seen it, and neither has a copy today. When Perutz' scientific papers were dispersed, the bound 1953 reprint volume with its typescript was lost. But in early 2003, a prominent rare book dealer in Los Angeles offered a large collection of early papers on molecular biology, mostly from either Cambridge or King's College. The asking price for the collection was $200,000! Richard Henderson, the present Director of the MRC (or the Laboratory of Molecular Biology or LMB as it is now known) intervened and persuaded the dealer to donate some of Perutz' material back to the laboratory, including the volume with the globlglobin paper. So 2003 was the golden anniversary of globlglobin as well as DNA.

How to Really Solve a Protein Structure

The principal player in this part of the story is Max F. Perutz (Plate I of the color signature). Max came from a middle class family in Vienna. His physical chemistry mentor at the University of Vienna, Hermann Mark, was one of the pioneers in fiber x-ray diffraction. He persuaded Perutz to pursue his doctorate in crystallography with J. D. Bernal in Cambridge, and Max joined that group in 1936. But after two years Bernal left Cambridge to take a chair at Birkbeck College in London. Hitler invaded Austria, meaning that not only were Max's funds cut off; he had to find some way of bringing his parents to England and supporting them. He plucked up his courage and approached W. L. Bragg (later Sir Lawrence), who had just replaced Rutherford as Cavendish professor. Armed only with an x-ray photograph of hemoglobin, he persuaded Bragg to take him on as a Ph.D. student and support him with an annual salary of £250 from the Rockefeller Foundation. Bragg became a strong mentor of Perutz, persuading him to remain at Cambridge after completing his doctorate, and to develop his hemoglobin x-ray structure project.

Max's status changed dramatically in May 1940. When Hitler annexed Austria in 1938, all the Austrian nationals living in Britain suddenly became Germans. And two years later, on the eve of the German invasion of the Netherlands, they were officially declared "Enemy Aliens." Max described his Odyssey in a piece of that title in a 1985 issue of the New Yorker (1), and later included it in two collections of essays: Nos. 12 and 13 of Appendix 2. He and many of his fellow countrymen were rounded up, shuttled from one temporary internment camp to another in England, and finally put aboard a ship bound for Canada. His friends in England were outraged, and Dorothy Crowfoot (Hodgkin) put the matter on public notice in a 1941 review of protein structure analysis (2):

> As Dr. Perutz is now interned and may be unable to send an account of his work himself, Professor Bragg has kindly given me permission to use his results.

By January 1941, friends had obtained Perutz' release, and he was back in Cambridge once again. But by a curious twist of fate, a year later he was recruited by the British Government for a top-secret military project, and sent back to Canada! The details are in his "Enemy Alien" article but the essence is this: Max had always been an avid skiier, and had taken part in expeditions in the Swiss Alps to study the crystal behavior of glacial ice. In 1958 his office wall had a photograph of himself and two students standing on the slope of a Swiss glacier. Two of the earliest thesis projects which he supervised had been on the structure of ice.

Shortly before World War II began, Perutz' former physical chemistry professor at the University of Vienna, Hermann Mark, had discovered that a mixture of water and wood pulp, when frozen, was stronger than concrete! Furthermore, it could be machined like wood and cast into shapes like metal. A copy of Mark's report had come into the hands a British government worker, Geoffrey Pyke. At a time when England was losing air bases on the European continent, Pyke urged the British government to build a fleet of floating aircraft carriers made of what later came to be called "pykrete," and Perutz was enlisted as an expert in ice. The project was given the code name "Habakkuk," after a minor Old Testament prophet who railed against attacks by evil enemies. The inspiration for naming this particular project almost certainly was derived from Habakkuk 1:5, which reads,

> Look among the nations, and see; wonder and be astounded. For I am doing a work in your days that you would not believe if told.

That certainly describes aircraft carriers built from synthetic icebergs!

Perutz, as an acknowledged expert on the physical chemistry of ice, was recruited for the project and sent to Canada to take part in the construction of a trial model in a northern lake. But the project foundered on the difficulty of freezing such a large mass of ice, and was made unnecessary by the increased range of land-based airplanes. By the end of 1943 the project was abandoned and Perutz sailed back to England, this time with the papers of a naturalized British subject. He plunged back into the research that would be his life's mission: the x-ray crystal structure analysis of the protein hemoglobin.

In 1958, while a postdoctoral fellow with John Kendrew, I asked Max, since he loved climbing and skiing so much, why he had abandoned mountainous Austria for Cambridge, one of the flattest places on Earth. His answer was simple: "Being in Cambridge, Gisela (his wife) and I can go back to Switzerland every Christmas and ski as much as we like. Had I stayed at the University of Vienna, we would not have been able to afford the train fare to Innsbruck!"

To understand the power of Perutz' methods, we must dig a little deeper into x-ray crystallography, again keeping mathematics down to an absolute minimum. If the repeating distances in three directions in the crystal are **a**, **b** and **c**, then the separations of the spots along the three axes of the diffraction pattern will be $1/\mathbf{a}$, $1/\mathbf{b}$ and $1/\mathbf{c}$. (To keep things simple we will always assume that these axes are at right angles to one another. If they are not, this just adds mathematical complication without changing the essential ideas.) The statement that *the diffraction pattern of a crystal is its Fourier transform* means that the three-dimensional image of the crystal has been broken down into component sinusoidal waves called Fourier waves. Each spot in the three-dimensional x-ray diffraction pattern (Figure 6.1) corresponds to one of these waves from which an image of the crystal can be reconstructed. The relationship between spots in the x-ray patterns and waves in the Fourier synthesis that will produce an image of the scattering molecules is:

- The *direction* of a given Fourier wave is determined by the vector from the center of the diffraction pattern to the spot in question.
- The *wavelength* of the wave is inversely proportional to the distance of the spot from the origin.
- The *amplitude* of the wave, F, is proportional to the square root of the intensity of the spot.
- The *phase* shift of the wave, ϕ, relative to a common origin *is not given directly by the x-ray pattern*. A Fourier synthesis that yields a picture of the molecule cannot be calculated until this "phase problem" is solved.

Perutz' great achievement was the first use of MIR (multiple isomorphous replacement) methods to solve the phase problem for a protein, hemoglobin.

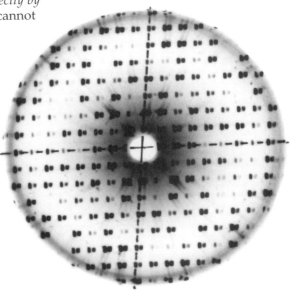

FIGURE 6.1   One section through the full three-dimensional diffraction pattern from a crystal of triclinic egg-white lysozyme. The spots are doubled because this actually is a superposition of two photos: one from native lysozyme, and the other from a heavy atom derivative in which mercury bromide has been diffused into the crystal and a single mercury atom has bound at the same position on every lysozyme molecule. These are the kinds of intensity changes that Perutz used in MIR phase analysis. This photograph contains the information needed to compute a projection of the structure down one axis of the crystal, to a resolution of 4 Å. For a three-dimensional image of the lysozyme molecule, the full three-dimensional diffraction pattern of spots must be used. From (7).

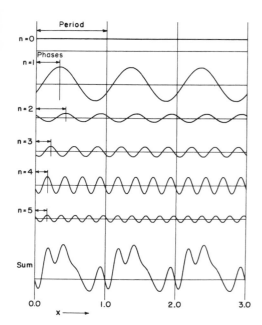

FIGURE 6.2   Addition of sinusoidal waves to generate an image of a "molecule". The one-dimensional "crystal" has a unit cell length or period as shown. Each wave is characterized by (a) a wavelength, or its inverse the wavenumber (the number of cycles of the wave within one period), (b) an amplitude (the height of the wave), and (c) a phase, or the amount by which the wave is shifted from a common origin before it is added into the synthesis. In this particular example, the phase angles of waves n=1 through n=5 are approximately 120°, 150°, 80°, 64° and 62° respectively. As Figure 5.1 showed, a phase shift of 360° displaces the wave by one period length, making it identical to an unshifted wave. The sum of the five waves at bottom is quite different from a sinusoidal wave, and could be imagined to describe the electron density in a one-dimensional molecule. Originally from (8), quoted in (9).

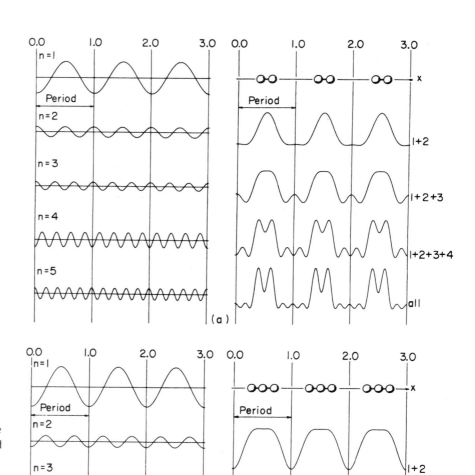

FIGURE 6.3   Two one-dimensional representations of the synthesis of an image of molecules by different additions of the same five waves. In each case, the waves to be added are shown at left, with their proper amplitudes and phase shifts relative to the origin. At right is the image that results from successive addition of waves down from the top of the list to that point. (a) Addition of five waves to build the image of a two-atom molecule. (b) Addition of these same five waves, but with different amplitudes and phase shifts, to build an image of a three-atom molecule. In both cases, use of long waves n=1 through n=3 builds what could be regarded as a "low-resolution map," which locates the molecule but does not reveal its structure. Addition of shorter waves 4 and 5 yields a "high-resolution map," in which the diatomic or triatomic structures of the molecules become clear. From (8) and (9).

The addition of Fourier waves with proper wavelength, amplitude and phase to yield a more complex periodic function is shown for one dimension in Figures 6.2 and 6.3. Figure 6.2 uses five waves which repeat once, twice, three, four or five times in the period of the function. Each wave has its own amplitude, and each is given a different phase shift relative to the origin. The result of adding all these waves together is the repeating sum function at the bottom, which can be regarded as the image of a molecule repeated throughout a one-dimensional crystal.

If the amplitudes of the component waves are changed, or if their phase shifts are altered, the sum function will appear quite different. Figure 6.3 shows how *the same five waves* can build images of two quite different molecules, diatomic at top and triatomic below. The same waves are used, only their amplitudes and phase shifts are changed. Figure 6.3 also illustrates the concept of *resolution* of the calculated electron density map. If only long waves are used, the picture is coarse and degraded. Only by adding short-wavelength terms can the fine details of the image emerge.

The resolution or amount of detail in a calculated electron density map depends on how many data are included in the synthesis, and on the wavelengths of the finest waves in the data set. Figure 6.4 illustrates the effect on the image of a small organic molecule of cutting off diffraction data at different points. The amount of information in the map is proportional to the number of terms in its Fourier series, or to the number of reflections included. The 1.54 Å map has twice the information of that at 2.2 Å, the 1.19 Å map has four times, and the 0.77 Å map has eight times the information. If all data out to a resolution of 0.77 Å are used, then individual atoms are clearly separated. If data are cut off at 1.19 Å, yielding half as many x-ray reflections, the atoms in the resulting map are still distinguishable although blurred. At 1.54 Å resolution the ring atoms have merged but the central hole in the ring is still visible. But at 2.20 Å the ring becomes only a flat disk with projections.

The process of x-ray crystal structure analysis would be quite straightforward but for one essential fact: *the phase angle of each Fourier wave is lost in the diffraction*

(a)

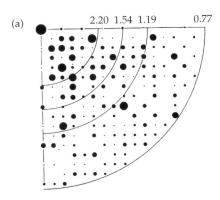

FIGURE 6.4   Illustration of the concept of resolution in an electron density map. (a) One quadrant of the x-ray diffraction pattern from an organic molecule, 4,5–diamino–2–chloropyrimidine. As one goes farther out from the center of the pattern, the Fourier wavelengths corresponding to individual reflections decrease. All reflections on the curve marked 2.20 correspond to Fourier wavelengths of 2.20 Å, but with waves running in different directions. All reflections on the 1.54 Å curve correspond to this wavelength, and the same is true for the 1.19Å and 0.77Å curves. (b) The images of the molecule that result when all data out to 0.77 Å are used, or when the diffraction pattern is truncated at 1.19, 1.54, or 2.20 Å. From (7).

(b)

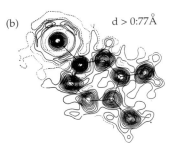 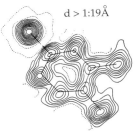 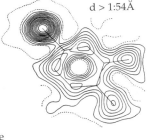 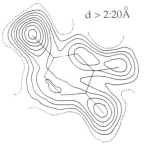

4,5–diamino–2–chloropyrimidine
Sharpened Fourier projections

pattern. Methods have been developed to get around this problem with smaller molecules, including:

(a) Guess a structure, calculate the expected amplitudes, F_c and phases, ϕ_c, for the guessed model, and then use these calculated phases with the real, observed amplitudes, F_o, to calculate a new hybrid image, hoping that the result will be something in between the trial model and reality. This is called trial-and-error molecular replacement, and works only when one's first guess is very close to the truth. Proteins are vastly too complicated for such an approach.

(b) Attach a heavy metal atom to the molecule, collect the complete diffraction pattern from crystals of these labeled molecules, locate the heavy atom position in the cell, calculate the phases for the heavy metal alone, and then use these phases with the measured amplitudes to calculate an image of the molecule. This is called heavy atom phasing, and although useful with organic molecules is of little help with proteins. Protein molecules are so large and complicated that the phases for an attached heavy atom bear little or no resemblance to those of the entire protein.

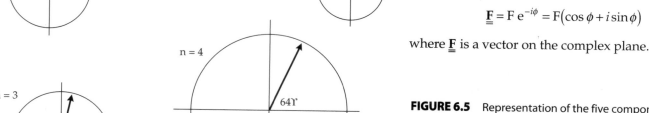

However, although an added heavy atom cannot *dominate* the scattering by a protein, it can visibly *alter* it. The electrons of the atoms are the entities that scatter x-rays. The difference between 80 electrons in a mercury atom and roughly 9,000 electrons in a myoglobin molecule led people to expect that binding a heavy metal to a protein would have a negligible effect on its diffraction pattern. But this, as Perutz was the first to demonstrate, is not the case. As Figure 6.1 shows, binding even a single mercury atom to every molecule of lysozyme in the crystal creates changes in diffraction intensities, sufficient to permit the kind of MIR phase analysis described below.

I Don't Recall the Name, but the Phase is Familiar

A very useful fact is that one can represent a sinusoidal wave such as we have been seeing, by a vector on the complex plane. Figure 6.5 represents the five waves shown in Figure 6.2. In each case the amplitude F and phase ϕ of the wave have been measured directly from Figure 6.2, and then used to draw its phase diagram in Figure 6.5. The mathematical expression for such a wave representation is:

$$\underline{\mathbf{F}} = F\,e^{-i\phi} = F\left(\cos\phi + i\sin\phi\right) \tag{1}$$

where $\underline{\mathbf{F}}$ is a vector on the complex plane.

FIGURE 6.5   Representation of the five component waves of Figure 6.2 as vectors in a phase diagram on the complex plane. Real axis horizontal, imaginary axis vertical. The length of the vector in each case equals the amplitude of the wave, F, and the orientation of the vector relative to the horizontal axis is the phase, ϕ. The wave then is represented by the complex number: $\underline{F} = Fe^{-i\phi}$.

Individual reflections in the three-dimensional diffraction pattern are labeled h, k and l along the three principal axes. (See the two-dimensional case in Figure 5.3.) If the phases of all reflections are known, then a map of the electron density $\phi(x,y,z)$ at every point (x,y,z) throughout the crystal can be calculated from the expression:

$$\rho(x,y,z) = \sum_h \sum_k \sum_l \underline{F}_{hkl}\, e^{-2\pi i(hx+ky+lz)} = \sum_h \sum_k \sum_l \left(F_{hkl}\, e^{-i\phi}\right) e^{-2\pi i(hx+ky+lz)} \qquad (2)$$

Conversely, the amplitude and phase of a particular reflection (h,k,l) can be calculated by knowing the positions (x, y, z) of all the atoms in the unit cell:

$$\underline{F}_{hkl} = F_{hkl}\, e^{-i\phi} = \sum_j f_j e^{2\pi i\left(hx_j + ky_j + lz_j\right)} \qquad (3)$$

The term f_j is the scattering power of atom j, and is related to the number of electrons in the atom.

So the scattering from a collection of atoms is just *the vector sum of scattering by each of the individual atoms*. The magnitude of an individual atom's vector is proportional to the number of electrons in the atom, and the phase of its vector depends on where the atom is. Figure 6.6a shows the phase diagram for a single carbon atom, and 6b shows how three carbon atoms in different locations might scatter. The addition is a vector sum, and the overall phase bears no particular resemblance to the phase of any of the three atoms. In Figure 6.6c, a bromine atom has been added to these same three carbon atoms. Bromine, with 35 electrons, scatters x-rays so strongly that to a good first approximation, the phases of bromine alone, ϕ_{Br}, and of the four-atom "C₃Br molecule," ϕ_{all}, are quite similar. If the electron density as in equation 2 is calculated using the true amplitudes but phases derived only from bromine, the resulting map will show the bromine atom strongly, and hopefully a reasonably accurate picture of the three carbon atoms. These then can be built into a trial model for another cycle of Fourier refinement. This is the heavy atom phasing method, and has been enormously successful for organic molecules.

One of the largest, if not the largest, crystal structures to have been solved by the heavy atom phasing method is vitamin B₁₂, which Dorothy Crowfoot Hodgkin and coworkers solved in 1954–6 (3). The staggering complexity of the molecule is evident from its schematic formula in Figure 6.7. With an empirical formula of $C_{63}H_{88}N_{14}O_{14}PCo$,

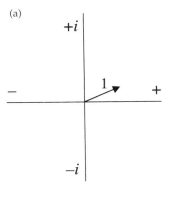
(a)

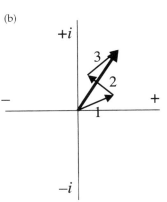
(b)

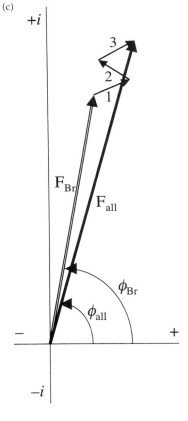
(c)

FIGURE 6.6 Addition of vectors from individual atoms, to yield the overall scattering vector from the entire molecule for a given x-ray reflection. The length of each individual vector, f_j, is proportional to the number of electrons in that atom. (a) A single carbon atom with 6 electrons. (b) Three carbon atoms at different locations (and hence having vectors of the same length but pointing in different directions). (c) These same three carbon atoms, plus one bromine atom with 35 electrons. The heavy bromine atom dominates, and the phase for the entire four-atom "molecule," ϕ_{all}, is not very different from that of bromine alone, ϕ_{Br}. If all the bromine-only phases are used with true measured amplitudes in equation (2), the result will be a map that shows many features of the full structure. This is the so-called heavy atom method of phase analysis.

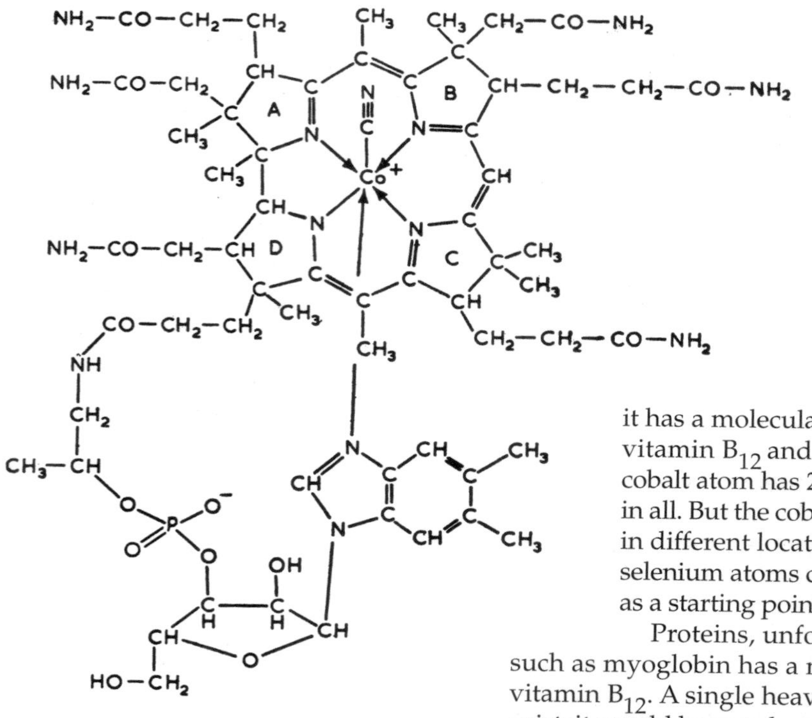

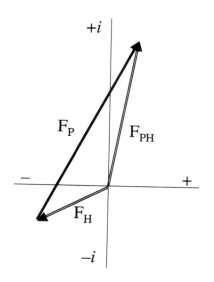

FIGURE 6.7   Empirical formula of vitamin B_{12}, $C_{63}H_{88}N_{14}O_{14}PCo$. Its three-dimensional structure, not shown here, was worked out by Dorothy Crowfoot Hodgkin (3), and is one of the most complex structures ever solved by heavy atom phase analysis. But even the smallest protein molecule is an order of magnitude more complex than this, and heavy atom phasing fails with proteins. From (6).

it has a molecular weight of 1356. Crystals were grown both of the native vitamin B_{12} and of a derivative containing selenium atoms. The central cobalt atom has 27 electrons, and the other 180 atoms contain 691 electrons in all. But the cobalt scattering is coherent, whereas that of the lighter atoms in different locations is incoherent. Hence the phases from the cobalt and selenium atoms can dominate the structure as in Figure 6.6c, and can serve as a starting point for structure analysis.

Proteins, unfortunately, are quite another story. Even a small protein such as myoglobin has a molecular weight of 17,184, nearly thirteen times that of vitamin B_{12}. A single heavy atom thirteen times the size of cobalt simply does not exist; it would have to have an atomic number of 350! Myoglobin has approximately 1200 non-hydrogen atoms, each with 6 or more electrons. This comes to something like 9000 electrons. How can even an 80-electron mercury atom dominate against such competition? But it does alter the diffraction pattern; Perutz established this for hemoglobin, and Figure 6.1 shows it for a different protein, lysozyme. The mercury atom phases themselves cannot be used directly as in the heavy atom method, since the mercury phases themselves do not even vaguely resemble the phases from the entire protein. But the mercury can cause measurable changes in intensity because all of its electrons are concentrated in a small region of space, whereas the electrons of the protein are spread over tens of Ångstroms. Scattering from the single mercury atom is coherent, whereas that from so many individual small atoms is not.

Figures 6.8 and 6.9 show graphically how multiple isomorphous replacement works. In Figure 6.8, for a particular x-ray reflection, $\underline{F}_P$ is the scattering vector for a protein molecule, with both an amplitude F_P and a phase, ϕ_P. $\underline{F}_H$ represents scattering by the heavy atom, and their vector sum, $\underline{F}_{PH} = \underline{F}_P + \underline{F}_H$, is the scattering by the heavy-atom-labeled protein. But only the magnitudes F_P and F_{PH} are known; they are the square roots of the measured intensities of the reflection in question in the native protein and in the heavy atom derivative. Figure 6.9a shows that, with only this information, there is a twofold uncertainty about the phase of $\underline{F}_P$. Either point a or point b on the phase circle could be correct. The dilemma is solved if another heavy atom derivative is available, Figure 6.9b. Then the second derivate also has a twofold ambiguity, with phase angles at a or c. But phase angle a is common to both derivatives and therefore is the true phase.

FIGURE 6.8   Phase diagram for a particular reflection in the x-ray pattern, illustrating the relationship: $\underline{F}_{PH} = \underline{F}_P + \underline{F}_H$, where $\underline{F}_P$ is the scattering vector (amplitude plus phase) for the protein, $\underline{F}_H$ is that for the heavy atom alone, and $\underline{F}_{PH}$ is the scattering for the protein labeled with the heavy atom. The dilemma in MIR phasing is that, although both the magnitude and phase of $\underline{F}_H$ are known once the heavy atom is located in the cell, only the magnitudes of $\underline{F}_P$ and $\underline{F}_{PH}$ are available, derived from the square roots of the intensities of reflections.

FIGURE 6.9   (a) Phase diagram for a single heavy atom derivative. Two points around the phase circle, a and b, satisfy the vector sum: $\underline{F}_{PH} = \underline{F}_P + \underline{F}_H$ or $\underline{F}_P = \underline{F}_{PH} - \underline{F}_H$. There is no way of choosing between them. (b) A second, different heavy atom derivative also yields a twofold ambiguity, with phase possibilities a and c. Hence the point common to both derivatives, a, must indicate the correct phase for $\underline{F}_P$. This graphic formulation of the phase problem was first developed by David Harker at the Polytechnic Institute of Brooklyn (see Key Papers list).

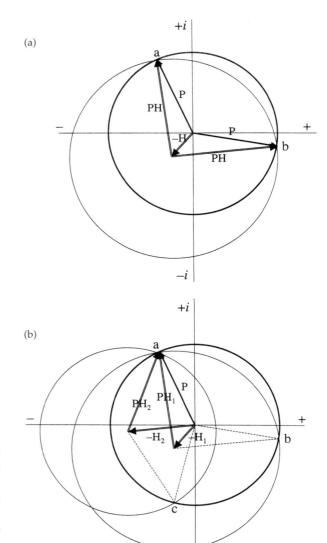

Of course phase diagrams are never as clean and unambiguous as Figure 6.9b. Slight errors in measured intensities from parent and derivative change the radii of the phase circles and can play havoc with their intersection points. A third derivative is almost always required to pin down the common crossing point, and various statistical methods are used to establish the probability of a given phase at different points around the phase circle. Hence the name "*Multiple* Isomorphous Replacement." And hence the folk wisdom about the importance of the third derivative. Many improvements have been made since Perutz' time in techniques of MIR and even SIR (Single Isomorphous Replacement) methods. But the basic idea remains the same.

So the strategy in MIR phase analysis is as follows:

- Prepare two or more isomorphous heavy atom derivatives of the protein, crystallize them, and collect a complete set of diffraction intensities from the native protein and each of the derivatives.
- Measure the intensities of all the reflections in each diffraction pattern.
- Locate the positions of the heavy atoms (usually not a difficult job using what are known as Patterson vector maps) and calculate the amplitude and phase of scattering by the heavy atom for each reflection in the diffraction pattern.
- Compute a phase diagram like Figure 6.9b for each reflection, and find the phase of that reflection.
- Use these phases with the observed amplitudes to construct a three-dimensional picture of the protein molecule.

For the first three-dimensional picture of any protein, myoglobin at 6 Å resolution, Kendrew drew phase circles like Figure 6.9b by hand, using ruler, compass and different colored pencils. Today all of this, of course, is done with a digital computer. It may sound difficult and tedious, and it is.

The Hemoglobin Saga

Perutz and coworkers published a series of nine massive papers on the structure of hemoglobin in the *Proceedings of the Royal Society, Series A*, between 1947 and 1962, with 228 pages in total. They began with crude first attempts to find the size and shape of the hemoglobin molecule, and ended fifteen years later with a complete molecular structure at a resolution of 5.5 Å (and a Nobel Prize). Because myoglobin is only one-quarter the size of hemoglobin, its structure analysis by Kendrew and coworkers went faster. A low resolution, 6 Å myoglobin structure appeared in 1958 (B, C), and by the time the hemoglobin structure had been solved at 5.5 Å, the myoglobin structure had progressed to the atomic level, 2 Å (next chapter). It took longer for the larger hemoglobin molecule to be brought to that level of detail, but even comparing 5.5 Å hemoglobin with 2 Å myoglobin, a striking reality was obvious: the two molecules obviously were closely related, suggesting a common evolutionary ancestor! That is the subject of the next chapter.

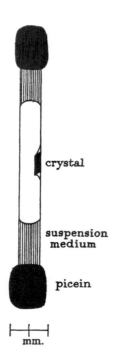

FIGURE 6.10 Sealing a wet protein crystal in a small glass capillary, to keep it from drying out during x-ray data collection. The technique keeps the crystal in equilibrium with its original crystallization liquid at the two ends of the capillary, without actually immersing it in liquid. If it were actually suspended in liquid, one could not keep the crystal stationary during x-ray photography. From (Hemoglobin I).

The nine *Proc. Roy. Soc.* papers on hemoglobin were important historically in establishing how a protein structure might be solved. They form a continuous narrative of 15 years of research. There is no point today in reprinting all 228 pages; there are too many blind alleys, and wrong ideas, and failed methods. But it is worth reading the first page of each of the papers, with its abstract summarizing what the authors had accomplished. These are identified by Roman numerals I through IX in the Key Papers list. If you want to see the full papers, they all are available at the previously mentioned JSTOR site on the internet.

Paper I (1947) laid the groundwork for hemoglobin analysis. Perutz established that wet crystals as first prepared from solution contained 52% liquid of crystallization or "mother liquor". This liquid surrounded the molecules, *but did not interpenetrate them*. Hemoglobin molecules possessed a structural integrity of their own, and remained unchanged when crystals were dried out or rehydrated. This was a critically important idea, for if protein molecules were sponge-like, as Perutz expressed it, and swelled and shrank with surrounding water content, then the x-ray crystallographic results might well have no bearing at all on the structure of the protein *in vivo*. Chargaff's attitude toward fiber diffraction in his essay "Gullible's Troubles" might have been equally relevant to crystalline proteins: "I set little trust in the biological relevance of X-ray photographs of stretched and pickled high-polymer preparations."

To keep crystals from drying out, and their diffraction pattern from deteriorating, they were sealed in glass capillaries as in Figure 6.10. (For an account of the trauma involved in learning how to mount crystals, see "A Little Ancient History" at the end of Chapter 7.) The horse hemoglobin molecules, with a molecular weight of 64,650, were arranged in flat layers or sheets as in Figure 6.11. Changes in water content caused the sheets to move closer together or farther apart, and to slide over one another. But the structure of each sheet was conserved.

Paper II from 1949 opens with a heartfelt remark. The diffraction pattern extended out to 2.8 Å resolution, and then faded away quickly. Even so, the complete data set contained 7840 independent x-ray reflections, all of which had to be measured from x-ray photographs. As Perutz expressed it:

The photographing, indexing, measuring, correcting and correlating of some 7000 reflexions was a task whose length and tediousness it will be better not to describe.

This sentence almost certainly was the inspiration for Donohue's remark in his globlglobin paper:

Complete intensity data were rapidly collected over a period of seven years with

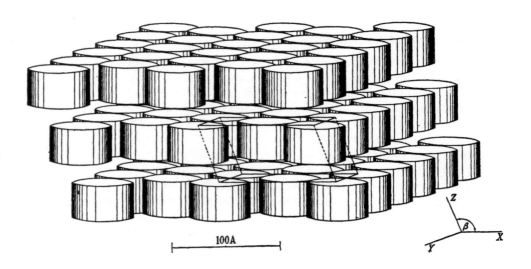

FIGURE 6.11 Approximate shape and arrangement of hemoglobin molecules in the crystal. The 64,650 molecular weight molecules pack into sheets, separated from neighboring sheets by an amount of water that can be decreased or increased as crystals are dried or rehydrated. But it is extremely significant that the hemoglobin molecules themselves are not changed by this process. From (Hemoglobin I).

an improved Weissenberg camera, the alterations in which had best not be described.

Paper II was devoted to attempts to learn something about hemoglobin structure by using what are called Patterson vector maps. The electron density expression :

$$\rho(x,y,z) = \sum_h \sum_k \sum_l \left(F_{hkl}e^{-i\phi}\right)e^{-2\pi i(hx+ky+lz)} \qquad (4)$$

cannot be calculated if one does not know the phase angles, ϕ_{hkl}, of all of the reflections. But A. L. Patterson (the man responsible for the "Amino acids in chains....." limerick in Chapter 3) discovered that if one simply deleted the phase terms and squared each amplitude:

$$P(x,y,z) = \sum_h \sum_k \sum_l \left(F_{hkl}\right)^2 e^{-2\pi i(hx+ky+lz)} \qquad (5)$$

the result would be a map displaying all the *vectors between all pairs of atoms* in the structure. For large molecules this is not as helpful as it might seem. The small organic molecule benzene has 6 carbon atoms, and if hydrogens are neglected, its Patterson map would contain $6 \times 5 = 30$ peaks (ignoring unit cell symmetry for simplicity). But a myoglobin molecule with ca. 1200 atoms would exhibit $1200 \times 1199 = 1,438,800$ interatomic vectors! Could any sense be made of such a map? Perutz hoped so. Even though individual vector peaks were not resolved, if some feature in the molecule repeated in a regular manner, then this would cause a ridge or peak of density in the map at that distance from the origin.

Paper II was an attempt to use the Patterson function to extract information about hemoglobin molecules, and is pretty much a lost cause. It was difficult to know what to make of the Patterson maps:

> **Arguing purely from considerations of packing there should be twenty such chains in the hemoglobin molecule. Porter and Sanger (1948), on the other hand, have shown the horse hemoglobin molecule to contain only six terminal α-amino groups. Hence the twenty chains cannot be independent, but must be combined into six bigger chains folded backwards and forwards through the molecule in long zigzags.**

Of course, we now know that the hemoglobin molecule contains only *four* terminal α-amino groups, one for each alpha or beta chain, not six. Figure 6.12 shows their final conclusions in Paper II: that one hemoglobin molecule has four layers of chains, each one of which is bent back and forth so as to fit in a cylindrical molecule. They caution that, "The details in the two pictures [Figure 6.12] are, of course, purely imaginative, but the general lay-out which they indicate follows from the vector structure." The packing between bends is 10 Å, and some feature, of nature unknown, appears to produce a repeat at intervals at 5 Å along the chains. One might think that the 5 Å spacing anticipated Pauling's alpha helix. Indeed, this is what they thought at first after the Pauling/Corey/Branson paper came out in 1951. But if these chains were alpha helices, their directions within the mol-

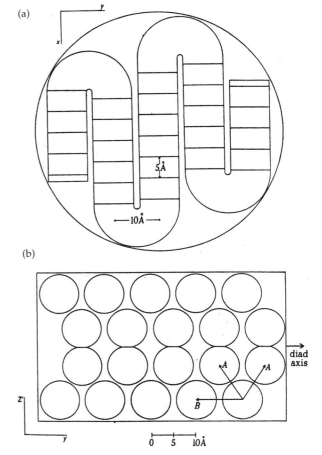

FIGURE 6.12 The incorrect picture of the hemoglobin molecule derived from interpretations of Patterson maps, which display all interatomic vectors. (a) Top view, (b) Side view in cross section. The "pillbox" seen in Figure 6.11 was thought to have four layers of polypeptide chain (b), each layer being bent back and forth as shown in (a). Tightly packed loops of one chain were 10 Å apart between centers, and some unknown structural feature repeated at intervals of 5 Å along the chain. From (Hemoglobin II).

FIGURE 6.13   A public lecture at the Royal Institution in London, as drawn by caricaturist James Gillray in 1802, a century and a half before Sir Lawrence Bragg became its Director. The founder of the Royal Institution, Count Rumford, stands at the right observing a public lecture in "Pneumaticks." Rumford, born Benjamin Thompson in pre-Revolutionary New England, is one of the most fascinating scoundrels (and geniuses) in the history of science. Space precludes recounting his sordid tale here, but have a look at references 10–12. In addition to discovering the First Law of Thermodynamics, he was nearly hanged as a spy by two governments (U. S. and Britain), fled a third (Bavaria) because of financial scandals, and along the way found time to invent the drip coffee pot! He married a wealthy New Hampshire widow at age 19, but deserted her and their daughter when he emigrated with the defeated British army at the end of the American Revolution. At age 52 after many vicissitudes he married the wealthy widow of the famous French chemist Antoine Lavoisier. In this cartoon Professor Thomas Garnett administers laughing gas to the Laboratory Manager, Sir John Hippisley, with odiferous consequences, while a sinister young Humphrey Davy stands behind holding a bellows. The lecture hall is still used today, and is little changed.

ecule had no connection at all with those established later in the true structure. Patterson methods had failed.

Papers III–VII chronicle the long and arduous road that Perutz and coworkers traveled, a road that ultimately turned into a highway when Perutz realized that MIR phase analysis would work with proteins. With financial support from the Medical Research Council, Perutz' protein structure project within the Cavendish Laboratory became known as the MRC Unit for the Study of the Molecular Structure of Biological Systems. William Astbury of Leeds is given credit for being one of the first, if not the first, people to use the expression "molecular biology" (4), but in

(a)

(b)

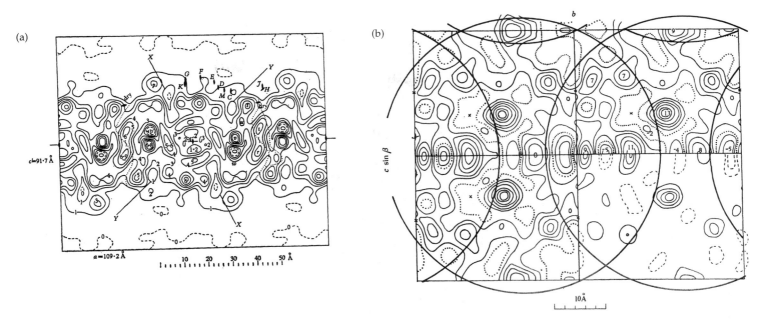

FIGURE 6.14   Projections of the hemoglobin molecule in two directions, using MIR phasing. (a) Centrosymmetric (h0l) projection down the b axis, at 6.5 Å resolution. (From Hemoglobin VI.) (b) Noncentrosymmetric (0kl) projection down the a axis, at 6 Å resolution. From (Hemoglobin VII). Neither projected map is interpretable in terms of a molecular structure.

1954 the term had not yet become fully accepted. Indeed, when John Kendrew started the *Journal of Molecular Biology* in 1957, there was debate as to whether this was a proper scientific term. But that same year, Perutz' laboratory was rechristened the MRC Laboratory of Molecular Biology, or simply the "MRC Lab".

Sir Lawrence Bragg retired as Director of the Cavendish Laboratory in 1953 at the age of 63. But he immediately succeeded his father, W. H. Bragg, as the head of the Royal Institution in London, and remained in active collaboration with Perutz' protein structure project. The Royal Institution or the "RI" was an interesting organization. It was created in 1801 by the expatriot American Benjamin Thompson, famous as the discoverer of the First Law of Thermodynamics. Thompson had achieved this distinction by demonstrating the equivalence of heat and work, and showing that heat was molecular motion rather than a substance, while boring cannons in Munich for the Elector of Bavaria. When knighted for his achievements there, he took the title of Count Rumford after his home town of Rumford (now Concord), New Hampshire. When his political situation in Munich became too controversial, he moved to London and founded the Royal Institution as a means of bringing modern technology to the working classes. The RI became an important research center, home first to Humphrey Davy and then to Michael Faraday and others. Figure 6.13 depicts a public lecture in "Pneumaticks" at the RI, with Count Rumford standing and watching benignly at the right. The room shown is still the main lecture hall of the RI.

But back to MIR phase analysis. Because for mathematical simplicity Perutz chose to work with a projection (h0l) that possessed a center of symmetry, the phase diagrams of Figure 6.9 were simplified by projecting them onto the real axis, and the issue became one not of phase angles, but only signs, positive or negative. In Paper VI all of this work comes to a head, and the projected structure of the hemoglobin molecule is presented (Figure 6.14a). But what does it mean? What has been learned? Since molecules overlap in projection, how does one tell where one molecule ends and the next begins? The authors drew in what they thought the boundaries of one molecule might be, as faint dotted lines.

Four years later in paper VII, graduate student David Blow at the MRC Laboratory completed the structure of the 0kl projection of hemoglobin crystals, which was not centrosymmetric, and for which the elaborate phase diagrams of Figure 6.9 were necessary. But again, the new projection of the structure (Figure 6.14b) was frustratingly uninformative. All this work demonstrated that at last we knew *how* to

(a)

(b)

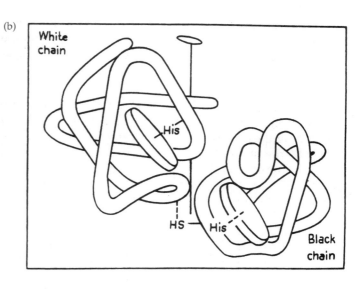

FIGURE 6.15   (a) Three-dimensional structure of the hemoglobin molecule at 5.5 Å resolution, using MIR phasing. Two alpha chains (white) sit at the top, related by a vertical twofold axis. Two beta chains (black) sit below, related by that same twofold axis. Two of the four heme groups are visible here as flat disks, each located in its own pocket protected by α-helices. Labels O_2 show where oxygen molecules bind. Two equivalent hemes are to be found on the back side of the molecule. The vertical column to left of the heme in the white alpha subunit is α-helix E. The polypeptide chain here runs from bottom to top through helix E, bends to the right, and then continues on a diagonal downward pathway as another alpha helix, F. (b) Schematic diagram of the two subunits nearest the viewer in (a). The straight cylinders are alpha helices. (From paper A in the following chapter.)

solve a protein structure. But the solution had to be done in three dimensions; projections were too cluttered to mean anything.

Papers VIII and IX, the last in our series, did finally present the complete, three-dimensional structure of the hemoglobin molecule at a resolution of 5.5 Å. The low resolution model in Figure 6.15 was built by cutting out contours on individual map sections and stacking them in three dimensions. But these results appeared three years after the smaller myoglobin had been solved at 6 Å resolution, and in fact at the same time that myoglobin was being extended to a near-atomic 2 Å. Although MIR phasing was first worked out and used in projection with hemoglobin, the honor for the first low resolution three-dimensional crystal structure analysis of any protein belongs to Kendrew with myoglobin. As we shall see in the next chapter, interpretation of the low resolution hemoglobin structure depended heavily on the high resolution myoglobin sructure.

Dr. Frankendrew's Monster

The 1958 Nature paper B on myoglobin was a true landmark; the first application of MIR phasing to a three-dimensional protein structure. By then it was known that sperm whale myoglobin had 152 (actually 153) amino acids and a single heme group, whereas hemoglobin had four chains and four hemes. Kendrew and colleagues made heavy atom derivatives by diffusing compounds containing silver, mercury and gold atoms into myoglobin crystals, and testing via the x-ray patterns to see which heavy atoms had bound uniformly to one or two positions on the molecule. They first tried the methodology in projections as Perutz had done, and like him, found their results uninterpretable. The leap to three dimensions demanded a vastly greater number of reflections, whose intensities had to be measured independently for each of the heavy atom derivatives. So a decision was made to look first at low resolution:

> In the present stage of the analysis the most urgent objective was an electron-density map detailed enough to show the general layout of the molecule—in other words, its tertiary structure. If the α-helix, or something like it, forms the basis of the structure, we need only work to a resolution sufficient to show up a helical chain as a rod of high electron density. For this purpose we require only reflections with spacings greater than about 6 Å; in all there are some 400 of these.... The [three-dimensional Fourier] synthesis was computed in 70 min. on the EDSAC Mark I electronic computer at Cambridge.....It is in the form of sixteen

sections perpendicular to *y* and spaced nearly 2 Å apart; these must be piled on top of one another to represent the electron density throughout the cell, containing two myoglobin molecules....

Eight sections through the electron density map are shown as Figure 18 of paper C; the other eight are produced by rotating these sections by 180°. Figure C.19 shows two views of the stacked Plexiglas sheets, each bearing the image of one of the 16 sections.

At 6 Å resolution Kendrew could follow a sausage-like winding of short segments of density through the myoglobin molecule. Two such segments formed a V-shaped pocket enclosing the heme group. Figure 2 of paper B shows photographs of a clay model, and Figure C.23 indicates how myoglobin molecules were packed within the crystal. Figure C.22 is an artist's sketch of front and back of the molecule. The front view exhibits one continuous chain of rods that could well be α-helices. The back view has a strange H-shaped crossbridge between rods, which eventually turned out to be an artifact. A slightly different drawing of low-resolution myoglobin, by a different artist, appears in Figure 6.16. Kendrew was strongly convinced that his rods did indeed represent α-helices, although he admitted in paper C, "There is no proof that this is the case..."

So the stage was set for a triumph. MIR phasing worked, and globular protein molecules had a structural integrity of their own that survived changes in local environment. In 1958–9 Kendrew worked as rapidly as possible toward a higher resolution map of myoglobin (at first planned at 2.5 Å, then changed to 2.0 Å), while Perutz pushed his much larger hemoglobin molecule toward a map comparable to that which Kendrew had just achieved. The journey toward Stockholm had begun.

In view of the subject of this chapter, Multiple Isomorphous Replacement phase analysis, it is appropriate to end with a quotation from reference 5 that is meaningful to protein crystallographers but may risk offending fans of William Butler Yeats' "The Second Coming":

FIGURE 6.16   John Kendrew and his "Frankendrew Monster" at 6 Å resolution. This drawing originated in Howard Schachman's laboratory at U. C. Berkeley shortly after the 6Å structure was published in 1958, but before the high-resolution 2 Å structure appeared in 1960. It was sent to me by Prof. Schachman in 1965. Regrettably, the name of the artist has been forgotten, even at Berkeley. If anyone recognizes this as his or her work, I would like to hear from you; please contact me at: red@mbi.ucla.edu.

THE THIRD DERIVATIVE

Turning and tumbling in the bubbling stream,
    The reactant cannot bear the reagent;
Things fall apart; the crystal cannot hold;
    Disorder is loosed upon the world;
The blood-red block is loosed, and everywhere
    The symmetry of solid state is drowned;
The best lack all cohesion, while the worst
    Are full of passionate intensity.
Surely some revelation is at hand (?);

....
....

....

And what rough protein, its hour come round at last,
    Slouches toward *Nature* to be born?

The Road to the Globins

In 1959, Max Perutz and John Kendrew achieved what Max had set out to obtain a quarter of a century earlier: the first molecular structures of any globular proteins. The single-chain myoglobin was at high resolution, and although the four-chain hemoglobin was then only at low resolution the structural and evolutionary relatedness of the two oxygen-binding proteins were obvious. Their work not only initiated structural molecular biology; it also launched the field of molecular evolution.

Plate I   Max Perutz in 1991 during a visit to UCLA. Taken during a garden party given for him by David and Lucy Eisenberg. Photo by Daniel H. Anderson.

Plate II   Comparison of three globin models in 1960. Right: Wire model of 2 Å myoglobin. Lower left: Clay model of the earlier 6 Å model. Upper left: Hemoglobin at 5.5 Å, turned so one of its α chains (white) has the same orientation. Photo by the author.

Plate III   Interaction of heme and protein helix F in deoxymyoglobin. The 5-coordinate iron atom is pulled into the heme plane when O_2 binds at W on the opposite side. In hemoglobin this induces a small shift in the F helix which propagates through the molecule and ultimately helps to make binding of oxygen an all-or-nothing situation. Drawing by Irving Geis, from Dickerson and Geis "Proteins," 1969. Courtesy of Geis Archives, Howard Hughes Medical Institute.

Plate IV   Front view of the horse methemoglobin or oxyhemoglobin molecule. Only the alpha carbon positions of amino acids are shown. Those whose side chains are involved in contacts with other subunits are numbered in bold face, as C3 and FG4. Drawing by Irving Geis, from Dickerson and Geis "Hemoglobin," 1983. Courtesy of Geis Archives, Howard Hughes Medical Institute.

Plate V   Top view of deoxyhemoglobin. This and the oxyhemoglobin view in Plate VI demonstrate the operation of the "molecular machine." Drawing by Irving Geis, from Dickerson and Geis "Hemoglobin," 1983. Courtesy of Geis Archives, Howard Hughes Medical Institute.

Plate VI   Top view of methemoglobin or oxyhemoglobin. Subunits α1/β1 at right and α2/β2 at left behave as rigid units. Oxygenation of one or two hemes leads to a shift of subunits which pushes iron atoms of unoxygenated hemes into their heme planes and encourages further O_2 binding. See reference 7 of Chapter 7 for details. Drawing by Irving Geis, from Dickerson and Geis "Hemoglobin," 1983. Courtesy of Geis Archives, Howard Hughes Medical Institute.

Plate VII   John Kendrew explaining his 2 Å myoglobin structure at the 1960 meeting of the International Union of Crystallography in Cambridge. The model was built in a forest of vertical wires in which electron density was represented by small colored clips. Photo by author.

Plate VIII   Kendrew explaining myoglobin sequence data at the 1960 IUCr meeting. Contrary to the custom today, the three-dimensional structures of myoglobin and hemoglobin were established *before* their amino acid sequences. Vertical support wires of the model are visible in the foreground. Photo by author.

Plate I

Plate II

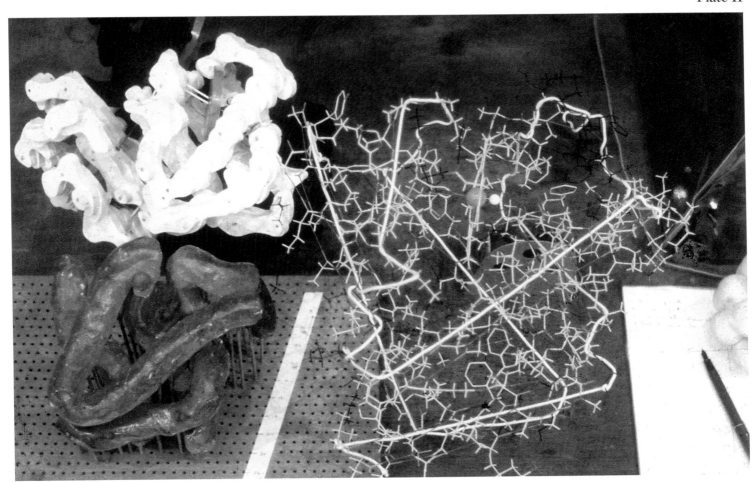

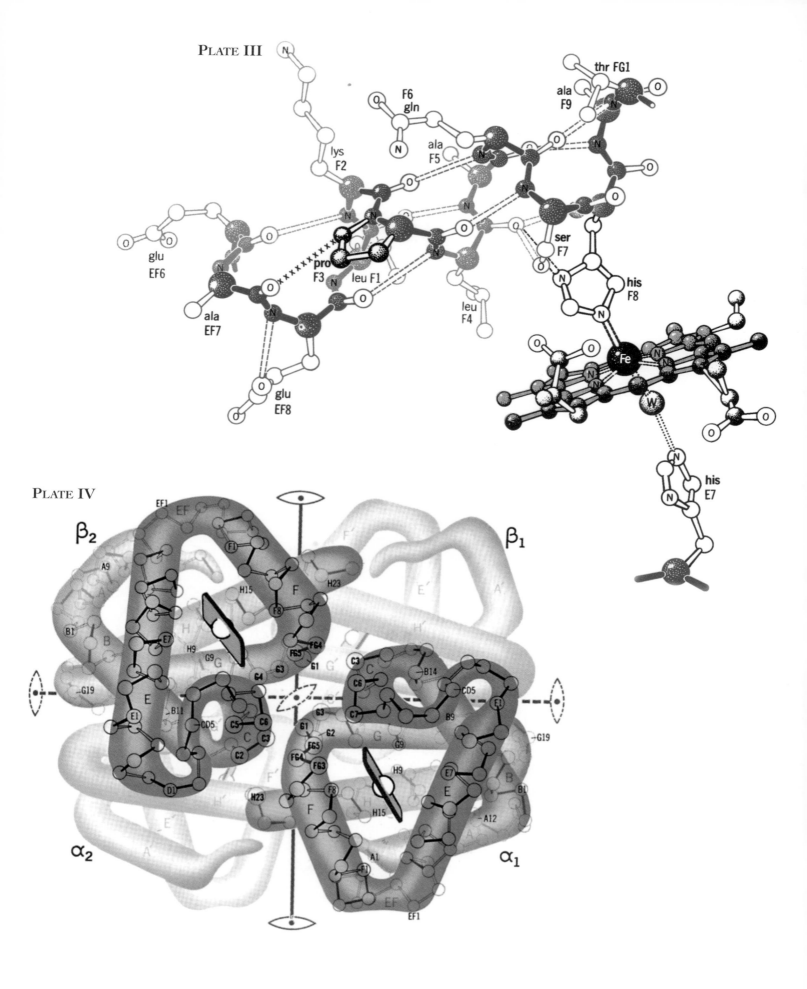

PLATE III

PLATE IV

PLATE V

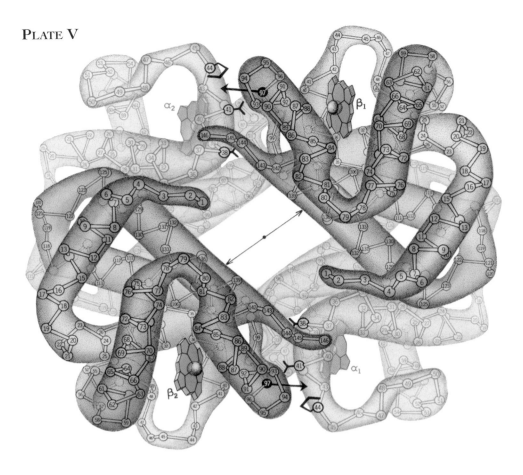

PLATE VI

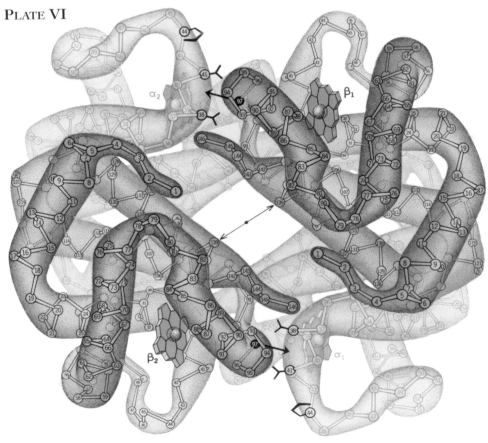

PLATE VII

PLATE VIII

PLATE IX

PLATE X

PLATE XI

PLATE XII

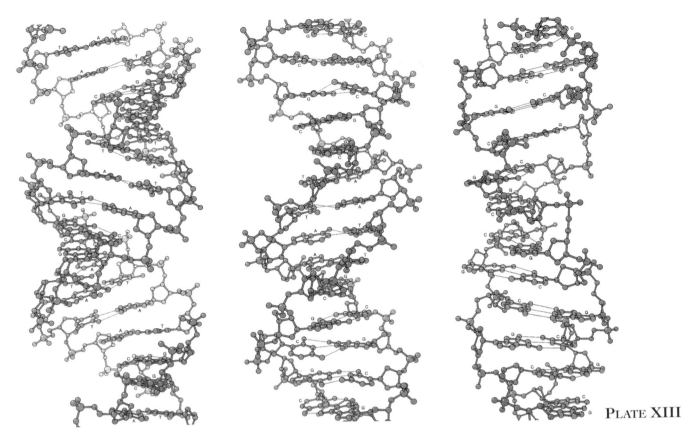

PLATE XIII

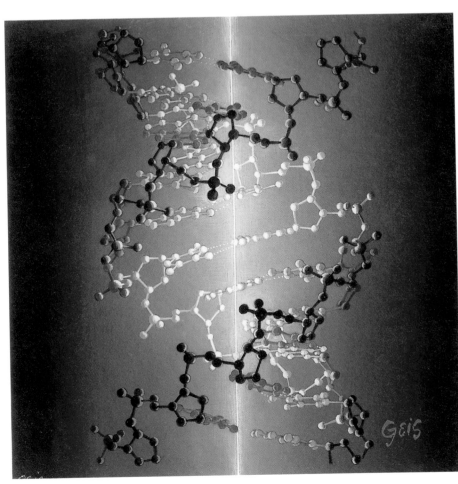

PLATE XIV

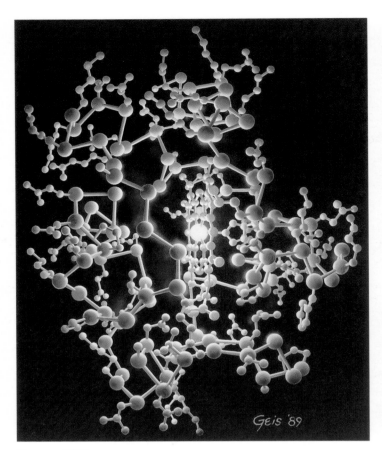
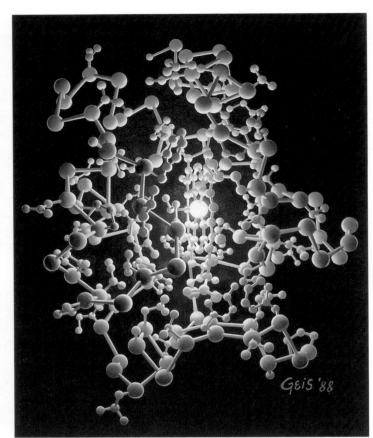

PLATE XV

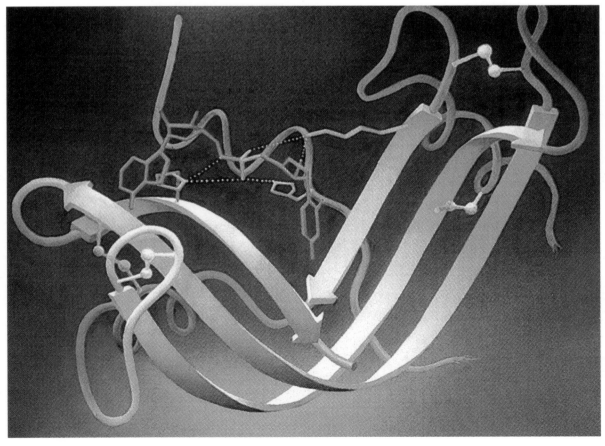

PLATE XVI

After the Flood

There was a pause after the 1960 Perutz/Kendrew thunderbolt, while other research groups set out to do the same thing with different proteins. Perutz became concerned with the slow pace of progress, and in early 1966 organized a protein structure workshop in Austria that was so successful that it was repeated three more times at two-year intervals. The man who set the standards for depicting macromolecular structures was Irving Geis, the subject of Appendix 3. Irv first drew the detailed picture of high-resolution myoglobin in our frontispiece in 1961, and then never stopped thereafter for thirty-six more years.

Plate IX The site of Max Perutz's 1966 first protein structure workshop at Hirschegg in the Kleines Walsertal, Austria. The meeting was carefully scheduled for March, when skiing was good. Looking southwest along the Gemstalbach, with the Elferkopf (2387 m) and Zwölferkopf (2208 m) at left and Widderstein (2536 m) at right. Photo by author.

Plate X View of Waldemar-Petersen Haus or Darmstadterhaus, the conference center owned by the Technical University of Darmstadt, where the 1966, 1968 and 1970 workshops were held. Typical morning activity of delegates shown. Photo by author.

Plate XI Inside Waldemar-Petersen Haus. Left to right: David Phillips (later Sir David and finally Lord Phillips of Ellesmere), Tony North, Max Perutz, Ken Holmes and Michael Rossmann. Photo by author.

Plate XII The conference room at Waldemar-Petersen Haus. Front row: (unknown), Herman Watson, Harold Wyckoff, Aaron Klug. Johan Jansonius and Gobinath Kartha are two rows behind Watson; Brian Matthews is half-hidden two rows behind Klug. Photo by author.

Plate XIII Drawings by Irving Geis of the three forms of DNA double helix: A, B and Z. A less colorful version of these drawings first appeared in the author's article on DNA structure in the December 1983 issue of *Scientific American*. Courtesy of Geis Archives, Howard Hughes Medical Institute.

Plate XIV Geis' "hot wire" painting of A-DNA, in which a glowing helix axis is the only source of illumination. Irv never found the B or Z forms especially inspiring artistically, but liked the angled bases and hollow core of the A form. From reference A of Appendix 3. Courtesy of Geis Archives, Howard Hughes Medical Institute.

Plate XV Geis' "hot heme" paintings of horse heart cytochrome *c*, with the protein backbone in blue. By analogy with Plate XIV, the sole source of illumination in these two paintings is an incandescent iron atom in the center of the heme group. Left: polar side chains in green. Right: nonpolar side chains in orange. From references A (left) and B (right) of Appendix 3. Courtesy of Geis Archives, Howard Hughes Medical Institute.

Plate XVI Geis' painting of the twisted β sheet structure in the enzyme bovine ribonuclease S, as determined by Wyckoff, Richards and coworkers in 1967 at Yale (*J. Biol. Chem.* 242, 3749 and 3984). That same year Harker and coworkers at Brooklyn Polytechnic Institute solved the structure of a slightly different form (*Nature* 213, 862). Ribonuclease was the second enzyme structure to be solved, preceded only by hen egg-white lysozyme, by Phillips and coworkers in 1965 at the Royal Institution (*Nature* 206, 575 and 761). From reference B of Appendix 3. Courtesy of Geis Archives, Howard Hughes Medical Institute.

References

1. Max Perutz. 1985. *New Yorker,* 12 August 1985, 35–54. "That Was the War: Enemy Alien."

2. Dorothy Crowfoot. 1941. *Chem. Reviews* 28, 215–228. "A Review of Some Recent X-Ray Work on Protein Crystals."

3. D. Crowfoot Hodgkin, J. Kamper, M. Mackay, J. Pickworth, K. N. Trueblood and J. G. White. 1956. *Nature* 178, 64–66. "Structure of Vitamin B_{12}."

4. W. T. Astbury. 1939. *Ann. Rev. Biochem.* 8, 113–132. "X-ray Studies of the Structure of Compounds of Biological Interest."

5. R. E. Dickerson, M. L. Kopka, C. L. Borders, Jr., Joan Varnum, Jr., J. E. Weinzerl and E. Margoliash. 1967. *J. Mol. Biol.* 29, 77–95. "A Centrosymmetric Projection at 4 Å of Horse Heart Oxidized Cytochrome *c*."

6. S. C. Nyburg. 1961. *X-Ray Analysis of Organic Structures,* Academic Press, New York.

7. R. E. Dickerson. 1964. In *The Proteins* (Hans Neurath, ed.), Academic Press, New York, 2nd edition, Vol. 2, pp. 603–778.

8. Jurg Waser. 1968. *J. Chem. Edu.* 45, 446–451. "Pictorial Representation of the Fourier Method of X-Ray Crystallography."

9. D. S. Eisenberg. 1970. In *Methods of Enzymology* (Paul D. Boyer, ed.), Academic Press, New York, 3rd edition, Vol. I, pp. 1–89. "X-Ray Crystallography and Enzyme Structure."

10. Sanborn C. Brown. 1954. *American Scientist* 42, 113-127. "Count Rumford: A Bicentennial Review."

11. S. C. Brown. 1962. *Count Rumford: Physicist Extraordinary.* Anchor Doubleday, New York.

12. S. C. Brown. 1979. *Benjamin Thompson, Count Rumford.* MIT Press, Cambridge MA.

Study Questions

1. What feature of protein structure, proposed first by Francis Crick, was Jerry Donohue mocking when he described the "paper chain paper chain" structure on page 4 of his parody report (A)?

2. What makes us think that the Donohue parody was written *after* the Watson/Crick B-DNA structure came out, but *before* Perutz had shown that Multiple Isomorphous Replacement phase analysis would work with proteins? What is this critical time frame?

3. When adding sine waves to reconstruct the electron density map of a protein structure using information derived from individual reflections in its diffraction pattern, how do you know in which direction a particular wave is going? How do you know its wavelength? How do you know how strong the wave is—that is, its amplitude?

4. In the process just described, what vital piece of information for each reflection is missing from the diffraction pattern?

5. In the MIR or Multiple Isomorphous Replacement method of phase analysis, why are at least two isomorphous derivatives needed? Why is a third derivative such an immense help? (For the curious, a SIR or Single Isomorphous Replacement approach does exist, which needs only one derivative. But it involves severe approximations and is less powerful.)

6. The two two-dimensional projections of the hemoglobin structure which Perutz calculated using MIR phasing (Figure 6.14) were correct, but were biologically useless. Can you think of two reasons why?

7. In view of the inability of Perutz to see any significant information in his projections at 6 Å resolution, why did Kendrew think that his three-dimensional map at the same resolution would tell him anything (B, C)?

8. Why was Kendrew fortunate in choosing a protein built from alpha helices rather than one only with beta sheets, such as an antibody?

9. How did Kendrew recognize the heme group in his 6 Å structure?

THE X-RAY STRUCTURE ANALYSIS OF α-GLOBLGLOBIN[*]

by

J. Donohue and J. Briekopf

Institute of Astrobotany and Metabolic Diseases

(Communicated June, 1953, and in revised form December, 1967,

by R. E. Dickerson)

---

[*] This paper originally appeared in the Collected Papers of the Medical Research Council Laboratory of Molecular Biology, Cambridge, 1953. The long delay in journal publication is a result of its rejection by the Grain and Feed Journal, Acta Retracta, and the Journal of Unverifiable Knowledge (J. UN. K.). The paper in its present form was referred by Mr. Harold Austin, but unfortunately the referee's comments could neither be followed nor reproduced here. We are grateful to the National Football League for making the referee's time available.

1

We wish to report the first complete structure determination of any protein. As is well known, proteins are very interesting substances, and because of their wide occurence they are sometimes rather important. It will be remembered that they are giant molecules containing hundreds, sometimes even thousands, of atoms. Work on the structure of globlglobin has been underway in this laboratory for the past twenty-three years, eight months and two weeks, except for a short lacuna four years ago caused by the epidemic of fowlpest. The elucidation of the detailed structure of globlglobin may well be of interest since it is now known that it is an important constituent of not only the tears of the Sahara crocodile from which it was first isolated, but is also, for example, found in the lateral extensor muscle of the tail of the Manx cat, the eyelid membrane of the snowy owl, and in the lining of the appendix of the great marsh tit. It is from this last source that most of our material was isolated. We wish to thank the Southwestern East Anglian Bird Watchers Association for their invaluable aid in maintaining a constant supply of tits. It has also been recently discovered that globlglobin is directly responsible for the reddish coloration of certain eyeballs; we are indebted to our colleague Dr. F. Bonamie, who is an expert in these matters, for this observation.

The only previous work on globlglobin is that of Sorey and co-workers, whose brilliant and painstaking researches were able to establish unequivocally beyond any doubt that it was not possible with the data at hand to come to any definite conclusions concerning the size and shape of the molecules, or their position in the unit cell.

2

We have discovered that avian globlglobin crystallizes in a variety of forms which differ only in cell dimensions, space group, and number of molecules in the asymmetric unit. This as yet unexplained phenomenon was of no use to us. The form chosen as most suitable for exhaustive examination was triclinic, with a = 735 Å, b = 75 Å, c = 74 Å, α = 103°, β = 91°, and γ = 1°37'. The space group is P1, and there are three molecules in the cell, if we asume that the molecular weight is best represented by an average of the values 169,325 and 21,768 as given by the ultracentrifuge and osmotic pressure, respectively. Moreover, the published data concerning the density are inadequate.

The crystals were prepared with the aid of Dr. O. N. Bearish, who was of great assistance with the washing up. The first attemps yielded small specimens only, but eventually crystals were obtained which could be measured with ease by use of the Standard British Yard. Complete intensity data were rapidly collected over a period of seven years with an improved Weissenberg camera, the alterations in which had best not be described. A new type of x-ray tube with a spinning cathode enables the exposure times to be reduced to the order of 10^{-3} microseconds. We have found that these short times entirely eliminate errors due to fluctuations in the power supply. Systematic absences observed are: (hk0) absent if k = 2n+7, and (h0l) absent if h = 3n+17. It is therefore obvious that each molecule consists of three identical halves.

All recent work points to the fact that the w-helix is the basic structure for proteins. We have accordingly adopted it in the present study, since globlglobin shows the familiar strong reflection at 0.063 Å, as predicted for the w-helix by all earlier workers. One difficulty is, however, that globl-globin is known to contain no end groups, whereas the light scattering data clearly show that the molecule is 120 Å long. We were therefore led to formulate what we call the paper chain model, which consists of interpene-trating rings of w-helices. This model is schematically illustrated in Figure 1.* The axis of the paper chains must coincide with the a axis of the unit cell in order to account (within 200%) for the measured pleochroism.

Figure 1

Excellent agreement between the observed and calculated intensities of the reflections (100) and (200) could be obtained with this model. The calculation of the intensities of these meridional reflections is given by the trivial standard expression. In order to test additional reflections, we have, in collaboration with Mr. I. M. A. Crock, evaluated the form factor for a paper chain. For reflections not on the meridian, and if the usual approxi-mations are made, it is found that:

$$F = \sum_i \int_o^\pi \sum_j f_i \, (o) \, \Gamma \, (x_i) \int_\pi^{\frac{\pi}{2}} B_m \, (Y_i) \int_{-e}^{+e} J_n \, (z_i) \, d\xi \, dn \, d\zeta \qquad (1)$$

using an obvious notation. We may therefore expect reflections at values

---

* The presence of the paper chain w-helix has recently been demonstrated in the DNA of HeLa cell mitochondria (Hudson and Vinograd, Nature, 216, 647 (1967). This finding illustrates once more the truism that the fundamental elements of living matter are the proteins, and that they must serve as patterns for nucleic acid structure.

4

of $5 \sin \xi / \pi \lambda$ given by the formula $p/T + q/E$, where T and E are, respectively, the sets of Tschebycheff polynomials and Eulerian numbers, and p and q are relatively prime perfect squares. This useful expression predicts that large values of F will be found approximately 18° from a line bisecting the angle between the meridian and c^*, as is observed on the photographs. This point is more easily grasped by referring to Figure 1, where it is easily seen that most of the atoms lie on surfaces which, in reciprocal space, lead inexorably to the above condition.[†]

A comparison between observed and calculated intensities for a few selected planes is made in Table 1. It is seen that the agreement is satisfactory, the average discrepancy, R, being 54%, significantly less than for a random structure. Preliminary considerations tend to indicate that the agreement could be improved by modifying the structure so that each paper chain bends back and joins itself to form, together with additional such units, what we call the paper chain paper chain. We intend to explore this

---

[†] Since the above was written, Mr. Crock has made a more rigorous proof of these relations. He has shown that large values of F will be expected to be uniformly distributed in the $a^* b^*$ plane, instead of 18° from the bisector. We do not consider this to be a significant discrepancy. In a private communication, Professor V. Cobbler informs us that he is attempting its explanation. We are awaiting his results with keen interest.

5

hypothesis more fully in a later publication.   In this regard, it has been emphasized to us by our colleague, Dr. S. Holmes, that it is not necessarily uncanonical to anticipate a certain degree of factitious teleology in these procedures, providing that the neostochastic approach is avoided.   We wish to thank him for his unwonted clarity.

At this stage it was discovered that additional data could be obtained from peroxygloblglobin prepared by extracting transversely cut sections of wild rabbits.   The entire laboratory staff gave most generously of their time to assist in the splitting of hares.   With this additional data, we proceeded to a Fourier analysis of the three principal zones.   Although none of these contain centers of symmetry, the statistical distribution of the first two orders in each tends to indicate that a pseudo-center is present.   Following J. D. Lenin, who foresaw all, we assume that the intensities fall naturally into groups which are logically called constellations.   Some confirmation of this is found in the observation that the reflection (390) is quite diffuse, corresponding to the nebula in Andromeda.   We have made a Fourier analysis of the various constellations of the (hk0) zone, but, instead of randomly permuting the phases, we have permuted the values for the intensities with a fixed set of signs.   This procedure allows many more interesting variations than mere sign permutation.   In order to achieve a truly random set of numbers, we chose to use for this purpose the number of lines of the successive sonnets of Shakespeare, taking them in the order in which they were written; we wish to thank the Bacon Society for advice on this point. The final electron density is shown in Figure 2.   It is seen that there is striking agreement with the paper chain hypothesis.   Additional projections

6

Figure 2

are being evaluated by Mr. E. Fouls, who is using a cunning method of calculation in which the sine waves are superimposed, by appropriate vibrations, on the surface of a trayful of jelly at the moment it sets. This method is very rapid and has the advantage that the unwanted results are edible.

Our proposed structure accounts for the observed rapidity with which owls blink. Simple entropy considerations of the twisting of the paper chains give 14 blinks per minute, in exact agreement with the values of 3 to 25 (average 14) reported in the current issue of Nature. It is also interesting that, although globlglobin has yet to be demonstrated as an important factor in hereditary mechanisms, the intertwining of the chains is very similar, and indeed, identical, with the published microphotographs of neurospora.

It has not escaped our notice that the paper chain structure provides a possible method of impulse transfer in the adrenal cortex.

We are now conducting a three-dimensional high resolution analysis of avian α-globlglobin, and the completed structure analysis will undoubtedly be published in detail in the very near future.

We wish to thank kindly all of our colleagues for their interminable discussion of various points related to the punctuation of the manuscript. One of us was aided by a grant from the Ministry of Fish.

7

Table I

hkl	$I_{obs.}$	$I_{calc.}$
1, 0, 0	2^a	350
2, 0, 0	110	17^b
3, 0, 0	(c)	415
0, 0, 1	12^d	1738
0, 0, 2	36	37^e
36, 7, 23	3495.73	3495.73

(a)   This reflection is probably subject to extinction.

(b)   The discrepancy here is not serious, and may be explained by an abnormality in the background.

(c)   Not observed because of the experimental arrangment.

(d)   Part of this reflection is cut off by the beam stop. The actual intensity is therefore probably much greater.

(e)   Agreement could probably be improved in this case by taking into account the scattering of the hydrogen.

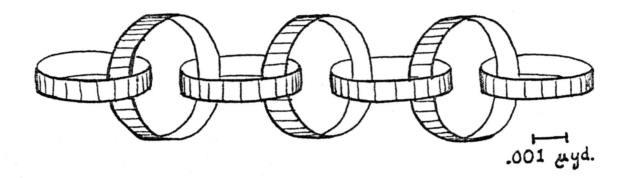

Figure 1.   The w—helix paper chain of α-globlglobin and HeLa mitochondrial DNA.

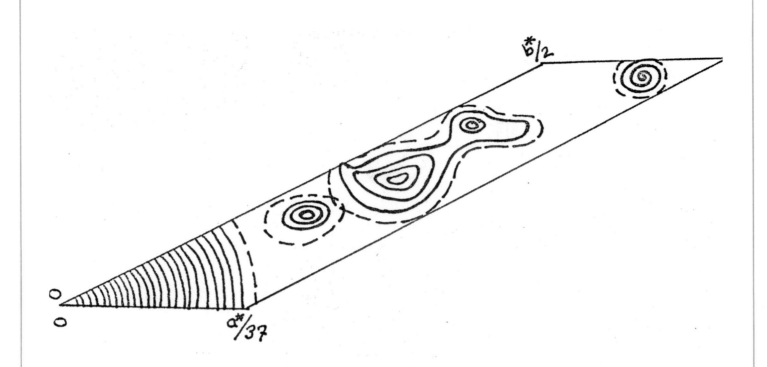

Figure 2.   hk0 projection of triclinic α–globlglobin using the data of
Table 1. According to Dr. A. L. Patterson, this should properly
be called a Fourier projection. M. Joseph Fourier insists with
equal firmness that it be referred to as a Patterson projection.

An X-ray study of horse methaemoglobin. I

By Joy Boyes-Watson, Edna Davidson and M. F. Perutz

Cavendish Laboratory and Molteno Institute, University of Cambridge

(*Communicated by Sir Lawrence Bragg, F.R.S.—Received* 3 *February* 1947)

[Plates 5 and 6]

The paper describes a detailed study of horse methaemoglobin by single crystal X-ray diffraction methods. The results give information on the arrangement of the molecules in the crystal, their shape and dimensions, and certain features of their internal structure.

Horse methaemoglobin crystallizes in the monoclinic space group $C2$ with two molecules of weight 66,700 per unit cell. In addition, the wet crystals contain liquid of crystallization which fills 52·4% of the unit cell volume. Deliberate variations in the amount and composition of the liquid of crystallization, and the study of the effects of such variations on the X-ray diffraction pattern, form the basis of the entire analysis.

The composition of the liquid of crystallization can be varied by allowing heavy ions to diffuse into the crystals. This increases the scattering contribution of the liquid relative to that of the protein molecules and renders it possible to distinguish the one from the other. The method is analogous to that of isomorphous replacement commonly used in X-ray analysis. It yielded valuable information on the shape and character of the haemoglobin molecules and also led to the determination of the phase angles of certain reflexions.

The amount of liquid of crystallization was varied by swelling and shrinkage of the crystals. This involves stepwise, reversible transitions between different well-defined lattices, each being stable in a particular environment of the crystal. The lattice changes were utilized in two different ways: the first involved comparison of Patterson projections at different stages of swelling and shrinkage, and the second an attempt to trace the molecular scattering curve as a function of the diffraction angle.

The results of the analysis can be summarized as follows. The methaemoglobin molecules resemble cylinders of an average height of 34 A and a diameter of 57A. In the crystal these cylinders form close-packed layers which alternate with layers of liquid of crystallization. The layers of haemoglobin molecules themselves do not swell or shrink, either in thickness or in area, except on complete drying, and lattice changes merely involve a shearing of the haemoglobin layers relative to each other, combined with changes in the thickness of the liquid layer. Thus the molecules do not seem to be penetrated by the liquid of crystallization, and their structure is unaffected by swelling and shrinkage of the crystal.

Space-group symmetry requires that each molecule consists of two chemically and structurally identical halves. Evidence concerning the internal structure of the molecules comes both from two-dimensional Patterson projections and one-dimensional Fourier projections. The former indicate that interatomic vectors of 9 to 11 A occur frequently in many directions, and the latter show four prominent concentrations of scattering matter just under 9 A apart along a line normal to the layers of haemoglobin molecules. No structural interpretation of these features is as yet attempted.

The liquid of crystallization consists of two distinct components: water 'bound' to the protein and not available as solvent to diffusing ions, and 'free' water in dynamic equilibrium with the suspension medium. An estimate of the 'frictional ratio' based on the molecular shape and hydration found in this analysis is in good agreement with the frictional ratio calculated from the sedimentation constant.

1. Introduction

(a) *Background and scope of research*

The molecular structure of the crystalline proteins is one of the major unsolved problems in biology to-day. During the last 25 years the recognition of their ubiquity and paramount importance in plant and animal metabolism has set in

An X-ray study of horse methaemoglobin. II

By M. F. PERUTZ

Cavendish Laboratory and Molteno Institute, University of Cambridge

(*Communicated by Sir Lawrence Bragg, F.R.S.—Received* 8 *June* 1948—
Read 16 *December* 1948)

A complete three-dimensional Patterson synthesis of haemoglobin has been calculated, giving the distribution of vector density in thirty-one sections through the unit cell. The sections show certain concentrations of vector density which can be interpreted in terms of polypeptide chain structure. The following are the conclusions tentatively arrived at on the evidence described in this paper.

The haemoglobin molecule resembles a cylinder of 57 Å diameter and 34 Å height, which consists of an assembly of polypeptide chains running parallel to the base of the cylinder. The chains show a short-range fold, with a prominent vector of 5 Å parallel to the chain direction. In addition to this the chains also contain a longer fold which may extend through the whole width of the molecule. This long fold may be due either to open chains folded backwards and forwards through the molecule or to closed loops of polypeptide chains. The average distance between neighbouring chains, or neighbouring portions of the same chain folded back on itself, is 10·5 Å. The chains are arranged in four layers which are about 9 Å apart and correspond to the four layers of scattering matter described in a previous paper. The haem groups lie with their flat sides approximately normal to the chain direction.

1. INTRODUCTION

A previous paper described how the arrangement of the haemoglobin molecules in the crystal, their shape and dimensions and certain features of their internal structure can be deduced from a detailed analysis of the X-ray diffraction pattern (Boyes-Watson, Davidson & Perutz 1947, henceforth referred to as I). Of necessity only a small part of the total diffraction pattern was used (i.e. the reflexions from the three principal crystal zones), while the intensities of the vast majority of reflexions had to be disregarded for lack of any method of interpretation.

At present the three-dimensional Patterson synthesis represents the only way of translating the information contained in the complete diffraction pattern into a form which is at least potentially capable of being interpreted in terms of molecular structure. Such a three-dimensional synthesis gives the vector density in a series of sections through the unit cell and is intrinsically much more likely to lend itself to reasoned interpretation than the two-dimensional projections described in I, because the overlapping of peaks is reduced and the resolution much increased. The actual chances of interpretation depend largely on the kind of molecular structure which the protein may be supposed to possess. For instance, if the globin molecule consisted of a complex interlocking system of coiled polypeptide chains where interatomic vectors occur with equal frequency in all possible directions, the Patterson synthesis would be unlikely to provide a clue to the structure. On the other hand, if the polypeptide chains were arranged in layers or parallel bundles, interatomic vectors within the layer plane or in the chain direction should occur particularly frequently and should give rise to a vector structure showing a corresponding system of layers or chains, which could then be interpreted without difficulty. All the more plausible hypotheses of globular protein structure put forward in recent years have

The structure of haemoglobin
III. Direct determination of the molecular transform

By M. F. Perutz*

Medical Research Council Unit for the Study of the Molecular Structure of Biological Systems, Cavendish Laboratory, University of Cambridge

(*Communicated by Sir Lawrence Bragg, F.R.S.—Received* 5 *March* 1954)

Horse methaemoglobin crystallizes with two molecules in a face-centred monoclinic unit cell (space group $C2$), in which rigid layers of molecules parallel to (001) alternate with layers of liquid. The crystals can be made to swell and shrink in a series of steps involving changes in $d(001)$ and in the angle β. It appears that only the distances between the molecular layers change, but not their internal structure. The lattice changes allow the modulus of the molecular Fourier transform to be sampled along lines of constant h and k. When $k=0$ the transform is real and the sampled values of $|F|$ describe a series of loops and nodes. Part I of this series dealt with the principles of deciphering these and established the absolute signs of the $00l$ reflexions. In part II the absolute signs of certain $20l$ reflexions were derived from the changes in intensity produced by the substitution of salt solution for water as the liquid of crystallization. In this paper the transform is measured for all values of h and l up to $\lambda/d = 0.24$, comprising nine layer lines in all. The absolute signs of layer lines with $h > 2$ are left in doubt, but many sign relations are established within each of them. It is difficult to assess exactly the number of sign relations found by the transform method, but it is estimated that the number of alternative sign combinations is reduced from 2^{96} to 2^{13}. The remaining uncertainties are cleared up by the isomorphous replacement method described in part IV.

1. Introduction

This paper is one of a series, representing an attack on a far more difficult problem of X-ray analysis than has been successfully tackled up to now. Part I (Bragg & Perutz 1952b) dealt with the principles of deciphering the nodes and loops of the molecular transform, an example being given of their application to the $00l$ reflexions. At the end of part I a further paper was promised giving an extension of the same principles to the $h0l$ reflexions. This is the promised extension, dealing with all reflexions of $\lambda/d < 0.24$. Ideally, the method should give complete sign relations along each line of constant h, but practical difficulties prevented this, leaving a number of points where the presence or absence of a node could not be established with certainty. In addition, it proved impossible to find the absolute signs of layer lines with $h > 2$.

The uncertainties left by the transform method have now been cleared up with the help of two further methods of sign determination. The first is based on a comparison of several isomorphous forms, one being pure haemoglobin and the other compounds of haemoglobin with heavy metals (part IV, Green, Ingram & Perutz 1954). The second method uses an apparently orthorhombic compound of haemoglobin with imidazole in which $F(h0l)$ equals either $F(h0l) + F(h0\bar{l})$ or $F(h0l) - F(h0\bar{l})$ of the normal monoclinic form; this provides a useful check for the sign distribution along some of the layer lines (part V, Howells & Perutz

* Elected F.R.S. on 18 March 1954.

The structure of haemoglobin
IV. Sign determination by the isomorphous replacement method

By D. W. Green, V. M. Ingram and M. F. Perutz*

Medical Research Council Unit for the Study of the Molecular Structure of Biological Systems, Cavendish Laboratory, University of Cambridge

(Communicated by Sir Lawrence Bragg, F.R.S.—Received 5 March 1954)

Native horse haemoglobin contains free sulphydryl groups and forms crystalline compounds with *para*-mercuribenzoate groups and with silver ions. Crystals in which two of the four available SH groups are so combined are exactly isomorphous with normal monoclinic methaemoglobin, but exhibit significant changes in the intensities of many reflexions. The changes in $F(h0l)$ were used to determine the x and z parameters of the pair of heavy atoms attached to each haemoglobin molecule; this was done both for the normal wet lattice and for one of the acid-expanded lattices. The positions of the heavy atoms proved to be slightly different in each case, giving rise to three sets of diffraction fringes, each set making measurable contributions in different areas of the reciprocal net.

In each case the isomorphous substitution allowed the signs of just over two-thirds of the reflexions to be found with certainty. Between them the three sets of diffraction fringes determined the signs over the entire area of the $h0l$ plane so far investigated. These signs were then superimposed on the waves of the transform described in previous papers of this series. All the sign relations established by the transform method were confirmed and the remaining uncertainties cleared up. Comparison of the transform with the three sets of isomorphous replacement results allowed the consistency of the signs to be rigorously checked; not a single inconsistent sign was found.

In the normal wet lattice the mercury and the silver compounds between them allowed the signs of 87 out of 94 reflexions to be found with certainty. This suggests that the isomorphous replacement method may offer a way of finding the phases in protein crystals even when practical difficulties preclude the use of the transform method.

1. Introduction

1·1. *Outline of analysis*

The previous papers of this series have described an attempt to interpret the diffraction by a haemoglobin molecule, and in part III (the preceding paper) the molecular transform was plotted along nine layer lines of constant index h. It has been shown that absolute signs could be established for the $00l$ reflexions and, with a lesser degree of certainty, for some of the $20l$ reflexions. The absolute signs of the remaining seven layer lines were left in doubt, but a large number of sign relations were established within each of them. In this paper it will be shown how the isomorphous replacement method determines the absolute signs of all significant loops in the transform.

Not counting the hydrogen, haemoglobin contains about 5000 atoms. It is not immediately obvious that the replacement of one or two light atoms by heavier ones would allow one to determine many signs. According to statistical arguments (Wilson 1942) the average intensity per lattice point (i.e. per molecule) should be

$$\bar{I} = N \times f^{-2} = 5000 \times 7^2 = 250\,000,$$

* Elected F.R.S. on 18 March 1954.

The structure of haemoglobin
V. Imidazole-methaemoglobin: a further check of the signs

By E. R. Howells and M. F. Perutz[*]

Medical Research Council Unit for the Study of the Molecular Structure of Biological Systems, Cavendish Laboratory, University of Cambridge

(*Communicated by Sir Lawrence Bragg, F.R.S.—Received 5 March* 1954)

The crystals are apparently orthorhombic; their unit-cell dimensions and the intensities of the $00l$ reflexions suggest a structure closely related to monoclinic methaemoglobin. Each structure appears to contain the same molecular layers parallel to (001), but in the monoclinic form the molecules are tilted the same way in each layer, while in the new form the tilt is alternately left and right in successive layers. Precession pictures show an interesting sequence of sharp and diffuse layer lines, those with certain indices h containing streaks parallel to c^*. In other layer lines the spots are broadened in varying degrees.

The amplitudes of the $h0l$ reflexions are related to the molecular Fourier transform described in previous papers of this series. Each value of $F(h0l)$ is compounded from the sum or the difference of the amplitude at two points on the transform having the co-ordinates $ha^* \, 0 \, lc^*$ and $h a^* \, 0 \, \bar{l}c^*$. The rules for addition and subtraction of amplitudes follow a simple scheme which is related to the sequence of sharp and diffuse layers. The scheme accounts for all observed values of $F(h0l)$, a reliability factor of 0·21 being obtained. This correlation provides an independent check for the sequence of signs along certain layer lines of the transform. A preliminary analysis of the structure is made, and molecules in neighbouring layers are shown to be displaced by 11·2 Å in the a direction.

1. Introduction

St George & Pauling (1951) suggested that the four haem groups in haemoglobin are embedded in a gap between the globin moieties, and that this gap might be prized open by the combination of haem with a large substituent group. This idea stimulated Kendrew & Parrish (1953, unpublished) to try the effect of combination with imidazole on the structure of metmyoglobin; they observed no change in unit-cell dimensions, but found that the intensities of many reflexions had changed. Encouraged by this observation we crystallized imidazole-methaemoglobin of horse, a compound first prepared by Russell & Pauling (1939), and analyzed its X-ray photographs. The results have no bearing as yet on the question which had led to the experiment, but they unexpectedly provide an independent check for the sequence of signs along several layer lines of the transform discussed in previous papers in this series. They also pose an interesting new problem in X-ray optics which will be dealt with elsewhere by Bragg & Howells (1954) and by Cochran & Howells (1954).

2. Morphology, unit cell, space group and intensities

Crystals were prepared by Dr V. M. Ingram, following the procedure described by Boyes-Watson, Davidson & Perutz (1947), except that imidazole was added in slight excess to the haemoglobin solution before dialysis against ammonium

---

[*] Elected F.R.S. on 18 March 1954.

The structure of haemoglobin
VI. Fourier projections on the 010 plane

By Sir Lawrence Bragg, F.R.S.

Cavendish Laboratory, University of Cambridge

and M. F. Perutz*

Medical Research Council Unit for the Study of the Molecular Structure of Biological Systems, Cavendish Laboratory, University of Cambridge

(*Received 5 March* 1954)

In the preceding papers the molecular Fourier transform was measured as a continuous function of c^* along nine lines of constant h in the plane of $k = 0$. Fourier inversion of this transform gives an electron density map of a single row of molecules in projection on the 010 plane. This map shows molecules of irregular outline and complex internal structure. Because of lack of resolution and the great depth of material projected neither polypeptide chains nor haem groups can as yet be recognized. Along the c^* direction the molecular dimensions can be measured with accuracy; along the a direction this is not yet possible, owing to overlap of neighbouring molecules in projection. Indications of the length of the molecules in that direction are obtained from the Fourier projection of a lattice slightly shrunk along a.

At low resolution the molecular dimensions can be estimated from a Fourier projection derived by inversion of the 'salt-water' transform described in part III, which corresponds to the diffraction by a row of 'ghost' molecules having the same shape as haemoglobin, but a uniform density throughout. The tilted spheroid of 71×54 Å, proposed by us in part II, is a good approximation to the outline of the molecule shown on this projection. Its only un-expected feature is a hollow at the central dyad where the thickness of the molecule is apparently reduced to 32 Å. It is not yet clear what this hollow represents.

1. Introduction

The structure of haemoglobin is a problem of much greater complexity than any other yet attacked by X-ray analysis. Some tentative solutions of the structure have been proposed in the past, but it has been impossible to prove them either right or wrong. What is novel in the present attack on the problem is the certainty of the results. The proof of their correctness, however, is different from that offered in the crystal-structure analysis of compounds where single atoms can be resolved. In those simpler structures proof rests on agreement of the atomic positions with the known facts of stereo-chemistry and with the observed intensities of the diffracted rays. The picture of the haemoglobin molecule which now emerges from the Fourier projections cannot yet be interpreted and contains only few features that can be recognized as intrinsically right. Its proof rests entirely on the agreement between the different sign determinations described in the preceding papers in this series (see References). The triple series of signs found by the heavy-atom method alone, the combination of these signs with the transform found by swelling and shrinkage, and the further check of the transform with the $|F|$'s of imidazole haemoglobin make us confident that all the terms included in the Fourier series have been given their correct signs.

* Elected F.R.S. on 18 March 1954.

The structure of haemoglobin

VII. Determination of phase angles in the non-centrosymmetric [100] zone

By D. M. Blow†

*Medical Research Council Unit for Molecular Biology,
Cavendish Laboratory, University of Cambridge*

(*Communicated by M. F. Perutz, F.R.S.—Received* 21 *December* 1957—
Revised 26 *March* 1958)

In the last paper in this series a Fourier projection down the [010] axis of horse haemoglobin was given (Bragg & Perutz 1954). This projection was centrosymmetric. As a first step towards the three-dimensional analysis, the projection down [100] has now been attacked. This projection is non-centrosymmetric, and arbitrary phase angles have had to be determined. All the fundamental problems of a three-dimensional study are met, but only a small number of reflexions need be dealt with.

The isomorphous replacement method has been used successfully with three mercury derivatives of haemoglobin. This provided a test of new methods for finding the vectors relating heavy atoms. Particular attention has been given to estimation of errors, and to their effect on the results. Further information about the phases has been derived from anomalous scattering by the mercury atoms, using $CrK\alpha$ and $CuK\alpha$ radiation.

By combining these results, the phases of most reflexions out to a spacing of about 6 Å have been determined with a standard error of about 25°. Ambiguous results are obtained for a few reflexions. The resulting electron density projection shows peaks up to four times the estimated standard error.

The prospects for three-dimensional structure analysis at 6 Å resolution are favourable. If the polypeptide chain is coiled in the α-form, the contrast should be sufficient for it to show up throughout its length.

1. Introduction

Using the isomorphous replacement method, Green, Ingram & Perutz (1954) were able to determine the signs of almost all the reflexions in the centrosymmetric [010] zone of horse haemoglobin, to a spacing of about 6 Å. Cullis, Dintzis & Perutz (1957) have since been able to extend the method to find most of the signs out to 2·8 Å spacing. From these results, projections of the electron density along the [010] axis have been calculated. The projection at 2·8 Å resolution shows a great amount of detail, but this cannot as yet be interpreted and it conveys little information about the internal organization of the haemoglobin molecule. This is not surprising, since the projection passes through a thickness of about 50 Å.

It is now obvious that a three-dimensional analysis will be needed to solve the structure of haemoglobin. In a centrosymmetric structure the phases of the reflexions are limited to 0 or π, as they are in a centrosymmetric zone. Proteins, however, consist largely of L-amino acids, and protein crystals are inevitably non-centrosymmetric. The phase angles of the general reflexions may take any value. In practical terms, this means that a method needs to be developed for determining the arbitrary phases of the general *hkl* reflexions.

† Now at Dept. of Biology, Massachusetts Institute of Technology, Cambridge, Mass. U.S.A.

The structure of haemoglobin

VIII. A three-dimensional Fourier synthesis at 5·5 Å resolution: determination of the phase angles

By Ann F. Cullis, Hilary Muirhead, M. F. Perutz, F.R.S.
and M. G. Rossmann

*Medical Research Council Unit for Molecular Biology, Cavendish Laboratory,
University of Cambridge*

and A. C. T. North

*Medical Research Council External Staff, Davy Faraday Research Laboratory,
Royal Institution, London W. 1*

(*Received* 21 *March* 1961)

Determination of the phase angles of a crystalline protein requires a series of isomorphous heavy-atom compounds, with heavy atoms attached to different sites on the protein molecule. The asymmetric unit of horse oxyhaemoglobin was found to combine with heavy atoms at two different sites which are now known to be sulphydryl groups. Altogether six different heavy-atom compounds of haemoglobin were made which proved isomorphous on X-ray analysis.

The positions of the heavy atoms were determined first by difference Patterson and Fourier projections on the centrosymmetric plane of the monoclinic crystals, and later by three-dimensional correlation functions, $(|F_{H_1}| - |F_{H_2}|)^2$ being used as coefficients, where F_{H_1} and F_{H_2} are the structure factors of the two different heavy-atom compounds. The parameters and anisotropic shape factors of the heavy atoms were refined by a three-dimensional least-squares method.

For each of the 1200 reflexions in the limiting sphere of $(5\cdot5 \text{ Å})^{-1}$ the structure amplitudes of all seven compounds were combined in an Argand diagram and the probability of the phase angle having a value α was calculated for $\alpha = 0, 5, 10, ..., 355°$. The coefficients for the final Fourier summation were then calculated in two different ways. In one method the vector from the origin to the centroid of the probability distribution, plotted around a circle of radius $|F|$, was chosen as the 'best $\mathbf{F}$'. The alternative set of coefficients was calculated, using the full, observed, value of $|F|$ and the most probable value of the phase angle α. The most probable error in phase angle was found to be 23°, and the standard error in electron density to be expected in the final results 0·12 $e/\text{Å}^3$.

1. Introduction

The present series of papers started with the electron density distribution in the haemoglobin molecule seen in projection on a line (I)†. This was obtained by comparing the intensities of the 00*l* reflexions from crystals at various degrees of swelling and shrinkage, and at various salt concentrations. Next, the analysis was extended to two dimensions. Using isomorphous replacement by heavy atoms, we determined the signs in the centrosymmetric zone of the monoclinic crystals to a limiting spacing of 6 Å, and later, of 2·8 Å, but the resulting Fourier projections proved uninterpretable (IV, VI and Cullis, Dintzis & Perutz 1958). A three-dimensional analysis requires determination of the phase angles of the general *hkl* reflexions. The

† For references to previous papers of this series see p. 38.

[15]

The structure of haemoglobin

IX. A three-dimensional Fourier synthesis at 5·5 Å resolution: description of the structure

By Ann F. Cullis, Hilary Muirhead, M. F. Perutz, F.R.S.
and M. G. Rossmann

*Medical Research Council Unit for Molecular Biology, Cavendish Laboratory,
University of Cambridge*

and A. C. T. North

*Medical Research Council External Staff, Davy Faraday Laboratory,
Royal Institution, London W. 1*

[Plates 13 to 19]

The electron density distribution in the unit cell is calculated at intervals of approximately 2Å and plotted in a series of sections parallel to (010). The contour maps show that haemoglobin consists of four subunits in a tetrahedral array. The subunits are identical in pairs in accordance with the twofold symmetry of the molecule. The two pairs are very similar in structure, and the members of each pair closely resemble the molecule of sperm-whale myoglobin. The four haem groups lie in separate pockets at the surface of the molecule. The positions of the iron atoms are confirmed by comparison of observed and calculated anomalous scattering effects, which also serve to determine the absolute configuration of the molecule.

The four subunits found by X-ray analysis correspond to the four polypeptide chains into which haemoglobin can be divided by chemical methods. In horse haemoglobin the amino acid sequence within these chains is still partly unknown, but in human haemoglobin it has already been determined. Comparison of this sequence with the tertiary structure of the chains as now revealed in horse haemoglobin and with the atomic model of sperm-whale myoglobin recently obtained by Kendrew and his collaborators shows many interesting relations. Prolines appear to come where the chains turn corners or where their configuration is known to be non-helical. On the other hand, the chains also have corners which contain no proline. Certain residues appear to be structurally vital, because they appear in identical positions in myoglobin and in the two chains of haemoglobin, while in other parts of the molecule a wide variety of different side-chains appears to be allowed.

1. Introduction

The preceding part (VIII, Cullis *et al.* 1961) dealt with the heavy-atom parameters and the calculation of the phase angles in the limiting sphere of $5 \cdot 5 \text{ Å}^{-1}$. Two alternative methods were used: one ignored anomalous scattering and determined the 'best F' from multiple isomorphous replacement by the centroid method of Blow & Crick (1959). In the second method the most probable value of the phase angle was found by isomorphous replacement and checked by comparison with anomalous scattering, and the full values of the observed structure amplitudes were used as coefficients of the terms. The two alternative Fourier syntheses calculated with these two sets of coefficients will now be described. The 'centroid Fourier', which gives the clearer results, will be discussed at length and the 'most probable Fourier' will be briefly shown for comparison. Various criteria will then be applied to test the reliability of the results.

[161]

662   **N A T U R E**   **March 8, 1958**   VOL. 181

A THREE-DIMENSIONAL MODEL OF THE MYOGLOBIN MOLECULE OBTAINED BY X-RAY ANALYSIS

By Drs. J. C. KENDREW, G. BODO, H. M. DINTZIS, R. G. PARRISH and H. WYCKOFF

Medical Research Council Unit for Molecular Biology, Cavendish Laboratory, Cambridge

AND

D. C. PHILLIPS

Davy Faraday Laboratory, The Royal Institution, London

MYOGLOBIN is a typical globular protein, and is found in many animal cells. Like hæmoglobin, it combines reversibly with molecular oxygen; but whereas the role of hæmoglobin is to transport oxygen in the blood stream, that of myoglobin is to store it temporarily within the cells (a function particularly important in diving animals such as whales, seals and penguins, the dark red tissues of which contain large amounts of myoglobin, and which have been our principal sources of the protein). Both molecules include a non-protein moiety, consisting of an iron–porphyrin complex known as the hæm group, and it is this group which actually combines with oxygen; hæmoglobin, with a molecular weight of 67,000, contains four hæm groups, whereas myoglobin has only one. This, together with about 152 amino-acid residues, makes up a molecular weight of 17,000, so that myoglobin is one of the smaller proteins. Its small size was one of the main reasons for our choice of myoglobin as a subject for X-ray analysis.

In describing a protein it is now common to distinguish the primary, secondary and tertiary structures. The *primary structure* is simply the order, or sequence, of the amino-acid residues along the polypeptide chains. This was first determined by Sanger using chemical techniques for the protein insulin[1], and has since been elucidated for a number of peptides and, in part, for one or two other small proteins. The *secondary structure* is the type of folding, coiling or puckering adopted by the polypeptide chain: the α-helix and the pleated sheet are examples. Secondary structure has been assigned in broad outline to a number of fibrous proteins such as silk, keratin and collagen; but we are ignorant of the nature of the secondary structure of any globular protein. True, there is suggestive evidence, though as yet no proof, that α-helices occur in globular proteins, to an extent which is difficult to gauge quantitatively in any particular case. The *tertiary structure* is the way in which the folded or coiled polypeptide chains are disposed to form the protein molecule as a three-dimensional object, in space. The chemical and physical properties of a protein cannot be fully interpreted until all three levels of structure are understood, for these properties depend on the spatial relationships between the amino-acids, and these in turn depend on the tertiary and secondary structures as much as on the primary.

Only X-ray diffraction methods seem capable, even in principle, of unravelling the tertiary and secondary structures. But the great efforts which have been devoted to the study of proteins by X-rays, while achieving successes in clarifying the secondary (though not yet the tertiary) structures of fibrous proteins, have hitherto paid small dividends among the metabolically more important globular, or crystalline, proteins. Progress here has been slow because globular proteins are much more complicated then the organic molecules which are the normal objects of X-ray analysis (not counting hydrogens, myoglobin contains 1,200 atoms, whereas the most complicated molecule the structure of which has been completely determined by X-rays, vitamin B_{12}, contains 93). Until five years ago, no one knew how, in practice, the complete structure of a crystalline protein might be found by X-rays, and it was realized that the methods then in vogue among protein crystallographers could at best give the most sketchy indications about the structure of the molecule. This situation was transformed by the discovery, made by Perutz and his colleagues[2], that heavy atoms could be attached to protein molecules in specific sites and that the resulting complexes gave diffraction patterns sufficiently different from normal to enable a classical method of structure analysis, the so-called 'method of isomorphous replacement', to be used to determine the relative phases of the reflexions. This method can most easily be applied in two dimensions, giving a projection of the contents of the unit cell along one of its axes. Perutz attached a *p*-chloro-mercuri-benzoate molecule to each of two free sulphydryl groups in hæmoglobin and used the resulting changes in certain of the reflexions to prepare a projection along the *y*-axis of the unit cell[3]. Disappointingly, the projection was largely uninterpretable. This was because the thickness of the molecule along the axis of projection was 63 A. (corresponding to some 40 atomic diameters), so that the various features of the molecule were superposed in inextricable confusion, and even at the increased resolution of 2·7 A. it has proved impossible to disentangle them[4]. It was clear that further progress could only be made if the analysis were extended to three dimensions. As we shall see, this involves the collection of many more observations and the production of three or four different isomorphous replacements of the same unit cell, a requirement which presents great technical difficulties in most proteins.

The present article describes the application, at low resolution, of the isomorphous replacement method in three dimensions to type *A* crystals of sperm whale myoglobin[5]. The result is a three-dimensional Fourier, or electron-density, map of the unit cell, which for the first time reveals the general nature of the tertiary structure of a protein molecule.

Isomorphous Replacement in Myoglobin

No type of myoglobin has yet been found to contain free sulphydryl groups, so that the method of attaching heavy atoms used by Perutz for hæmoglobin could not be employed. Eventually, we were able to attach several heavy atoms to the myoglobin molecule at different specific sites by crystallizing it with a variety of heavy ions chosen because they might be expected, on general chemical grounds, to possess affinity for protein side-chains. X-ray, rather than chemical, methods were used to determine whether combination had taken place, and, if so, whether the ligand was situated predominantly at a single site on the surface of the molecule. Among others, the following ligands were found to combine in a way suitable for the present purpose : (i) potassium mercuri-iodide and auri-iodide ; (ii) silver nitrate, potassium auri-chloride ; (iii) p-chloromercuri-benzene sulphonate ; (iv) mercury diammine $(Hg(NH_3)^{2+}$, prepared by dissolving mercuric oxide in hot strong ammonium sulphate), p-chloro-aniline ; (v) p-iodo-phenylhydroxylamine. Each group of ligands combined specifically at a particular site, five distinct sites being found in all. The substituted phenylhydroxylamine is a specific reagent for the iron atom of the hæm group[6], and may be assumed to combine with that group ; in none of the other ligands have we any certain knowledge of the mechanism of attachment or of the chemical nature of the site involved.

Methods of X-ray Analysis

Type *A* crystals of myoglobin are monoclinic (space group $P2_1$) and contain two protein molecules per unit cell. Only the $h0l$ reflexions are 'real', that is, can be regarded as having relative phase angles limited to 0 or π, or positive or negative signs, rather than general phases ; when introduced into a Fourier synthesis, these reflexions give a projection of the contents of the cell along its *y*-axis. In two dimensions the analysis followed lines[7] similar to that of hæmoglobin. First, the heavy atom was located by carrying out a so-called difference-Patterson synthesis ; if all the heavy atoms are located at the same site on every molecule in the crystal, this synthesis will contain only one peak, from the position of which the *x*- and *z*-co-ordinates of the heavy atom can be deduced, and the signs of the $h0l$ reflexions determined. These signs were cross-checked by repeating the analysis for each separate isomorphous replacement in turn ; we are sure of almost all of them to a resolution of 4 A., and of most to 1·9 A. Using the signs, together with the measured amplitudes, we may, finally, compute an electron-density projection of the contents of the unit cell along *y* ; but, as in hæmoglobin and for the same reasons, the projection is in most respects uninterpretable (even though here the axis of projection is only 31 A.). On the other hand, knowledge of the signs of the $h0l$ reflexions to high resolution enabled us to determine the *x*- and *z*-co-ordinates of all the heavy atoms with some precision. This was the starting point for the three-dimensional analysis now to be described.

In three dimensions the procedure is much more lengthy because all the general reflexions hkl must be included in the synthesis, and more complicated because these reflexions may have any relative phase angles, not only 0 or π. Furthermore, we need to know all three co-ordinates of the heavy atoms ; the two-dimensional analysis gives *x* and *z*, but to find *y* is more difficult, and details of the methods used will be published elsewhere, including among others two

proposed by Perutz[8] and one proposed by Bragg[9]. Finally, a formal ambiguity enters into the deduction of general phase angles if only one isomorphous replacement is available ; this can be resolved by using several replacements[10], such as are available in the present case. Once the phases of the general reflexions have been determined, one can carry out a three-dimensional Fourier synthesis which will be a representation of the electron density at every point in the unit cell.

Before such a programme is embarked upon, however, the resolution to be aimed at must be decided. The number of reflexions needed, and hence the amount of labour, is proportional to the cube of the resolution. To resolve individual atoms it would be necessary to include at least all terms of the series with spacings greater than 1·5 A.—some 20,000 in all ; and it is to be remembered that the intensities of all the reflexions would have to be measured for *each* isomorphous derivative. Besides this, introduction of a heavy group may cause slight distortion of the crystal lattice ; as the resolution is increased, this distortion has an increasingly serious effect on the accuracy of phase determination. In the present stage of the analysis the most urgent objective was an electron-density map detailed enough to show the general layout of the molecule—in other words, its tertiary structure. If the α-helix, or something like it, forms the basis of the structure, we need only work to a resolution sufficient to show up a helical chain as a rod of high electron density. For this purpose we require only reflexions with spacings greater than about 6 A. ; in all there are some 400 of these, of which about 100 are $h0l$'s already investigated in the two-dimensional study. The Fourier synthesis described here is computed from these 400 reflexions only, and is in consequence blurred ; besides this, it is distorted by an unknown amount of experimental error, believed to be small but at the moment difficult to estimate. Thus while the general features of the synthesis are undoubtedly correct, there may be some spurious detail which will require correction at a later stage.

The Three-dimensional Fourier Synthesis

The synthesis was computed in 70 min. on the EDSAC Mark I electronic computer at Cambridge (as a check, parts of the computation were repeated on DEUCE at the National Physical Laboratory). It is in the form of sixteen sections perpendicular to *y* and spaced nearly 2 A. apart ; these must be piled on top of one another to represent the electron density throughout the cell, containing two myoglobin molecules together with associated mother liquor (which amounts to nearly half the whole). Unfortunately, the synthesis cannot be so represented within the two-dimensional pages of a journal ; furthermore, if the sections are displayed side by side, they give no useful idea of the structure they represent. The examples reproduced in Fig. 1 illustrate some of the more striking features.

A first glance at the synthesis shows that it contains a number of prominent rods of high electron density ; these usually run fairly straight for distances of 20, 30 or 40 A., though there are some curved ones as well. Their cross-section is generally nearly circular, their diameter about 5 A., and they tend to lie at distances from their neighbours of 8–10 A. (axis to axis). In some instances two segments of rod are joined by fairly sharp corners. Fig. 1a

664   N A T U R E   **March 8, 1958** VOL. 181

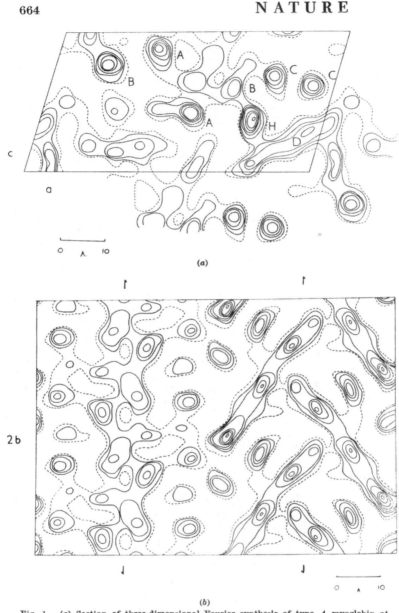

Fig. 1. (a) Section of three-dimensional Fourier synthesis of type A myoglobin at $y = -1/8\ b$. A–D, polypeptide chains; H, hæm group. (b) Section parallel to $[20\bar{1}]$ at $x = 0$, showing polypeptide chain A (on the right)

value greater than at any other point in the cell. A section through this disk is shown at H in Fig. 1a. We identify this feature as the hæm group itself, for the following reasons : (i) the hæm group is a flat disk of about the same size ; (ii) its centre is occupied by an iron atom and therefore has a higher electron density than any other point in the whole molecule ; (iii) a difference-Fourier projection of the p-iodo-phenylhydroxylamine derivative shows that, at least in y-projection, the position of the iodine atom is near that of our group ; this is what we should expect, since this reagent specifically combines with the hæm group ; (iv) the orientation of the disk corresponds, as closely as the limited resolution of the map allows one to determine it, with the orientation of the hæm group deduced from measurements of electron spin resonance[5,11].

We cannot understand the structure of the molecules in the crystal unless we can decide where one ends and its neighbours begin. In a protein crystal the interstices are occupied by mother liquor, in this case strong ammonium sulphate, the electron density of which is nearly equal to the average for the whole cell. Hence it is to be expected that in the intermolecular regions the electron density will be near average (the density of coiled polypeptide chains is much above average, and that of side-chains well below). It should also be fairly uniform ; these regions should not be crossed by major features such as polypeptide chains. Using these criteria, it is possible to outline the whole molecule with minor uncertainties. It was gratifying to find that the result agreed very well, in projection, with a salt-water difference-Fourier projection made as part of the two-dimensional programme (for the principles involved, see ref. 12). Moreover, the dimensions of the molecule agreed closely with those deduced from packing considerations in various types of unit cell.

shows several rods—three of them (A, B and C) cross the plane of the section almost at right angles, while one (D) lies nearly in that plane. D is part of a nearly straight segment of chain about 40 A. long, of which some 20 A. is visible in this section. It seems virtually certain that these rods of high density are the polypeptide chains themselves—indeed, it is hard to think of any other features of the structure which they could possibly be. Their circular cross-section is what would be expected if the configuration were helical, and the electron density along their axes is of the right order for a helical arrangement such as the α-helix. The various rods in the structure are intertwined in a very complex manner, the nature of which we shall describe later.

Another prominent feature is a single disk-shaped region of high electron density which reaches a peak

The Myoglobin Molecule

We are now in a position to study the tertiary structure of a single myoglobin molecule separated from its neighbours. Fig. 2 illustrates various views of a three-dimensional model constructed to show the regions of high electron density in the isolated molecule. Several points must be noticed. First, the model shows only the general distribution of dense regions. The core of a helical polypeptide chain would be such a region ; but if the chain were pulled out, into a β-configuration, for example, its mean density would drop to near the average for the cell and the chain would fade out at this resolution.

Similarly, side-chains should, in general, scarcely show up, so that the polypeptide rods in the model must be imagined as clothed in an invisible integument of side-chains, so thick that neighbouring chains in reality touch. Third, features other than polypeptide chains may be responsible for some of the regions of high density; patches of adsorbed salt, for example. Fourth, the surface chosen to demarcate a molecule cannot be traced everywhere with certainty, so it is possible that the molecule shown contains parts of its neighbours, and correspondingly lacks part of its own substance.

Making due allowance for these difficulties, we may note the main features. It is known[13] that myoglobin has only one terminal amino-group: it is simplest to suppose that it consists of a single poly-peptide chain. This chain is folded to form a flat disk of dimensions about 43 A. × 35 A. × 23 A. Within the disk chains pursue a complicated course, turning through large angles and generally behaving so irregularly that it is difficult to describe the arrangement in simple terms; but we note the strong tendency for neighbouring chains to lie 8–10 A. apart in spite of the irregularity. One might loosely say that the molecule consists of two layers of chains, the predominant directions of which are nearly at right angles in the two layers. If we attempt to trace a single continuous chain throughout the model, we soon run into difficulties and ambiguities, because we must follow it around corners, and it is precisely at corners that the chain must lose the tightly packed configuration which alone makes it visible at this resolution (an α-helix, for example, cannot turn corners without its helical configuration being disrupted). Also, there are several apparent bridges between neighbouring chains, perhaps due to the apposition of bulky side-chains. The model is certainly compatible with a single continuous chain, but there are at least two alternative ways of tracing it through the molecule, and it will not be possible to ascertain which (if either) is correct until the resolution has been improved. Of the secondary structure we can see virtually nothing directly at this stage. Owing to the corners, the chain cannot be in helical configuration throughout; in fact, the total length of chain in the model is 300 A., whereas an α-helix of 152 residues would be only 228 A. long. The 300 A. might correspond, for example, to 70 per cent α-helix and 30 per cent fully extended chain, but of course intermediate configurations are probably present, too. The hæm group is held in the structure by links to at least four neighbouring chains; nevertheless, one side of it is readily accessible from the environment to oxygen and to larger reagents such as p-iodo-phenylhydroxylamine (in the difference-Fourier projection of this complex, referred to above, the position of the iodine atom indicates that the

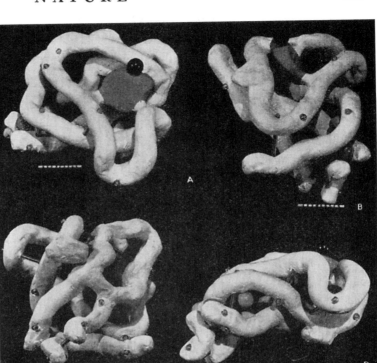

Fig. 2. Photographs of a model of the myoglobin molecule. Polypeptide chains are white; the grey disk is the hæm group. The three spheres show positions at which heavy atoms were attached to the molecule (black: Hg of p-chloro-mercuri-benzene-sulphonate; dark grey: Hg of mercury diammine; light grey: Au of auri-chloride). The marks on the scale are 1 A. apart

ligand is attached to the outside of the group). Clearly, however, the model cannot at present be correlated in detail with what we know of the chemistry of myoglobin; this must await further refinement.

Perhaps the most remarkable features of the molecule are its complexity and its lack of symmetry. The arrangement seems to be almost totally lacking in the kind of regularities which one instinctively anticipates, and it is more complicated than has been predicated by any theory of protein structure. Though the detailed principles of construction do not yet emerge, we may hope that they will do so at a later stage of the analysis. We are at present engaged in extending the resolution to 3 A., which should show us something of the secondary structure; we anticipate that still further extensions will later be possible—eventually, perhaps, to the point of revealing even the primary structure.

Full details of this work will be published elsewhere. We wish to record our debt to Miss Mary Pinkerton for assistance of all kinds; to the Mathematical Laboratory, University of Cambridge, for computing facilities on the EDSAC; to Dr. J. S. Rollett and the National Physical Laboratory for similar facilities on the DEUCE; to Mrs. Joan Blows and Miss Ann Mansfield for assistance in computing; for fellowships to the U.S. Public Health Service (H. W.), the Merck Fellowship Board (R. G. P.), the U.S. National Science Foundation (R. G. P. and H. M. D.), and the Rockefeller Foundation (H. M. D.); and to Sir Lawrence Bragg for his interest and encouragement. Finally, we wish to

666   NATURE   March 8, 1958 VOL. 181

express our profound gratitude to the Rockefeller Foundation, which has actively supported this research from its earliest beginnings.

[1] Sanger, F., and Tuppy, H., *Biochem. J.*, **49**, 481 (1951). Sanger, F., and Thompson, E. O. P., *ibid.*, **53**, 353, 366 (1953).

[2] Green, D. W., Ingram, V. M., and Perutz, M. F., *Proc. Roy. Soc.*, A, **225**, 287 (1954).

[3] Bragg, W. L., and Perutz, M. F., *Proc. Roy. Soc.*, A, **225**, 315 (1954).

[4] Dintzis, H. M., Cullis, A. F., and Perutz, M. F. (in the press).

[5] Kendrew, J. C., and Parrish, R. G., *Proc. Roy. Soc.*, A, **238**, 305 (1956).

[6] Jung, F., *Naturwiss.*, **28**, 264 (1940). Keilin, D., and Hartree, E. F., *Nature*, **151**, 390 (1943).

[7] Bluhm, M. M., Bodo, G., Dintzis, H. M., and Kendrew, J. C. (in the press).

[8] Perutz, M. F., *Acta Cryst.*, **9**, 867 (1956).

[9] Bragg, W. L. (in the press).

[10] Bokhoven, C., Schoone, J. C., and Bijvoet, J. M., *Acta Cryst.*, **4**, 275 (1951).

[11] Ingram, D. J. E., and Kendrew, J. C., *Nature*, **178**, 905 (1956).

[12] Bragg, W. L., and Perutz, M. F., *Acta Cryst.*, **5**, 277 (1952).

[13] Schmid, K., *Helv. Chim. Acta*, **32**, 105 (1949). Ingram, V. M. (unpublished work).

RECENT CLIMATIC CHANGES

By Dr. E. B. KRAUS

Snowy Mountains Hydro-Electric Authority, Cooma North, New South Wales, Australia

CLIMATE is nowhere invariant. The existence of a never-ending change is most apparent along the shifting boundaries of the various climatic provinces. During the past few hundred years the boundaries of the Alpine and Sub-Arctic regions have experienced some striking changes in glaciation and considerable variations in the extent and the seasonal duration of sea ice. It has now been established from the analysis of rainfall and stream-flow records that fluctuations of corresponding or even larger amplitude occurred in the sub-tropics at the boundary of the arid zone. During the same time the climates of the Antarctic ice cap, the central Sahara or the Amazon rain forest may not have altered a great deal.

Changes in a time series can be illustrated to advantage by graphs of cumulative percentual deviations from a mean. Graphs of this type accentuate changes in a time series, while running averages or orthodox statistical analyses tend to smooth them out. They demonstrate, at a glance, the existence of a trend and the time when it changes.

In the following diagrams (Figs. 1–6), the ordinate :

$$y(n) = 100 \sum_{l=1881}^{n} \left(\frac{r_l}{\bar{r}} - 1 \right)$$

where n represents the running calendar year, r_l the rainfall or stream discharge for the year l, and $\bar{r}$ the mean for the years 1881–1940. It is easily seen that y must be zero at the beginning of 1881 and the end of 1940. The graph of y rises for $r_l > \bar{r}$ and vice versa. It will be concave upwards during periods when r is increasing with time and convex when it is decreasing. The percentual deviation from $\bar{r}$ of the mean for any other period, say the years n to $(n + m)$, is given simply by $[y(n + m) - y(m)]/m$.

The accompanying six diagrams have been brought together to demonstrate how fluctuations of the same type occur at about the same time all around the globe. Fig. 1 illustrates the changes which occurred in the rainfall regime of semi-arid, north-western New South Wales (Australian rainfall district No. 48). It can be deduced from the curve that the mean rainfall for the years 1879–97 was 26 per cent above the 1881–1940 average. It was 11 per cent below that average during the period 1898–1946. In absolute terms, the mean annual rainfall from the beginning of records to 1897 was 16·90 in ; during the following forty-nine years it was 12·02 in. During the past ten years conditions were again more similar to those of the nineteenth century.

Fig. 2 deals with the east coast of Queensland (Australian rainfall district No. 40). The rainfall there is more seasonal but also more than three times as heavy as in north-western New South Wales. The onset of the dry period occurred two years later. It was interrupted by wetter conditions after about 1923, though a second severe drought occurred in the late '30's and early 1940's.

Conditions on the east coast of North America are illustrated by Fig. 3, which gives the mean rainfall for Charleston and Cape Hatteras. In an earlier paper[4] it had been shown that fluctuations there were paralleled by the records of other stations along the American east coast, and that the records suggest the persistence of relatively wet conditions all through the nineteenth and the later eighteenth centuries.

Fig. 4 shows the changes which occurred in the space-average of a large number of rainfall stations used by the Indian Meteorological Service to evaluate and forecast monsoon rainfall over the Indian peninsula. The change in the rainfall regime there was comparatively small and the vertical scale of the diagram has therefore been doubled. The same pattern of change with somewhat larger amplitude occurred in the more arid districts of north-west India and West Pakistan. All over this region the dry period which began at the end of the past century was ameliorated or terminated by about 1910.

Fig. 5 treats the discharge of the Nile at Aswan in a similar way. The difference in the mean discharge before and after 1898 amounts, in this case, to more than 30 per cent or 27×10^9 m.³/year.

The changes which occurred in parts of South Africa are illustrated by Fig. 6, which represents the South African rainfall district No. 16A in the Central Cape Province. Unfortunately, data after 1946 were not available in Australia for the two African cases.

Arakawa[1] and I have shown that changes similar to those demonstrated here have occurred in maritime parts of East Asia and in the West Indies, Hawaii and other tropical stations. They can also be found in Rhodesian rainfall records. Together, this material would seem to prove the existence of parallel and rather abrupt climatic fluctuations over a large part of the Earth.

The amplitude of the change in rainfall regime was largest at the fringes of the arid zone. In the extreme case of Aden, the rainfall record shows a difference of 84 per cent between the means for the thirty years after 1894 and the preceding thirteen years of avail-

The crystal structure of myoglobin
V. A low-resolution three-dimensional Fourier synthesis of sperm-whale myoglobin crystals

By G. Bodo, H. M. Dintzis, J. C. Kendrew and H. W. Wyckoff

Medical Research Council Unit for Molecular Biology, Cavendish Laboratory, University of Cambridge

(*Communicated by Sir Lawrence Bragg, F.R.S.—Received* 22 *April* 1959)

[Plate 8]

The study of type A crystals of sperm-whale has now been extended to three dimensions by using the method of isomorphous replacement to determine the phases of all the general X-ray reflexions having $d > 6$ Å, and a three-dimensional Fourier synthesis of the electron density in the unit cell has been computed. Data were obtained from the same derivatives which had been used in the previous two-dimensional study (Bluhm, Bodo, Dintzis & Kendrew 1958), in the course of which the x and z co-ordinates of the heavy atoms had been determined. Several methods were used to determine the y co-ordinates from the three-dimensional data; with a knowledge of all three co-ordinates of each heavy atom it was possible to establish the phases of nearly all the reflexions by a graphical method. The three-dimensional Fourier synthesis was evaluated on a high-speed computer from these phases and from the observed amplitudes of the reflexions.

A resolution of 6 Å was chosen because it should clearly reveal polypeptide chains having a compact configuration such as a helix. The electron-density map was in fact found to contain a large number of dense rod-like features which are considered to be polypeptide chains, probably helically coiled. In addition, a very dense flattened disk is believed to be the haem group with its central iron atom. Finally it was possible to identify the boundaries of the protein molecules by locating the intermolecular regions containing salt solution.

An isolated myoglobin molecule has dimensions about $45 \times 35 \times 25$ Å and within it the polypeptide chain is folded in a complex and irregular manner. For the most part the course of the chain can be followed, but there are some doubtful stretches, presumably where the helical configuration breaks down; a crude measurement of the total visible length of chain suggests that about 70 % of it may be in a helical or some similarly compact configuration. The haem group is near the surface of the molecule.

1. Introduction

In the previous paper of this series (Bluhm, Bodo, Dintzis & Kendrew 1958; hereafter referred to as IV) it was shown how the method of isomorphous replacement could be used to determine the signs of the $h0l$ reflexions from monoclinic type A crystals of sperm-whale myoglobin. A Fourier projection was computed from all such reflexions having spacings greater than 4 Å, but this was found to be uninterpretable owing to the large degree of overlapping, the axis of projection being 31 Å, or some twenty atomic diameters. It was concluded that the analysis should be extended to three dimensions, which would involve the determination of the general phases of the hkl reflexions. For this purpose the method of isomorphous replacement would again be used, the ambiguities inherent in the determination of the phase of general reflexions being resolved by using several different heavy-atom derivatives. As a preliminary to this three-dimensional analysis it is necessary to determine as accurately as possible the x and z co-ordinates of the heavy atoms, already known

[70]

5. THE COLLECTION OF THREE-DIMENSIONAL DATA

(a) *Choice of resolution*

The labour involved in collecting the data for a three-dimensional Fourier synthesis increases rapidly with the resolution desired, because the number of reflexions to which phases must be assigned is proportional to the cube of the radius of the sphere in reciprocal space over which the summation is to be carried out. Furthermore there are real (though slight) departures from strict isomorphism between the various derivatives studied, generally of the order $0\cdot1$–$0\cdot3\%$ of the cell axis, so that the accuracy of phase determination must decrease as one goes farther out in reciprocal space. For these reasons we decided to test the method of isomorphous replacement on a small scale, by working in the first instance at the lowest resolution which might be expected to give useful information about the structure of the myoglobin molecule.

We adopted the working hypothesis that a considerable proportion of the polypeptide chain has a configuration similar to the α-helix of Pauling, Corey & Branson (1951). There is no proof that this is the case, but there is some direct evidence of it in haemoglobin (Perutz 1951), as well as indirect evidence in myoglobin from studies of optical rotation (P. M. Doty, unpublished) and of rates of deuterium exchange (E. E. Benson & K. Linderstrøm-Lang, unpublished). At low resolution the α-helix would appear as a solid rod with axial electron density about $1\cdot0$ electrons/Å^3, embedded in a matrix of side chains of mean electron density about $0\cdot3$ electron/Å^3 (the mean overall electron density of the myoglobin molecule is about $0\cdot4$ electron/Å^3, and in type A crystals the electron density of the liquid regions has about the same value). Neighbouring α-helices would pack together with axial separations of 9 to 10 Å. We reached the conclusion that helices, if indeed they exist in myoglobin, would be clearly resolved if the Fourier synthesis included all terms having $d > 6$ Å; they should appear in such a synthesis as solid rods, since the region of reciprocal space being scanned includes only the first maximum of the Fourier transform of an α-helix. In fact there are about 400 reflexions having $d > 6$ Å, of which about 100 are $h0l$ reflexions with real phases which had already been determined in the two-dimensional work.

The choice of a 6 Å limit has a further advantage which can be appreciated from a study of the radial distribution of intensities plotted against reciprocal spacing (figure 11). As in many other protein crystals this function has a minimum at about 6 Å^{-1}, so pronounced that if the data are cut off sharply at this point a Fourier synthesis can be calculated without fear of serious series termination errors even though no artificial temperature factor is applied to the data.

(b) *X-ray methods*

In all we measured the $|F|$ values of the hkl reflexions from six different types of myoglobin crystal, namely unsubstituted met-myoglobin, the PCMBS, $HgAm_2$ and Au derivatives, and the double derivatives PCMBS/$HgAm_2$ and PCMBS/Au. Nearly all these data were obtained with the Buerger precession camera and $CuK\alpha$ radiation from the rotating-anode X-ray tube developed by Mr D. A. G. Broad. The method

The crystal structure of myoglobin. V 93

by himself. The results were identical apart from trivial differences nowhere exceeding, and generally much less than, one-fifth of the contour intervals used in figure 18; these discrepancies were probably due to minor errors in preparing the data for one or other computation, and are certainly not significant.

The eight sections of the Fourier synthesis are shown in figure 18. Contours have been drawn at convenient arbitrary intervals, which actually correspond to 9.08×10^{-2} electron/Å³. Since no $F(000)$ term was included in the synthesis, the zero contour corresponds to the mean electron density of the unit cell, which is 0.395 electron/Å³. To convert the contours shown to absolute electron densities it would be necessary to add 4.32 contour intervals at every point of the synthesis; this would have the effect of making the electron density almost everywhere positive, with only a few minor negative excursions. This result is a satisfactory check of the absolute scale of intensities used, though precise values of the electron densities could not be expected in a synthesis calculated from limited data with a sharp cut-off.

To study the synthesis a three-dimensional model was made by drawing the sections on thin Perspex sheets and piling them up at appropriate intervals; figure 19, plate 8, shows two photographs of this model. In reading the discussion of the synthesis in the next section it must be borne in mind that it is difficult to appreciate its features from the individual sections of figure 18; unfortunately it is impossible to give an adequate representation of the three-dimensional synthesis within the pages of a journal.

The diagrams used to determine the phase of each protein reflexion also indicate directly the phase angles appropriate to each of the heavy-atom derivatives; and from these, together with the observed structure amplitudes, Fourier syntheses were calculated (by Dr W. Hoppe) for some of the derivatives (PCMBS, $HgAm_2$, Au and PCMBS/$HgAm_2$). In general these syntheses resemble that of the unsubstituted protein very closely; there is no evidence that the protein molecules are appreciably displaced, and individual density values differ by less than one contour interval (on average by half a contour interval). The heavy atoms appear as high peaks in the expected positions; their co-ordinates are listed in table 1, and correspond within experimental error to those assumed in the structure factor calculations, a result which suggests that the latter were substantially correct for each of the heavy atoms, since the phases are based on an average of all the derivatives (a synthesis carried out with phases deduced from one derivative only would of course reproduce the co-ordinates assumed, however much these were in error). Examples of sections through the heavy-atom peaks are shown in figure 20, together with the corresponding difference-sections obtained by subtracting the Fourier synthesis of the protein from that of the heavy-atom derivative. The additional scattering matter in the heavy-atom peaks may be obtained by integrating the change in electron density over the whole peak; for PCMBS, $HgAm_2$ and Au the results are 81, 74 and 52 electrons respectively. These are all of the right order of magnitude, but higher than the assumed figures ($= \frac{1}{2} f(000)$), which were 65, 55 and 39 electrons respectively. However, the heavy-atom peaks are surrounded by diffraction fringes owing to the sharp termination of the series, and exact numerical

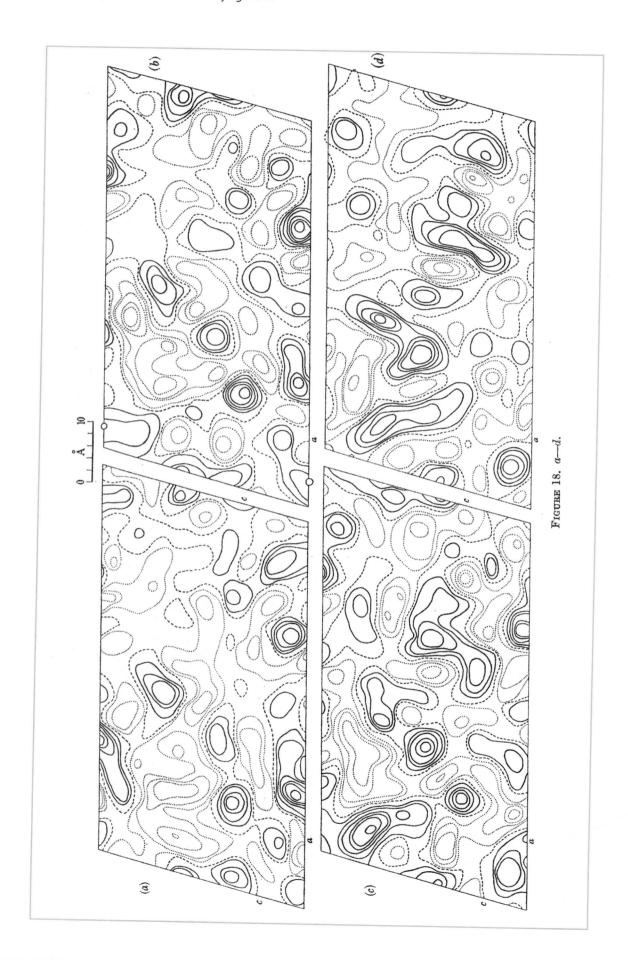

FIGURE 18. *a—d.*

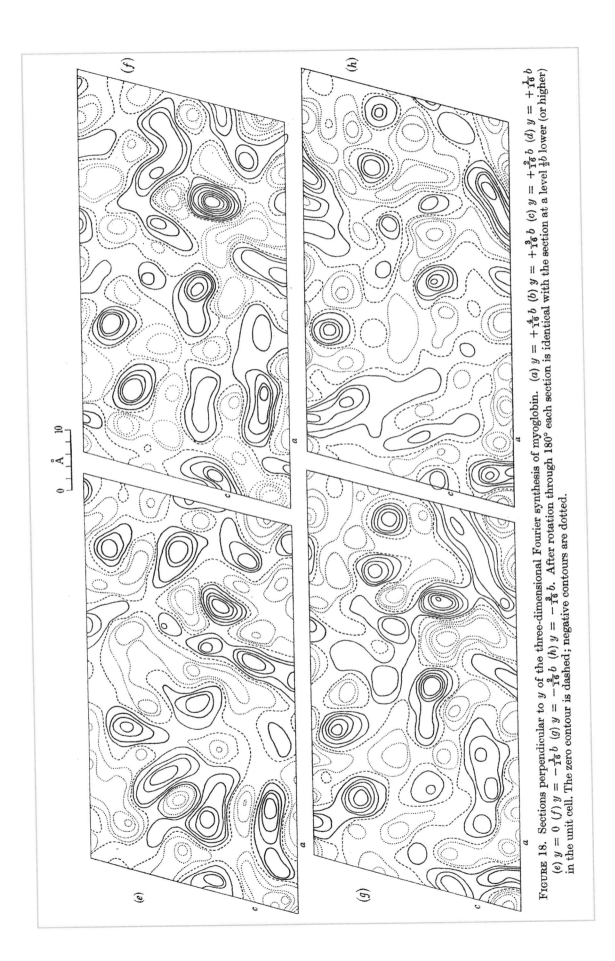

FIGURE 18. Sections perpendicular to y of the three-dimensional Fourier synthesis of myoglobin. (a) $y = +\frac{4}{16}b$ (b) $y = +\frac{3}{16}b$ (c) $y = +\frac{2}{16}b$ (d) $y = +\frac{1}{16}b$ (e) $y = 0$ (f) $y = -\frac{1}{16}b$ (g) $y = -\frac{2}{16}b$ (h) $y = -\frac{3}{16}b$. After rotation through 180° each section is identical with the section at a level $\frac{1}{2}b$ lower (or higher) in the unit cell. The zero contour is dashed; negative contours are dotted.

Bodo et al.   *Proc. Roy. Soc. A, volume 253, plate 8*

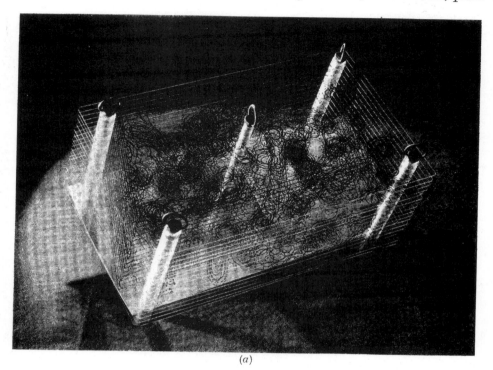

(a)

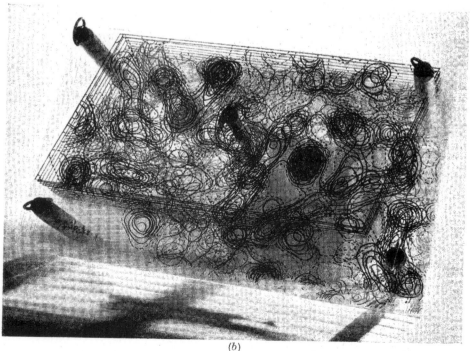

(b)

FIGURE 19. Photographs of the three-dimensional Fourier synthesis of myoglobin, constructed from the sections illustrated in figure 18.

(*Facing p.* 93)

100   G. Bodo, H. M. Dintzis, J. C. Kendrew and H. W. Wyckoff

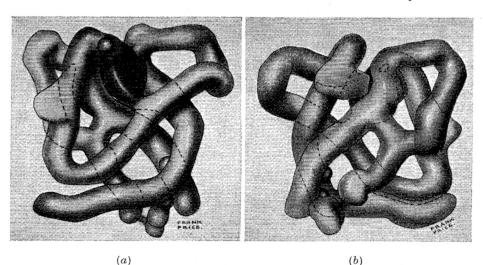

(a)   (b)

FIGURE 22. Drawings of a model of the myoglobin molecule. The haem group is shown in
(a) as a foreshortened dark grey disk. The small spheres are the heavy atoms used in phase
determination; in (a) the upper one is mercury of PCMBS. The two lower spheres visible
in both drawings, but partly concealed by polypeptide chains, are gold of $AuCl_4'$ (upper)
and mercury of $HgAm_2$ (lower).

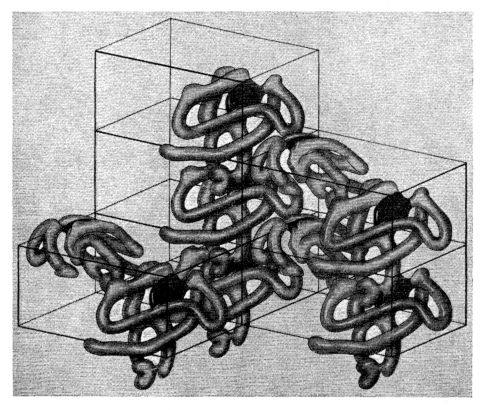

FIGURE 23. Drawing showing the arrangement of myoglobin molecules in the crystal lattice.
The edges of the unit cell are indicated by lines, the y axis being vertical.

HIGH-RESOLUTION
PROTEIN STRUCTURE ANALYSIS

What immortal hand or eye
could frame thy fearful symmetry?

WILLIAM BLAKE, "THE TYGER"

KEY PAPERS

(A) 1960   M. F. Perutz, M. G. Rossmann, A. F. Cullis, H. Muirhead, G. Will and A. C. T. North, *Nature* 185, 416–422. "Structure of Haemoglobin. A Three-Dimensional Fourier Synthesis at 5.5 Å Resolution, Obtained by X-Ray Analysis."

(B) 1960   J. C. Kendrew, R. E. Dickerson, B. E. Strandberg, R. G. Hart, D. R. Davies, D. C. Phillips and V. C. Shore, *Nature* 185, 422–427. "Structure of Myoglobin. A Three-Dimensional Fourier Synthesis at 2 Å Resolution."

     1961   J. C. Kendrew, H. C. Watson, B. E. Strandberg, R. E. Dickerson, D. C. Phillips and V. C. Shore. *Nature* 190, 666–670. "The Amino-acid Sequence of Sperm Whale Myoglobin: A Partial Determination by X-Ray Methods, and its Correlation with Chemical Data."

(C) 1992   R. E. Dickerson, *Protein Science* 1, 182–186. "A Little Ancient History."

     1961   J. C. Kendrew, *Scientific American*, December, pp. 96–110. "The Three-Dimensional Structure of a Protein Molecule."

     1964   M. F. Perutz, *Scientific American*, November, pp. 64–76. "The Haemoglobin Molecule."

The Cambridge MRC protein structure project finally came to fruition in 1959, when Kendrew, using Perutz' discovery of MIR phasing, solved the structure of met-myoglobin at 2 Å resolution (B), while Perutz produced a 5.5 Å map of the larger oxyhemoglobin (A). (Both myoglobin and hemoglobin normally function with their heme iron in the Fe^{+2} oxidation state, and these are termed oxy- when O_2 is bound and deoxy- when it is not. The prefix met- signifies that the iron is in the Fe^{+3} state. Molecular structures of met- and oxy- globins are generally the same, except that the O_2 molecule of oxy- is replaced by a water molecule in met-.) Paper C is an informal account of how the 2 Å myoglobin map was attained. The two *Scientific American* papers also are well-written, well-illustrated, and informative. That on myoglobin was illustrated by Irving Geis, who would go on to set new standards for illustration of complex protein and DNA structures (see Appendix 3).

The most striking new feature of these two globins was that the backbone chain of myoglobin, and the α and β chains of hemoglobin, were folded in an identical manner (Plate II). The amino acid sequence of myoglobin was currently being determined in parallel with the x-ray analysis (1), and the sequences of the hemoglobin chains were still unknown. But the common folding pattern of the three chains made it obvious that their sequences must be related in some manner. Perutz drew the correct conclusion when he remarked in paper A that "....their structural similarity suggests that they have developed from a common genetic precursor." The globins gave us our first look at what today is called *molecular evolution*.

Myoglobin was solved by MIR phasing using three heavy atom derivatives (or more accurately, 2.75. See paper C). Phase angles at the 6 Å stage had been found from hand-drawn diagrams like Figure 9 of the previous chapter. Since the diffraction pattern is in three dimensions, the number of data points that must be collected rises as the inverse cube of the boundary resolution. A 6 Å map requires the correct phasing of 400 reflections, but a 2 Å map requires $(6/2)^3 = 3^3 = 27$ times as many data, or 10,800 reflections. Hence phase analysis was computerized, and a weighted phase angle was used based on a profile of probabilities of phases all around each 360° phase circle (2). As Kendrew described in paper B, for each data set 22 separate precession camera photographs were taken, each using a fresh crystal, and the data from these photos were measured and merged. This process was carried out for the native myoglobin crystals, and separately for each of the heavy atom derivatives. The process required nearly a year of work by Bror Strandberg and myself at Cambridge, and our collaborator David Phillips at the Royal Institution in London. As Kendrew acknowledged, hundreds of hours of computer time were required on the EDSAC 2 computer, a marvel in its time but less powerful than a typical desktop computer today. Calculation of the final electron density map required an all-night session at the computer by all the principals, including Phillips who came down from London for the occasion (C). As the only American participant in the process, I was moved to a Key thought:

> **Oh say, can we see, at the dawn's early light,**
> **What we anxiously sought at the twilight's last gleaming?**

We could, indeed.

But how could one display a three-dimensional electron density map of something so large as a protein, in that pre-computer-graphics era? The initial attempt was simply to transfer electron density contours to Plexiglas sheets and stack these sheets over a light box, as had been done at 6 Å (Figure B.3). But this was cumbersome; 96 sections were required rather than 16. And one could not build an atomic model of a protein through stacked sheets of plastic.

Kendrew's solution was to obtain two six-foot square pieces of 1" plywood, and drill a rectangular array of holes corresponding to the lattice points of the crystal. Steel rods then were inserted in the holes, and electron density at each lattice point up one rod was represented by colored metal clips: white for the highest density, then yellow, orange, red, green, blue and black. This gave an immediate overall picture of the electron density map, and high-value white and yellow clips could be followed along the chains through the open forest of vertical rods. Kendrew then invented accurate wire framework models of atoms, held together by small cylindrical clamps and screws. With these he laboriously built the protein backbone through the high-density regions of the model, as shown in his Figure B.4. That figure illustrates the heme in its heme pocket, bounded by α-helix E to the right and F to the left, with histidine F8 coordinated directly to the heme iron. Kendrew model parts were later sold commercially, and were the standard for protein model building before they were made obsolete by computer graphics. Pauling and Corey independently invented a different kind of accurate atomic model for proteins, in which the planar amide was a single metal "dogbone" unit. These never became widely known, and were never distributed outside Caltech, probably because Pauling and Corey never solved a globular protein structure together. The only protein structure built with Pauling/Corey metal model parts was horse heart cytochrome *c* in 1968 (see Figure 8.2), and that model was disassembled as soon as commercial Kendrew model parts arrived.

An issue of primary importance during the high-resolution analysis was whether the straight rods of density visible at 6 Å would turn out to be α-helices, as had been assumed. The stage following 6 Å originally was planned to be at 2.5 Å resolution, but after discussions this was changed to 2.0 Å in order to increase chances of learning the structures of the rods. But what would be seen at 2.0 Å resolution? A hollow "garden hose"? And would the garden hose have spiral windings of density around its circumference that suggested an α-helix? Paper C relates how, at an evening party on Peterhouse lawn to celebrate the new structure, Sir Lawrence Bragg kept dragging visitors over to the Plexiglas sheets stacked on a light box, pointing down the inclined axis of one of the helices, and saying excitedly, "Look! See, it's hollow!"

Reality exceeded all expectations. Figure B.1 (paper B) shows the first direct proof that the myoglobin molecule was indeed built from α-helices. A cylindrical projection of one helix, when unrolled, not only had a zig-zag backbone chain of the proper dimensions, it also showed the C=O groups, allowing the direction of the chain to be determined.

The entire α-helical backbone pathway in myoglobin is depicted schematically in Figure 7.1. The molecule is a bundle of eight helices, lettered A through H from N-terminus to C-terminus. Helices E and F form a pocket into which the heme

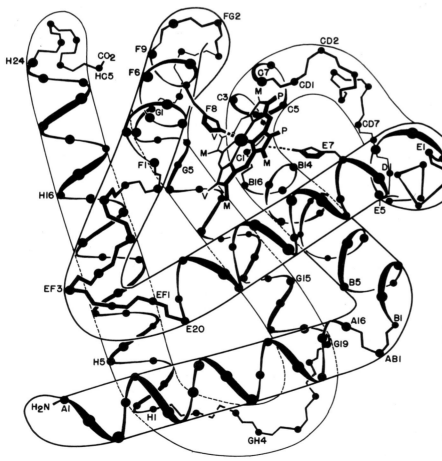

FIGURE 7.1   Alpha helix diagram of myoglobin at 2 Å. Black dots locate alpha carbon positions. Side chains are not shown except for the two histidines which interact with the heme. Amide linkages between alpha carbons are represented as smooth helical curves within an alpha helix, and three-bond zig-zags in nonhelical regions. Helices A through H are marked, and their amino acids are numbered. The heme sits in a pocket between E and F helices. Histidine F8 coordinates directly to the heme iron (Plate III). Histidine E7 is farther removed, with the oxygen-binding site between it and the heme iron. In metmyoglobin with an oxidized iron, this O_2 site is occupied by a water molecule. Compare this with the low-resolution view from the same direction in Figure 16 of the previous chapter. From (5).

group is fitted. The heme is "trapped" between two histidine side chains (Plate III). His F8 from helix F coordinates directly to the iron atom, whereas His E7 from helix E is farther removed from the iron. In fact, the binding pocket for the O_2 molecule lies between the heme iron and His E7 (occupied in metmyoglobin by a water molecule, W).

The 2 Å metmyoglobin structure made the 5.5 Å oxyhemoglobin analysis (A) far more interpretable than would have been the case by itself. Plate II shows how well a white alpha chain of hemoglobin matches the myoglobin structure. The arrangement of all four hemoglobin chains is as in Plate IV. The molecule is a tetramer, with two identical α chains of 141 amino acids each, and two identical β chains of 146 residues. The chains marked α_1 and β_1 at the right of Plate IV move as a unit in the hemoglobin mechanism to be described below, as do α_2 and β_2 to the left. The α_1/β_1 pair and the α_2/β_2 pair are related by a true twofold axis of symmetry, vertical in Plate IV. Approximate twofold axes (dashed lines) are horizontal from right to left, and from front to back. They would be true twofold axes only if α and β subunits were identical.

During the subsequent decade and a half, the horse methemoglobin analysis was pushed to high resolution (3), as were those of horse deoxyhemoglobin (4), human hemoglobin and those from other species, and many derivatives with various molecules or ions bound at the O_2 site between the heme and His E7. This is reviewed in Appendix table 2.1 of reference 7. Undoubtedly the most striking feature to come out of the hemoglobin comparisons was that *hemoglobin is a molecular machine*. It was known from kinetic measurements that the binding of O_2 tended to be a cooperative all-or-nothing process; attachment of one or two oxygens greatly facilitated binding the third and fourth. Conversely, loss of one or two oxygens induced some kind of molecular change that weakened its hold on the remaining two. This meant that the binding curve of oxygen for hemoglobin was sigmoid: under high-oxygen conditions at the lungs, a hemoglobin molecule would saturate with O_2 easily; and at oxygen-depleted conditions at the tissues, it would quickly hand over all of its oxygens to the myoglobin storage molecules.

The mechanical explanation of this kinetic behavior followed from Perutz' x-ray structure analyses. A top view of deoxyhemoglobin is seen in Plate V, and of oxyhemoglobin (methemoglobin) in Plate VI. The most noticeable difference is that the gap between H helices of subunits β_1 and β_2 in deoxyhemoglobin is narrowed when oxygen binds. This occurs because the pair of subunits α_1/β_1 on the right rotates downward while α_2/β_2 on the left rotates upward, about a horizontal axis at the back of the drawing. Perutz observed this motion between his oxy- and deoxyhemoglobin structures at high resolution. Even more, he observed a molecular click-switch that makes this motion a one-step event rather than a continuous process. In deoxyhemoglobin the side chain of His 97 on each beta chain sits in between those of Thr 41 and Phe 44 on its neighboring alpha chain. When the α_1/β_1 unit rotates relative to the α_2/β_2 unit, the histidine side chain jumps over Thr 41, to fit between Thr 41 and Thr 38, one turn of helix away. Hence the hemoglobin molecule is always either in one extreme state or the other; an intermediate conformation is forbidden by steric clash between His 97 and Thr 41.

But why does this happen, and what purpose does it serve? The answer lies in Plate III. In deoxyhemoglobin, the heme iron is 5-coordinate, and extends out of the heme plane in a pyrimidal manner toward the ring N of His F8. When O_2 binds at site W on the other side of the heme, the heme iron becomes 6-coordinate and is pulled back into the plane of the heme. This pulls helix F slightly closer to the heme, and the helix F displacement induces His β 97 to slip from α44/α41, to α41/α38. The distortion is propagated throughout the entire molecule, and pushes His F8 closer to the heme group even in those subunits that have not yet picked up O_2. The shift, consequently, increases the binding constant for subsequent oxygen molecules, and encourages the as-yet-unfilled hemes to take up oxygen. Conversely, when the loss of one or two oxygens allows the oxyhemoglobin molecule to slip back into its deoxy- subunit conformation, the pull of the His F8 side chains weakens oxygen binding and encourages those hemes which still have O_2, to release it.

In summary, myoglobin with its single chain is merely a passive "box" for oxygen. When oxygen is available (from hemoglobin) it is accepted; when oxygen is needed for metabolic processes in the cell, it is given up. Hemoglobin, in contrast, is a molecular machine. Its job is to saturate itself with oxygen where plentiful, and to hand over all four of its O_2 molecules to myoglobin in regions where oxygen is scarce. It needs to be a multi-subunit molecule in order to exhibit cooperativity in binding and releasing oxygen.

The announcement in early 1960 of the first high-resolution protein structure, and its relationship with an intermediate-resolution relative, had a major impact. By a stroke of good fortune, the meeting of the International Union of Crystallography that fall was scheduled for Cambridge. One of the prime attractions was Kendrew's new myoglobin structure, in its forest of wires and colored clips (Plates VII and VIII). This was on display in a room occupied by Electrical Engineering. Kendrew was told that he could use the room during the meeting, but that he had to "get his stuff out of there" before the fall term commenced. Parts of the model reside today in the Science Museum in South Kensington. The members of the Nobel Committee in Stockholm were somewhat more prescient, and awarded Perutz and Kendrew jointly the Nobel Prize in Chemistry in 1962. Since Watson, Crick and Wilkins also received the Prize in Medicine and Physiology that same year, quite a bit of champagne flowed at Cambridge that December.

There is an interesting afterthought to this story. Kendrew's myoglobin structure beautifully confirmed the central importance of Pauling's α-helix. There was joy on both sides of the Atlantic. But what would have been the reaction in Pasadena, had Kendrew been working instead on gamma globulin, which has not a single α-helix? The α-helix is indeed one of the most important substructures of globular proteins; but it is by no means the only such.

Postscript

It is strange that, just at the moment when Kendrew and coworkers gave the world its first picture of any globular protein molecule at the atomic level, the cyclol theory raised its ugly head for one last time. In 1954 Jacob Segal and A. Wolf at Humboldt University in East Berlin had proposed what they termed the "Faltentrommel" or Drum-Fold model for protein structures. In this model, a globular protein is built from a large number of short parallel polypeptide chains, linked laterally by hydrogen bonds like those in Pauling and Corey's parallel pleated β-sheet. A given chain is connected top and bottom to two different neighboring strands by cyclol rings as in Figure 7.2a. The result is a barrel of parallel polypeptide strands, linked at their ends to form a continuous chain (Figure 7.2b), rather like the zig-zag lacing around the sides of a snare drum. Hence the name "Drum-Fold." The drum is not rigid; chains can pack tightly together in a bundle, or can open up into a large hollow cylinder (Figure 7.2c). Segal claimed to see evidence for this bundle of parallel chains in Perutz' 1949 paper on the Patterson vector map of hemoglobin; the same evidence, in fact, that led Perutz at the time erroneously to invoke a compact pillbox of folded antiparallel chains (see Figure 6.12). But Perutz abandoned his tentative model as soon as isomorphous replacement phase analysis became possible.

When news of Kendrew's high-resolution map became known, Segal and his collaborators Käte Dornberger-Schiff and Angel Kalaidjiev were just finishing a book that would lay out the Faltentrommel case in detail (8). They came to visit Kendrew in Cambridge and to inspect his new map. They were convinced that they could explain the electron density map just as well with Faltentrommels as Kendrew could with α-helices. Dornberger-Schiff specifically asked whether they

FIGURE 7.2 Cyclol's last gasp: The Segal/Wolf "Faltentrommel" or Drum-Fold model of polypeptide chain folding.
(a) Short parallel polypeptide chains, linked to neighbors at their ends via diketopiperazine rings. (b) A continuous-chain barrel formed by 30 parallel linked polypeptides. (c) Schematic top view of the barrel in various stages of packing. Circles are top views of polypeptide chains; short and long bonds between them are diketopiperazine linkages as in Figure 8a. From (8).

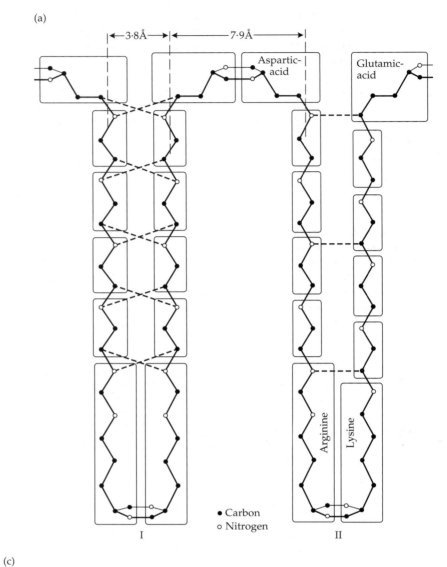

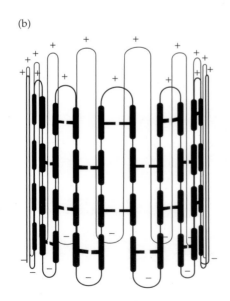

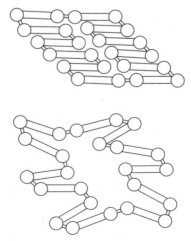

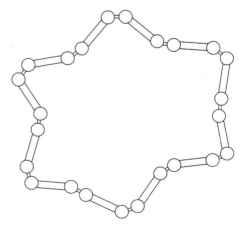

might take the electron density sections back with them to East Berlin for study. Kendrew replied that they were welcome to study the maps in Cambridge for as long as they liked, but that the maps would not leave the premises until he had finished interpreting them.

The book appeared in 1960 as planned, but with an added Appendix that briefly summarized Perutz' Patterson map analysis and Kendrew's MIR analysis and concluded that the case against the Faltentrommel was not yet proven. Some rather strange arguments had been presented in earlier chapters in support of the Faltentrommel:

> We wish to warn emphatically against such an uncritical acceptance of the conclusions drawn from X-ray work, as far as it concerns proteins.
>
> Proteins are obviously produced in organisms as copies of molecules that exist already. The simplest hypothesis is that of direct reduplication......A prerequisite of such a process is a molecular structure in which all the structural elements are arranged on the surface of the molecule. If some of the groups were inside the molecule, it could be duplicated only if the peptide chain were completely unwound, i.e., if the molecule were totally denatured. The results of such an operation would thus be, that instead of one native protein molecule, there would be two irreversibly denatured and therefore biologically worthless molecules. All the structural elements as well as all the folds must therefore lie on the surface of the molecule.

The ability of a Faltentrommel to open up as in Figure 7.2c, according to Segal, would provide just this mechanism for bringing all the side chains to the surface of the molecule. What came to be termed the Central Dogma of protein synthesis as espoused by Watson and Crick was: "DNA makes RNA makes protein." Pauling, by suggesting that amino acids might polymerize along the surface of his inside-out triple helix, in effect simplified this to: "DNA makes protein." Segal appears, even as late as 1960, to have arrived at the ultimate simplification: "Protein makes protein."

Segal's book quickly fell into well-deserved obscurity. Copies are hard to find today, and have become collectors' items. At the end of their visit to Kendrew's laboratory in 1960, Segal lamented, "It's a pity that your myoglobin structure came out when it did; had you published it six months earlier, we could possibly have rewritten parts of our book; had you published it six months later, we might have sold a few copies!" But Segal continued to defend his Faltentrommel theory vigorously even in 1963, the year after Kendrew and Perutz received the Nobel Prize. David Phillips attended a conference in Madras, India at which Segal spoke, and wrote me to say:

> Segal's intervention at Madras was quite embarrassing. Before he gave his paper I had a private conversation with him—and quite failed to make contact.....In his talk he carefully avoided mentioning myoglobin but did discuss his drum model of haemoglobin, presenting what he said was a theoretical Patterson projection derived from it. It looked like a copy of the true Patterson. At the end I wondered what I could say, and was saved by Edsall *[John Edsall of Harvard]* who asked him very mildly how his model agreed with the later work, and whether he had taken the new x-ray evidence into account at all. This was met by a semi-hysterical outburst about computing difficulties and lack of facilities, and lack of detail in the published work, at the end of which everyone was silent.

Ideas, good or bad, die hard.

References

1. A. B. Edmundson and C. H. W. Hirs. 1961. *Nature* 190, 663–665. "The Amino-Acid Sequence of Sperm Whale Myoglobin: Chemical Studies."

2. R. E. Dickerson, J. C. Kendrew and B. E. Strandberg. 1961. *Acta Cryst.* 14, 1188–1195. "The Crystal Structure of Myoglobin: Phase Determination to a Resolution of 2 Å by the Method of Isomorphous Replacement."

3. M. F. Perutz, H. Muirhead, J. M. Cox and L. C. G. Goaman. 1968. *Nature* 219, 131–139. "Three-dimensional Fourier Synthesis of Horse Oxyhaemoglobin at 2.8 Å Resolution: The Atomic Model."

4. W. Bolton and M. F. Perutz. 1970. *Nature* 228, 551–552. "Three-dimensional Fourier Synthesis of Horse Deoxyhaemoglobin at 2.8 Å Resolution."

5. R. E. Dickerson. 1964. In *The Proteins* (Hans Neurath, ed.), Academic Press, New York, 2nd edition, Vol. 2, pp. 603–778. "X-Ray Analysis and Protein Structure."

6. R. E. Dickerson and I. Geis. 1969. *The Structure and Action of Proteins*, Benjamin/Cummings, Menlo Park, CA.

7. R. E. Dickerson and I. Geis. 1983. *Hemoglobin: Structure, Function, Evolution and Pathology*, Benjamin/Cummings, Menlo Park, CA.

8. J. Segal, K. Dornberger-Schiff and A. Kalaidjiev. 1960. *Globular Protein Molecules: their Structure and Dynamic Properties.* VEB Deutscher Verlag der Wissenschaften, Berlin, DDR.

Study Questions

1. Why were myoglobin and hemoglobin so important in developing the concept of molecular evolution, even before their amino acid sequences were known (A)?

2. How was the image of an alpha helix improved in going from 6 Å to 2 Å resolution? Why did Sir Lawrence Bragg exclaim excitedly, "Look! It's hollow!" How else was it improved (B, C)?

3. Compared to a 6 Å map, does a 2Å map use: (a) one-third as many, (b) three times, (c) nine times, (d) 27 times, or (e) 81 times as many reflections in its diffraction pattern?

4. Why was a forest of steel rods better than the customary stack of transparent sheets in displaying and interpreting the 2 Å myoglobin molecule (B)?

5. Why did Perutz understand so much more about hemoglobin structure from his low resolution 5.5 Å map (A) than Kendrew was able to understand about myoglobin from its low resolution 6 Å map (B of previous chapter)? (The difference in detail between 6 Å and 5.5 Å maps is not the primary answer.)

6. What error did Kendrew make in his 6 Å structure concerning the orientation of one of the main components of the myoglobin molecule? Why was such an error possible at 6 Å, and why was it instantly cleared up at 2 Å (B)?

7. Why does the iron of the heme group sit *in the plane* of the heme when oxygen is bound, but slightly *out of plane* when oxygen is not present? How, in hemoglobin, is this translated into a signal that influences oxygen binding in other hemes of the molecule?

8. Why is the behavior just described advantageous for hemoglobin function?

416   N A T U R E   February 13, 1960 VOL. 185

STRUCTURE OF HÆMOGLOBIN

A THREE-DIMENSIONAL FOURIER SYNTHESIS AT 5·5-Å. RESOLUTION, OBTAINED BY X-RAY ANALYSIS

By Dr. M. F. PERUTZ, F.R.S., Dr. M. G. ROSSMANN, ANN F. CULLIS, HILARY MUIRHEAD and Dr. GEORG WILL

Medical Research Council Unit for Molecular Biology, Cavendish Laboratory, University of Cambridge

AND

Dr. A. C. T. NORTH

Medical Research Council External Staff, Davy Faraday Research Laboratory, Royal Institution, London, W.1

VERTEBRATE hæmoglobin is a protein of molecular weight 67,000. Four of its 10,000 atoms are iron atoms which are combined with protoporphyrin to form four hæm groups. The remaining atoms are in four polypeptide chains of roughly equal size, which are identical in pairs[1-3]. Their amino-acid sequence is still largely unknown.

We have used horse oxy- or met-hæmoglobin because it crystallizes in a form especially suited for X-ray analysis, and employed the method of isomorphous replacement with heavy atoms to determine the phase angles of the diffracted rays[4-7]. The Fourier synthesis which we have calculated shows that hæmoglobin consists of four sub-units in a tetrahedral array and that each sub-unit closely resembles Kendrew's model of sperm whale myoglobin[6]. The four hæm groups lie in separate pockets on the surface of the molecule.

Method of Analysis

Horse oxyhæmoglobin, crystallized from 1·9 M ammonium sulphate solution at pH 7, has the space group $C2$ with two molecules in the unit cell which lie on dyad axes[8]. In order to determine the phase angles of the 1,200 reflexions contained in the limiting sphere of 5·5 Å.$^{-1}$, six different isomorphous heavy-atom compounds were used (ref. 9 and unpublished work). Intensities were measured photographically and by counter spectrometer (Arndt, U. W., and Phillips, D. C., unpublished work). The relative positions and shapes of the heavy-atom replacement groups were found by correlation functions based on Patterson methods[10] and refined by least squares

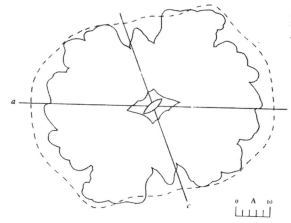

Fig. 2. External shape of the molecule. Full lines indicate the boundary derived from the contour at 0·54 electron/Å.^{3}. The broken line shows the boundary derived by Bragg and Perutz from two-dimensional data (ref. 13). Note the hole in the middle

procedures. For each reflexion the structure amplitudes of all seven compounds were combined in an Argand diagram[11], and the probability of the phase angle having a value α was calculated for $\alpha = 0, 5, 10, \ldots 355°$. The centroid of the probability distribution, plotted around a circle, was then chosen as the best vector F in the Fourier synthesis[12]. The results were finally plotted on 32 contour maps showing the distribution of electron density in sections spaced 2 Å. apart normal to b (Fig. 1). The absolute configuration of the molecule was determined from anomalous dispersion[4].

External Shape of the Molecule

More than half the volume of the crystals is taken up by liquid of crystallization, which mainly fills the spaces between the molecules and shows up on the contour maps in the form of flat, featureless regions (Fig. 1). The outlines of the two molecules in the unit cell can be traced by following the boundaries between these regions and the continuous electron-dense regions described below. In Fig. 2 the outline of one molecule, traced from the periphery of the 0·54

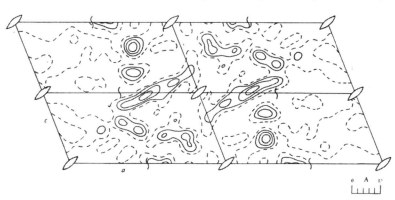

Fig. 1. Section at $y = 1/32b$. This cuts through the middle of the molecule on which the diagram is centred. 'Flat' areas indicating liquid appear on the left and right. Contours are drawn at intervals of 0·14 electron/Å.^{3}. The broken line marks 0·4 electron/Å.^{3}. Contours at lower levels are omitted

No. 4711   **February 13, 1960**   N A T U R E   417

Hæm 1 and 2

Fig. 3. Section at $z = 1/4\,c$ showing one of the two hæm groups in the asymmetric unit. The straight line indicates the hæm orientation derived from electron spin resonance (ref. 14). The lowest contour shown is at 0·4 electron/Å.³

electron/Å.³ contour, is seen in projection on the b-plane. To a first approximation it can be regarded as a spheroid with a length of 64 Å., a width of 55 Å. and a height of 50 Å. normal to the plane of the paper. Except for a slight shortening along a, this shape agrees with the earlier picture obtained from two-dimensional data, even including the dimple in the centre of the molecule (see Fig. 2 of ref. 12).

Positions of the Hæm and Sulphydryl Groups

Four peaks stand out from the rest, clearly representing the iron atoms with their surrounding porphyrin rings. A flattening of the peaks indicates the approximate orientation of the rings (Fig. 3). Fortunately, their exact orientation was already known from electron spin resonance[14], and it only remained to assign the correct one of the four alternative orientations of the hæms to each of the electron density peaks. The results are shown in Fig. 4. The iron atoms lie at the corners of an irregular tetrahedron with distances of 33·4 and 36·0 Å. between symmetrically related pairs. The closest approach between symmetrically unrelated iron atoms is 25·2 Å. (Table 1).

Table 1

	Fe₁	Fe₃	S₁
x	−6·6 Å.	12·3 Å.	5 Å.
y	7·3 Å.	−10·7 Å.	10 Å.
z	13·1 Å.	18·2 Å.	16 Å.
Fe₁ − Fe₂ = 33·4 Å.			
Fe₃ − Fe₄ = 36·0 Å.			
Fe₁ − Fe₃ = 25·2 Å.			
Fe₁ − S₁	13 Å.		
Fe₂ − S₁	21 Å.		

Horse hæmoglobin contains four cysteine residues, but only two sulphydryls combine with mercury in the native protein[5]. From the positions of the mercury atoms we inferred that each of the two

sulphydryl groups is about 13 Å. away from one iron atom and 21 Å. away from another. (Fig. 4 and Table 1). The significance of this situation is discussed below.

Configuration of the Polypeptide Chains

The most prominent feature of the Fourier synthesis consists of more or less cylindrical clouds of high density, like the vapour trails of an aeroplane; they are curved to form intricate three-dimensional figures. Sections through various parts of these appear in Fig. 1. To build a model of the figures, we rolled out sheets of a thermo-setting plastic to the thickness of our sections on a scale of 2 Å. = 1 cm., and cut out the shape of each region on the contour maps where the density exceeds 0·54 electron/Å.³ (this corresponds to the first full contour line in Fig. 1. From this section, for example, 14 shapes would be cut). The shapes were then assembled in accordance with their positions and heights in the different sections, and the hæm groups were attached in the appropriate orientation. The model was then baked to set it permanently. For comparison, a Fourier synthesis of sperm whale myoglobin was calculated at a resolution of 5·5 Å., using the new X-ray data of Kendrew et al.[15], and a model of the electron density distribution was constructed by the method just described.

From the hæmoglobin Fourier synthesis there emerged four separate units which are identical in pairs. Fig. 5 shows one member of each pair together with the model of myoglobin on the left. In each unit the cloud of high density describes a complicated figure which, in the white unit (middle), can be traced from end to end by following the superimposed line. Except for two small gaps where the density sinks slightly below 0·54 electron/Å.³, the black unit (right) closely resembles the white one. There are several gaps interrupting the myoglobin model where, probably due to increased thermal motion, the electron density falls below the contour-level chosen for cutting the model. However, we know from the recent work of Kendrew and his collaborators[15] that

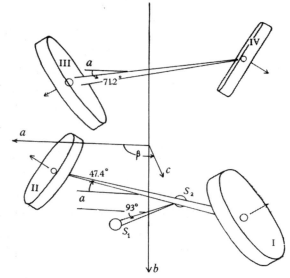

Fig. 4. Arrangement of hæm groups in hæmoglobin. Arrows indicate the reactive side of each group. The c-axis comes out of the paper towards the observer. This picture and all subsequent ones show the absolute configuration

418   N A T U R E   February 13, 1960   VOL. 185

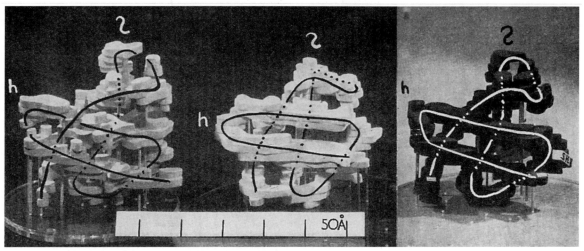

Fig. 5.   Two different polypeptide chains in the asymmetric unit of hæmoglobin compared with myoglobin (left).   The hæm groups are at the back of the chains

the gaps are bridged by a continuous polypeptide chain, and it is evident from Fig. 5 that apart from the gaps, the model has a configuration closely similar to the hæmoglobin units.

Clearly, the four tortuous clouds of high electron density in hæmoglobin represent the four polypeptide chains. The black and the white chains have similar, but not identical configurations. In the black chain the S-shaped bend at the top is more pronounced, the hæm group is lower, and the bend h sharper than in the white chain. These, however, are details. The

most important result is their resemblance to each other and to sperm whale myoglobin.

Arrangement of the Four Sub-units

The first step in the assembly of the molecule is the matching of each chain by its symmetrically related partner (Fig. 6). It will be noted that there is comparatively little contact between the members of each pair, suggesting rather tenuous linkages. In the next step the white pair is inverted and

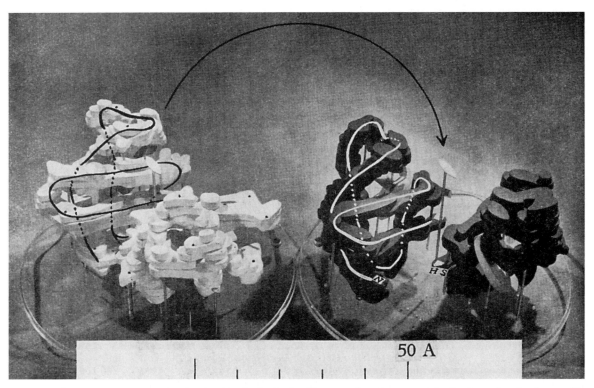

Fig. 6.   Two pairs of chains symmetrically related by the dyad axis. The arrow shows how one pair is placed over the other to assemble the complete molecule

No. 4711 **February 13, 1960** N A T U R E 419

placed over the black pair as indicated by the arrow. Fig. 7 shows one white chain placed over the pair of black ones and Fig. 8 shows the molecule completely assembled. The resulting arrangement is tetrahedral and has almost, but not quite, the orthorhombic point group symmetry 222. It contains two 'pseudo dyads' which lie approximately at right angles to each other and to the true dyad, one emerging from the centre of Fig. 8 and the other from the centre of Fig. 9. This means that, to a first approximation, each sub-unit can be generated by a rotation of 180° from any of its neighbours. Figs. 7, 8 and 9 also show that the surface contours of the white chains exactly fit those of the black, so that there is a large area of contact between them. This structural complementarity is one of the most striking features of the molecule. Fig. 10 is a view down the true dyad axis and reveals a hole going right through the centre of the molecule, as was to be expected from the Fourier projection[13]. However, the van der Waals radii of the chains are much bigger than appears in the model, and little room may, in fact, be left for water or electrolytes to pass through. Fig. 10 also reveals a dimple at the top where the white chains meet. This is matched by a similar, but larger, hollow at the bottom where the black chains meet.

The hæm groups are seen to lie in separate pockets on the surface of the molecule (Fig. 8). Each pocket is formed by the folds in one of the polypeptide chains, which appears to make contact with the hæm group at four different points at least. The iron atoms in the neighbouring pockets formed by the black and the white chains are 25 Å. apart.

Information from the Fourier Synthesis of Myoglobin at 2-Å. Resolution (ref. 15)

Thanks to the similarity with myoglobin, the interpretation can be carried further than would have been possible on the basis of our results alone. Kendrew *et al.* have found that the straight stretches of rod indicated in Fig. 5 are α-helices and that the N-terminal end of the chain is at the bottom left. The ends of the two hæmoglobin chains have been labelled N and C accordingly, as seen in Figs. 7 and 10. The myoglobin Fourier synthesis also reveals the hæm-linked amino-acid side-chain, probably histidine, on one side of the (ferric) iron atom and a small peak, probably representing a water molecule, on the other side. If this information is transferred to hæmoglobin, the reactive side of the hæm group is as indicated by the arrows in Fig. 4 and by the labels O_2 in Figs. 7 and 8a.

Figs. 7 and 8b show the reactive sulphydryl group (which is absent in myoglobin) to be attached to the portion of the black chain carrying the hæm-linked histidine. It is seen that the histidine and cysteine side-chains point in roughly opposite directions, one towards and the other away from the hæm group.

Fig. 7. Partially assembled molecule showing two black chains and one white

The sulphydryl group may possibly be in contact with the loop of the white chain which lies below the hæm group on the left. This is the situation of one pair of sulphydryl groups in the molecule. The second pair is probably attached to the white chains, but it is unreactive in the native protein and its position is still unknown.

Reliability of Results

The method of isomorphous replacement makes no assumptions about the structure of the protein, and its results suffer from none of the ambiguities which bedevil the interpretation of vector maps. Ideally, the parent protein, in combination with two different isomorphous heavy-atom compounds, should give accurate phase angles and electron density maps which are free from all except series termination errors. In practice, errors and uncertainties arise from several sources and must be minimized by using more than two heavy-atom compounds. In our case it was possible to estimate the accuracy of the phase angle for each reflexion from the measure of agreement between the angles indicated by the six different compounds. From the standard error in the vector F, averaged over all reflexions, the standard error in the final electron density was calculated as 0·12 electron/Å.³, which amounts to 0·85 of the interval between successive contours in Figs. 1 and 3 (0·14 electron/Å.³), or 0·15 of the difference in density between 'peaks' and 'valleys' (0·7 electron/Å.³) (ref. 12). This calculated error is borne out by the observed fluctuations in the liquid regions between the molecules and by the difference in height between the two iron peaks (0·13 electron/Å.³).

420   **NATURE**   **February 13, 1960**   VOL. 185

An error of at least five times this magnitude would be needed to turn a peak into a valley and thus to simulate the backbone of a polypeptide chain in a region actually occupied by side-chains.

Normally, the accuracy of a structure is checked by comparison of calculated and observed intensities. In hæmoglobin this is impossible so long as the light atoms are not resolved, but the positions of the iron atoms at least can be checked by their anomalous dispersion effect. Depending on the phase relationship between the structure factor of the four iron atoms by themselves and that of the entire molecule, the total intensity of any given reflexion $I(hkl)$ may be larger or smaller than $I(\bar{h}\bar{k}\bar{l})$. We selected 50 reflexions for which the effect of anomalous dispersion was expected to be largest and measured the intensity of each reflexion in the four symmetrically related quadrants on the counter spectrometer. Statistically significant differences between $I(hkl)+I(\bar{h}k\bar{l})$ and $I(\bar{h}\bar{k}l)+I(h\bar{k}l)$ were found in 36 reflexions. In 34 of them the signs of the differences agreed with prediction, indicating that the phase angles of the reflexions as well as the positions of the iron atoms are correct.

As a further check on the positions of the iron atoms, an attempt was made to label them with heavy atoms. This was done by allowing crystals of methæmoglobin, which are isomorphous with oxyhæmoglobin, to react with p-iodophenyl–hydroxylamine, which attaches itself to the iron atoms. A difference Fourier projection on the b-plane showed four prominent peaks which lie well within the calculated distances between the iodine and the iron atoms.

Finally, to avoid any possible bias in the interpretation of the Fourier synthesis, we have tried to construct objective models containing only those features actually found in the electron density maps. Due to series termination errors and inaccuracies in many of the phase angles, these maps must certainly contain errors of detail. Large errors, on the other hand, are unlikely because, quite apart from the checks just described, the results really prove themselves. No combination of errors could have led to the appearance of four distinct chains of roughly equal length, in agreement with chemical evidence, to the similarity between the two pairs of chains which are not related by crystal symmetry, and to the resemblance between hæmoglobin and myoglobin.

(a)

(b)

Fig. 8. (a) Hæmoglobin model viewed normal to a. The hæm groups are indicated by grey disks. (b) Chain configuration in the two sub-units facing the observer. The other two chains are produced by the operation of the dyad axis

Discussion

The polypeptide chain-fold which Kendrew and his collaborators first discovered in sperm whale myoglobin has since been found also in seal myoglobin[16].

No. 4711   February 13, 1960   N A T U R E   421

Fig. 9.   Hæmoglobin model viewed normal to *c*

structural changes detectable by X-ray analysis[17], both forms being isomorphous with the metmyoglobin normally studied. In the hæmoglobin of horse and of man, on the other hand, the oxygenated and reduced forms are crystallographically different[18-20]. The structure of reduced hæmoglobin is still unknown, but it would not be surprising if loss of oxygen caused the four sub-units to rearrange themselves relative to each other, rather than to change their individual structure to a marked degree.

Riggs has shown that blocking the sulphydryl groups reduces hæm–hæm interaction[21]. As Fig. 8 shows, these groups occupy key positions close to the hæm-linked histidines and to points of contact between two different sub-units. They may well play an important part in the transition between the oxygenated and reduced forms. Incidentally, the cysteine residue should provide a convenient marker for the peptide containing the hæm-linked histidine, and so help to determine the sequence in this important part of the chain.

A full account of this work will be published elsewhere.

We thank the Director and staff of the University of Cambridge Mathematics Laboratory for making their electronic computor *Edsac* II available to us for the many calculations involved in this work. We are

Its appearance in horse hæmoglobin suggests that all hæmoglobins and myoglobins of vertebrates follow the same pattern. How does this arise ? It is scarcely conceivable that a three-dimensional template forces the chain to take up this fold. More probably the chain, once it is synthesized and provided with a hæm group around which it can coil, takes up this configuration spontaneously, as the only one which satisfies the stereochemical requirements of its amino-acid sequence. This suggests the occurrence of similar sequences throughout this group of proteins, despite their marked differences in amino-acid content. This seems all the more likely, since their structural similarity suggests that they have developed from a common genetic precursor.

Little can be said as yet about the relation between structure and function. The hæm groups are much too far apart for the combination with oxygen of any one of them to affect the oxygen affinity of its neighbours directly. Whatever interaction between the hæm groups exists must be of a subtle and indirect kind that we cannot yet guess. A few observations of possible significance might be mentioned. Kendrew found that the combination of reduced myoglobin with oxygen involves no

Fig. 10.   A view down the *b*-axis. Note the proximity of the *C* and *N* terminal ends which could serve to form links between the two white chains

422   N A T U R E   February 13, 1960   VOL. 185

also deeply grateful to the Rockefeller Foundation for its long-continued financial support and to Sir Lawrence Bragg for his unfailing enthusiasm and encouragement. Finally, we wish to thank Dr. B. R. Baker, of the Stanford Research Institute, for a gift of organic mercurials, Dr. J. Chatt of Imperial Chemical Industries, Ltd., and Mr. A. R. Powell, of Johnson Matthey, Ltd., for gifts of heavy-atom compounds, and Mrs. Margaret Allen, Miss Ann Jury and Miss Brenda Davies for assistance.

¹ Rhinesmith, S. H., Schroeder, W. A., and Pauling, L., *J. Amer. Chem. Soc.*, **79**, 4682 (1957).

² Braunitzer, G., *Hoppe-Seyl. Z.*, **312**, 72 (1958).

³ Wilson, S., and Smith, D. B., *Can. J. Biochem. Physiol.*, **37**, 405 (1959).

⁴ Bokhoven, C., Schoone, J. C., and Bijvoet, J. B., *Acta Cryst.*, **4**, 275 (1951).

⁵ Green, D. W., Ingram, V. M., and Perutz, M. F., *Proc. Roy. Soc.*, A, **225**, 287 (1954).

⁶ Kendrew, J. C., Bodo, G., Dintzis, H. M., Parrish, R. G., Wyckoff, H. W., and Phillips, D. C., *Nature*, **181**, 662 (1958).

⁷ Blow, D. M., *Proc. Roy. Soc.*, A, **247**, 302 (1958).

⁸ Perutz, M. F., *Proc. Roy. Soc.*, A, **225**, 264 (1954).

⁹ Cullis, A. F., Dintzis, H. M., and Perutz, M. F., Conference on Hæmoglobin, National Academy of Sciences, NAS-NRC Publication 557, p. 50 (Washington, 1958).

¹⁰ Rossmann, M. G., *Acta Cryst.* (in the press).

¹¹ Bodo, G., Dintzis, H. M., Kendrew, J. C., and Wyckoff, H. W., *Proc. Roy. Soc.*, A, **253**, 70 (1959).

¹² Blow, D. M., and Crick, F. H. C., *Acta Cryst.*, **12**, 794 (1959).

¹³ Bragg, W. L., and Perutz, M. F., *Proc. Roy. Soc.*, A, **225**, 315 (1954).

¹⁴ Ingram, D. J. E., Gibson, J. F., and Perutz, M. F., *Nature*, **178**, 905 (1956).

¹⁵ Kendrew, J. C., Dickerson, R. E., Strandberg, B., Hart, R. G., Davies, D. R., Phill'ps, D. C., and Shore, V. C. (see following article).

¹⁶ Scouloudi, H., *Nature*, **183**, 374 (1959).

¹⁷ Kendrew, J. C. (private communication).

¹⁸ Haurowitz, F., *Hoppe-Seyl. Z.*, **254**, 266 (1938).

¹⁹ Jope, H. M., and O'Brien, J. R. P., "Hæmoglobin", 269 (Butterworths, London, 1949).

²⁰ Perutz, M. F., Trotter, I. F., Howells, E. R., and Green, D. W., *Acta Cryst.*, **8**, 241 (1955).

²¹ Riggs, A. F., *J. Gen. Physiol.*, **36**, 1 (1952).

STRUCTURE OF MYOGLOBIN

A THREE-DIMENSIONAL FOURIER SYNTHESIS AT 2 Å. RESOLUTION

By Drs. J. C. KENDREW, R. E. DICKERSON, B. E. STRANDBERG, R. G. HART
and D. R. DAVIES*

Medical Research Council Unit for Molecular Biology, Cavendish Laboratory, Cambridge

AND

D. C. PHILLIPS and V. C. SHORE

Davy Faraday Laboratory, The Royal Institution, London

MYOGLOBIN is a conjugated protein consisting of a single polypeptide chain of about 153 amino-acid residues associated with an iron–porphyrin complex, the hæm group ; its molecular weight is about 18,000, and the molecule contains some 1,200 atoms (excluding hydrogen). Two years ago a preliminary report of the first stage of an investigation of the three-dimensional structure of sperm-whale myoglobin was published in *Nature*¹ (a detailed account of this work has appeared recently²). Several isomorphous crystalline derivatives of myoglobin containing heavy atoms (mercury or gold) at single sites on the molecule were prepared, and by comparing the X-ray diffraction patterns of these crystals with those of the unsubstituted protein, it was possible to deduce the phases of all the reflexions in the X-ray pattern having spacings greater than 6 Å. These phases, together with the observed amplitudes, were used to compute a three-dimensional Fourier synthesis of the electron density in the unit cell (which contains two molecules) at a resolution of 6 Å. In this synthesis the polypeptide chain was visible as a rod of high electron density, folded in a complex pattern (Fig. 5a) ; in addition, the single hæm group with its iron atom, which is much more dense than any other atom in the molecule, could be identified as a disk of high electron density. The shape of the molecule could be determined with some confidence, as could most of the course of the single polypeptide chain, though there were several ambiguities where it turned through a large angle, so that the ends of the chain could not be located with certainty. Thus, the general nature of the tertiary

structure of the molecule was revealed, but not the secondary structure of the polypeptide chain, though the results were consistent with a helical configuration.

More recently, Scouloudi³ used similar methods to obtain a two-dimensional Fourier projection of the unit cell of seal myoglobin, in which the molecular arrangement is entirely different from that in sperm-whale crystals ; she was able to show that the myoglobins of these two species have essentially the same tertiary structure in spite of their different amino-acid compositions. Her work, by implication, confirmed the correctness of both analyses, as well as the deductions made about the shape of the molecule. In the accompanying article⁴ Perutz *et al.* now describe a three-dimensional analysis of the related protein hæmoglobin, at a slightly greater resolution, and show that each of the four sub-units of which this molecule is composed bears a close structural resemblance to myoglobin. It is apparent, therefore, that sperm-whale myoglobin possesses a structure the significance of which extends beyond a particular species and even beyond a particular protein.

We now present the results of a second stage in the analysis of sperm-whale myoglobin ; in this the resolution has been increased to 2 Å., that is to say, not far short of atomic resolution. The resulting Fourier synthesis is very complicated, and a detailed study of it will take many months ; in the meantime, our preliminary findings may be of interest.

Methods of X-ray Analysis

In this stage we have simply extended the methods which proved successful in the first stage of the analysis, comparing the diffraction pattern of unsub-

* Visiting scientist from the National Institute of Mental Health, Bethesda, Md.

No. 4711   **February 13, 1960**   N A T U R E   423

stituted myoglobin crystals with those of the *p*-chloro-mercuribenzene sulphonate, mercury diammine and aurichloride derivatives, together with a double derivative containing the first two substituents simultaneously. Whereas myoglobin crystals give 400 reflexions having spacings greater than 6 Å., the number of reflexions with spacings greater than 2 Å. is 9,600, each of which has to be measured not only for the unsubstituted protein but also for each of the derivatives. The very much greater number of data posed many problems, both in recording intensities and in computation, and in this stage we relied much more heavily than before on the use of a high-speed computer; it was fortunate that about the time the work began the *Edsac* Mark I computer used previously was superseded by the very much faster and more powerful Mark II.

The data for each derivative were recorded on twenty-two precession photographs; a separate crystal had to be used for each photograph to keep radiation damage within acceptable limits. The results from the different photographs were scaled together on the computer, the best set of scaling factors being determined by solving an appropriate 22×22 matrix[5]. The degree of isomorphism of each derivative was tested, and found adequate, by means of a computer programme which used the $h0l$ reflexions to refine the preliminary values of the heavy-atom parameters, temperature factor, etc., and then compared the values of $\delta F_{obs.}$ and $\delta F_{calc.}$ as a function of $\sin \theta$. The co-ordinates of the heavy atoms were further refined using correlation functions[6] computed by means of programmes devised by Dr. M. G. Rossmann, and finally refined again during the process of phase determination itself. The phases were determined by essentially the same method as before, but owing to the very large number of reflexions the determination was carried out on the computer rather than graphically. The 'best' phases and amplitudes[7] were computed and used in the final Fourier synthesis, to which a moderate degree of sharpening was applied. In all, some hundreds of hours of computer time were required, and the Fourier synthesis itself, which was calculated at intervals of about 2/3 Å., took about 12 hr.

The Fourier Synthesis

The electron density distribution was plotted in the form of 96 sections perpendicular to x^* and spaced 2/3 Å. apart, the density in each section being represented by a series of contours. For some purposes this method of representation is unsatisfactory, and we have also constructed models of parts of the structure on a scale of 5 cm. = 1 Å., by erecting vertical steel rods parallel to y in an array corresponding to the grid of points in the xz-plane at which densities were calculated, the value of electron density at points along the rods being indicated by coloured clips. On this scale the whole molecule is about 6 ft. cube, and about 2,500 rods each 6 ft. high are required (see Fig. 4).

The distance between singly bonded carbon atoms is 1·54 Å.; many of the common types of covalent bond present in protein molecules are shorter than this. We could not expect, therefore, that neighbouring covalently bonded atoms would be resolved in the present Fourier synthesis, and in fact such resolution was not achieved; on the other hand, the degree of resolution actually obtained is very nearly

Fig. 1. (*a*) Cylindrical projection of a helical segment of polypeptide chain, with the α-helix structure superposed: for explanation see text. (*b*) Key to the atomic arrangement in the α-helix. The points marked β and β′ are the two alternative projected positions of C$_\beta$; β is the position in a right-handed and β′ that in a left-handed helix of L-amino-acids

as great as theory would indicate, showing that the errors in the synthesis cannot be very large (see Fig. 2). The distances between neighbouring groups in Van der Waals contact is of the order of 3 Å., however, and as would be anticipated, we find that such groups appear as discrete regions of high density. One factor which complicates the interpretation is that similar groups (for example, peptide bonds) in different parts of the structure have substantially different densities. We attribute these differences to variation in the amplitude of thermal vibration in different parts of the molecule. Thus the terminal amino-end of the chain, which is on the outside of the molecule, has low density and is presumably relatively flexible; on the other hand, structures near the hæm group, which is surrounded by several segments of polypeptide chain running in different directions, possess high density, presumably because this part of the molecule is stabilized by many inter-chain bonds.

Configuration of the Polypeptide Chain

In the 6-Å. Fourier synthesis the polypeptide chain appeared as a solid rod of high density with approximately circular cross-section; segments of the chain were more or less straight, and these were joined by dense regions where the chain turned corners. In the 2-Å. synthesis all the straight segments of chain were found to be hollow cylindrical tubes of high density; projecting at intervals from the cylindrical core are dense regions of various shapes and sizes, which are clearly the amino-acid side-chains. More detailed

424 NATURE February 13, 1960 VOL. 185

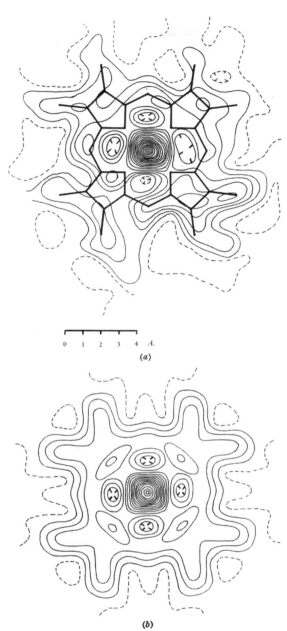

Fig. 2. (a) Observed electron density distribution in the plane of the hæm group, with the atomic arrangement in the group superposed. One contour interval = 0·5 electron/Å.³. (b) Calculated electron density distribution, computed from the known atomic arrangement as described in the text. One contour interval = 0·5 electron/Å.³

observed density follows the configuration of the α-helix with remarkable precision ; and it may confidently be asserted that not only this, but also several other lengths of polypeptide chain which have been analysed in the same way, possess the α-helical configuration. It is also found that the side-chains emerge at intervals of 100° around the periphery of the helix, and at axial intervals of 1·5 Å., in conformity with the parameters of the α-helix. Several indirect lines of evidence had led to the conclusion that parts of the polypeptide chain in globular proteins are in the form of α-helices, and in hæmoglobin Perutz[9] had found traces of an X-ray reflexion characteristic of this configuration ; but these results are the first direct proof that α-helices are present and, indeed, enable them to be seen directly for the first time.

The next question is whether the α-helices are left-handed or right-handed ; helices of either hand

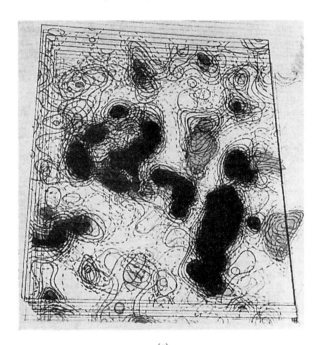

(a)

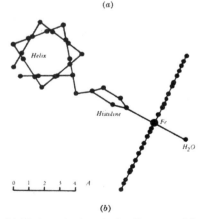

(b)

Fig. 3. (a) Photograph of a set of sections normal to the plane of the hæm groups showing, from left to right, a helix in cross-section, the histidine residue nearly edge-on, the hæm group edge-on, and a presumed water molecule. (b) Sketch showing the atomic arrangement in Fig. 3a

examination shows that, in fact, the cylindrical tubes are helices, consisting of a single strand of high density with axial repeat about 5·4 Å. The density in one such helix is shown in Fig. 1a. This was obtained by projecting all the density lying within a cylindrical shell of radii 1·6 and 2·3 Å. on to the surface of a cylinder of radius 1·95 Å. and co-axial with the helix ; the cylinder was then cut along a line parallel to its axis and unrolled. Superposed on the contours of electron density is the cylindrical projection of an α-helix, with the dimensions given by Pauling and Corey[8]. It will be seen that the

No. 4711 **February 13, 1960** N A T U R E 425

can be built, and although the right-handed form appears to be marginally the more stable, the difference is not so great as to make it certain that left-handed helices cannot exist. To decide this question, it is necessary to determine the absolute configuration of the Fourier synthesis; this function has no centre or plane of symmetry and might be plotted in two ways, each of which would be the mirror image of the other. The absolute configuration could be settled by making use of the anomalous dispersion of X-rays from the iron atoms or the introduced heavy atoms; we have not in fact made use of this effect, but instead have proceeded in the following way. In a right-handed α-helix composed of L-amino-acids, the first atom of the side-chain (C_β) projects from the main chain in a direction *opposite* to the C—O bonds of the carbonyl groups; in a left-handed helix of L-amino-acids C_β projects in the *same* direction as the carbonyl groups (see Fig. 1b). By constructing a second cylindrical projection (not illustrated) of the density near a radius of 3·34 Å., corresponding

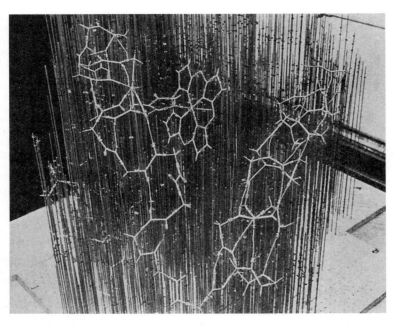

Fig. 4. Photograph of a model of part of the molecule near the hæm group, showing the vertical rods and coloured clips which indicate the electron density at each point of the grid, and atomic models of the hæm group and neighbouring helices. The terminal carboxyl end of the chain is on the extreme left

approximately to the radius at which C_β is found in either case, we found that C_β was systematically on the side of the main chain opposite the oxygen atoms of the carbonyl groups. This shows that the helix must be right-handed, and we were able to plot the Fourier synthesis with the correct absolute configuration by taking account of the known absolute configuration of a L-amino-acid[10]. The molecule was then found to be of the same hand as all the closely similar sub-units in hæmoglobin, the absolute configuration of which was determined by measurements of anomalous dispersion[4]. All the lengths of α-helix in the myoglobin molecule turn out to be right-handed. Finally, it will be clear from Fig. 1 that it is possible to determine by inspection the direction in which the C—O group points, and hence to see which is the terminal carboxyl end of each segment.

The α-helices can also be located by building atomic models into the model of the unit cell made with steel rods; this has been done for all the segments of helix in the molecule, and it is found that the total number of amino-acid residues contained in these segments is 100–110, whereas the number of residues contained in the whole molecule is believed to be 153 (Edmundson, A., unpublished results). Thus, 65–72 per cent of the molecule consists of regular right-handed α-helix, made up of about eight segments, each containing between seven and about twenty residues.

When the chain turns a corner, its regular helical configuration is necessarily disrupted. At the present resolution the precise arrangement of the residues at corners is difficult to determine; we have, nevertheless, built plausible models of several corners, though we do not yet claim to have established their configurations with certainty. Most of the corners take up two or three residues, and in addition there is one region (on the extreme right of Fig. 5a), consisting of about 13–18 residues, in which the arrangement is irregular (part of it is helical, but probably not α-helical). Further studies will be devoted to elucidating these problems.

No serious attempt has yet been made to identify the side-chains, and indeed it is doubtful whether this can be done systematically at the present resolution, partly because many of them appear to be subject to considerable thermal vibration, partly because there are often interactions between two side-chains on adjacent turns of the same helix or on adjacent helices, and it is difficult to tell where one ends and the other begins. In special cases, however, identification is easy; thus, in several places two helices approach one another closely, and one or more side-chains must for steric reasons be glycine. On the other hand, some side-chains are so long that they can only be arginine or lysine. It remains to be seen how many positive identifications can be made; but it is clear that if the whole amino-acid sequence of the protein were known, it would be possible at this stage to construct a model of the complete structure with fair precision. Unfortunately, this information is not yet available; but in the meantime it may be possible to correlate and check our findings with the partial results of the sequence determination now being undertaken by Mr. A. Edmundson, of the Rockefeller Institute.

The Hæm Group

The identification of a disk of high electron density in the 6-Å. synthesis as the hæm group is confirmed by the 2-Å. synthesis; but the iron atom is now just resolved from the nitrogens of the porphyrin ring (the iron–nitrogen distance is about 1·9 Å.) and the structure of the group as a whole corresponds closely with theoretical expectation. Fig. 2a shows a section through the density distribution, cut in the plane of the hæm group; superposed on the section is a model of the hæm group with the dimensions

426   N A T U R E   February 13, 1960 VOL. 185

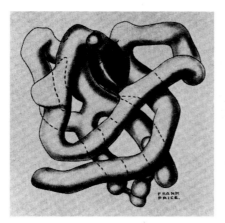

(a)

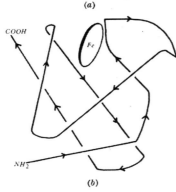

(b)

Fig. 5. (a) Drawing of the tertiary structure of myoglobin as
deduced from the 6-Å. Fourier synthesis. (Note: the plane of the
hæm group is indicated incorrectly in this drawing—see text).
(b) The course of the polypeptide chain, deduced from the 2-Å.
Fourier synthesis

found by Crute in his recent structure analysis of
nickel etioporphyrin[11]. In the model, only the carbon
atoms of the methyl groups and the first carbon atoms
of the vinyl and propionic acid groups have been
indicated; our analysis has not yet proceeded far
enough to identify the latter groups. Fig. 2b is a
Fourier synthesis made by calculating structure
factors for the model structure in a convenient unit
cell, applying to these the same temperature factor
as that observed in myoglobin, and using them as
terms of a Fourier synthesis cut off at 2-Å. spacings.
It will be seen that the calculated and observed
density distributions agree very closely; furthermore,
the observed peak density (8·8 electrons/Å.³) and half-
width (1·8 Å.) of the iron atom agree well with the
predicted values (8·2 electrons/Å.³ peak density, 1·8 Å.
half-width).   Errors are, of course, inherent in a
Fourier synthesis produced from experimental results;
the good agreement between the observed and calculated versions of the hæm group, and between the
helical polypeptide chain and the α-helix (Fig. 1),
indicate that in the present case these errors are
fairly small, certainly less than we had anticipated.
In one respect the 6-Å. model now requires correction. Measurements of electron spin resonance[12]
had shown that the two hæm groups in the unit cell
were tilted out of the *bc* plane through 21° about an
axis parallel to *c*, one in one direction and one in the
other, but from these measurements one could not
decide which direction of tilt was associated with the
hæm group of which molecule. In the 6-Å. analysis

the wrong choice was made; the new results show
that the angle of tilt is indeed very close to 21°, but
that it should be in the direction opposite to that
shown in earlier models (including Fig. 5a).

Chemical studies of hæmoglobin and myoglobin
have suggested that the hæm group is attached to
the protein by a bond from the iron atom to a
nitrogen atom of a histidine residue, and that the
sixth co-ordination position of the iron atom is
occupied in met-myoglobin (at *p*H's below 7) by a
water molecule. Our results are in conformity with
these suggestions. Fig. 3a is a photograph of a series
of sections cut through the hæm group normal to its
plane. The hæm group is seen edge-on; to its left is
the histidine residue, also nearly edge-on, and to the
left of that is the helix to which the histidine is
attached, seen in cross-section. To the right of the
hæm group is a small isolated peak which we take to
be the oxygen atom of the water molecule (see Fig. 3b).
When this structure is built with atomic models (see
Fig. 4), it is found to satisfy all known requirements
as to bond-lengths, angles, etc. Thus the iron–oxygen
distance is close to 2·1 Å.; that between the iron
and one of the histidine nitrogen atoms is 1·9 Å.,
while the second histidine nitrogen points with its
hydrogen directly at a carbonyl group on the helix,
presumably forming a hydrogen bond (forked at the
carbonyl end); and finally the bonds about the
C_α—C_β bond of histidine are almost exactly in the
staggered configuration.

General Configuration of the Molecule

In the 2-Å. synthesis the continuity of the polypeptide chain can be followed throughout the molecule; in addition, its direction can be ascertained in
each straight segment by examining the direction of
the carbonyl groups in the cylindrical projections or
by direct model-building. The 'run' of the chain so
deduced everywhere agrees with that derived from
the 6-Å. synthesis, as well as resolving several
ambiguities in that model, and the terminal amino-
and carboxyl-ends of the chain can now be identified.
The arrangement is shown in Fig. 5 (see also Fig. 5
in the accompanying article by Perutz *et al.*). The
overall shape of the molecule is as previously deduced.

Conservative estimates suggest that it may now
be possible to locate with some precision at least
43 per cent of all the atoms (excluding hydrogens) in
the molecule—namely, those making up the hæm
group and the backbone of the regular segments of
helix, together with the associated C_β's. We believe
that it may be possible, either by straightforward
refinement procedures or by heavy-atom methods, to
extend the resolution of the synthesis to a point near
the limit of the diffraction pattern of sperm-whale
myoglobin crystals (about 1·5 Å.), and thus to reveal
in any event most of the details of the atomic arrangement of the molecule. Further work is in hand with
this end in view.

We wish to thank the Rockefeller Foundation for
its generous financial support; the National Science
Foundation (U.S.A.) for providing a fellowship for
one of us (R. E. D.); to the National Institutes of
Health (U.S.A.) for fellowships (R. G. H., R. E. D.);
and to the Director and Staff of the Mathematics
Laboratory, University of Cambridge, for facilities
on the *Edsac* and for their enthusiastic co-operation.
We also wish to thank Miss Mary Pinkerton, who
was responsible for the organization of the very

No. 4711   **February 13, 1960**   NATURE   427

extensive programme of data processing, and who helped us in many other ways; a large number of computing assistants, especially Mrs. J. Blows, Mrs. W. Browne, Mrs. V. Cotterill, **Mrs.** A. Hartley, Miss J. Hughes, Mrs. J. Moss, Mr. P. Mulhall, Mrs. K. Parkes and Miss E. Rowland; Dr. M. G. Rossmann and Dr. E. Alver for making available several of their computer programmes; and Sir Lawrence Bragg for his encouragement at all stages of the work.

[1] Kendrew, J. C., Bodo, G., Dintzis, H. M., Parrish, R. G., Wyckoff, H. W., and Phillips, D. C., *Nature*, **181**, 662 (1958).

[2] Bodo, G., Dintzis, H. M., Kendrew, J. C., and Wyckoff, H. W., *Proc. Roy. Soc.*, A, **253**, 70 (1959).

[3] Scouloudi, H., *Nature*, **183**, 374 (1959).

[4] Perutz, M. F., Rossmann, M. G., Cullis, A. F., Muirhead, H., Will, G., and North, A. C. T. (see preceding article).

[5] Dickerson, R. E., *Acta Cryst.*, **12**, 610 (1959).

[6] Rossmann, M. G., *Acta Cryst.* (in the press).

[7] Blow, D. M., and Crick, F. H. C., *Acta Cryst.*, **12**, 794 (1959).

[8] Pauling, L., and Corey, R. B., *Proc. U.S. Nat. Acad. Sci.*, **37**, 235 (1951).

[9] Perutz, M. F., *Nature*, **167**, 1053 (1951).

[10] Trommel, J., and Bijvoet, J. M., *Acta Cryst.*, **7**, 703 (1954).

[11] Crute, M. B., *Acta Cryst.*, **12**, 24 (1959).

[12] Ingram, D. J. E., and Kendrew, J. C., *Nature*, **178**, 905 (1956).

DEVELOPMENTS IN GAMMA-RAY OPTICS

By Prof. P. B. MOON, F.R.S.

Physics Department, University of Birmingham

DURING the past decade, several phenomena that had previously been observed with light or with X-rays have been studied with nuclear γ-rays, where the quantum energy is typically 100 keV.–10 MeV. and the wave-length correspondingly short, 10^{-9} to 10^{-11} cm. Some new phenomena, specifically nuclear or specifically high-energy, have been added: but most of the experiments, though often technically more difficult, are extensions of those familiar at lower energies.

An excellent example of such an extension is the curved-crystal γ-ray spectroscopy of γ-rays, achieved with great elegance and precision by DuMond[1] and his collaborators. It will suffice to quote their value for the wave-length of electron–positron annihilation radiation, namely, $24 \cdot 262 \pm 0 \cdot 0033$ mÅ., and to mention that this radiation, when generated within a metal, has been observed to possess a spectral width which reflects the spread of energy of the conduction electrons.

A second example is the observation in the γ-ray region of Rayleigh scattering, a name taken over from visual optics to describe scattering of photons by the electronic structure of an atom without change of the atom's quantum state and therefore without any change of the photon's energy, except the small fraction taken by the recoil of the atom necessary to conserve momentum. This is the 'unmodified line' of X-ray scattering terminology, as opposed to the 'modified line' that arises from Compton scattering. The existence and (if it existed) the origin of elastic scattering at higher energies was a matter of controversy in 1930–33, but it is now clear that Meitner and Kösters observed it[2] and that its origin was electronic, not nuclear. In taking up the study of this scattering process with the wider range of monoenergetic lines provided by artificial radionuclides, Moon[3], Storruste[4] and Wilson[5] were able to verify semi-quantitatively the calculations of Franz[6] regarding its intensity and angular distribution, and many other workers have since made accurate experiments at energies above 1 MeV. where Franz's theory (which is essentially a Debye form factor) must be replaced by the more accurate relativistic treatment of Brown, Peierls and Woodward[7].

Two phenomena not observable in the X-ray region were predicted to be interwoven with the Rayleigh scattering; first, the classically calculable 'Thomson' scattering of photons by the nucleus, which oscillates in the electric field of the incident wave and radiates at the same frequency; secondly, scattering by virtual pair production in the field of the nucleus, which is a specifically quantum phenomenon.

The interest in the nuclear Thomson scattering seems to date from 1950, when it was remarked[3] that it should be coherent with Rayleigh scattering, but the process must surely have been considered long before then; the virtual pair production process was suggested as long ago as 1933 by Delbrück in a note to the paper of Meitner and Kösters[2].

Both processes were difficult to identify because the scattering radiation is of exactly the same frequency as the Rayleigh scattering but, at ordinary γ-ray energies, of much smaller intensity. The Thomson scattering was plausibly identified by the fact that the observed angular distribution of the total elastic scattering agreed much better with theory when this component was added into the calculation; its existence and its interference with the Rayleigh component were positively established by Sood's[8] measurements of the polarization of the elastic scattering at about 1 MeV., at which energy the Rayleigh and nuclear Thomson components have substantially different polarizations. The use of polarization for examining elastic scattering of γ-rays is pleasingly reminiscent of Lord Rayleigh and the blue of the sky.

The characteristic features of the Delbrück process are that its magnitude increases as a high power of the quantum energy and its angular distribution is strongly peaked around 0°. It has at last been unquestionably observed by Moffatt and Stringfellow[9] in Oxford, using 90-MeV. radiation from an electron synchrotron and working at angles of about 0·1°.

Another process of elastic scattering of γ-rays is that associated with the excitation of a nucleus from its ground-state to an excited state followed by the re-emission of a photon. This is the nuclear analogue of the optical resonance radiation so beautifully demonstrated by R. W. Wood in the early years of this century. It was sought experimentally at intervals over twenty years, the difficulty being not that the process is intrinsically weak (its cross-section, like that of any comparable resonant process, can be of the order of $\lambda^2/4\pi$) but that γ-ray lines are so narrow that the loss of energy to nuclear recoil ($h^2\nu^2/2Mc^2$ on emission and equal amount on absorption) destroys the resonance[2][5].

Protein Science (1992), *1*, 182–186. Cambridge University Press. Printed in the USA.
Copyright © 1992 The Protein Society 0961-8368/92 $5.00 + .00

RECOLLECTIONS

A little ancient history

RICHARD E. DICKERSON

Molecular Biology Institute, University of California, Los Angeles, Los Angeles, California 90024

(RECEIVED August 28, 1991; ACCEPTED August 29, 1991)

In September 1957, Peter J. Wheatley despaired of supporting a wife and two daughters on his Leeds University professorial salary of £900 per annum (then $2,500), resigned from the university, and prepared to move to Zurich to head up a new crystallographic laboratory for Monsanto. He was to take with him his graduate student John Daly, but with only 3 months left of my postdoctoral year at Leeds, I was told to find another supervisor. (John Daly was notable in my memory for two things: He brewed ginger beer in the closet of his apartment, having occasionally to endure the trauma of exploding beer bottles, and he was a tireless promoter of a series of fantasy novels that no one else had ever heard of, *The Lord of the Rings*, by an obscure British academic named Tolkien.)

Leeds University had a decent inorganic X-ray structure group for its day. It had oscillation cameras for data collection. An electronic computer was available at Manchester University, just over the Pennine Hills, and one could always go there to calculate a three-dimensional Patterson map from a new data set. After that you were on your own at Leeds, with Beevers-Lipson strips and a desk calculator. Understandably, we worked in projections. My project was a five-atom structure, dimethylsulfoximine, $(CH_3)_2SONH$, which to no one's surprise turned out to be tetrahedral. It was somewhat of a comedown from my nine-atom boron hydride thesis project with Bill Lipscomb at Minnesota. But my wife Lola and I thoroughly enjoyed Yorkshire, and we hated to cut short our time in England.

Fortunately, just at that juncture, Max Perutz and John Kendrew at Cambridge University began advertising worldwide for postdoctoral fellows. John's low-resolution map (6 Å) of myoglobin showed solid cylinders that everyone fully expected would turn out to be Linus Pauling's α-helices. John was gearing up for a high-resolution analysis of myoglobin, while Max was working toward an initial low-resolution picture of hemoglobin. Peter Wheatley, perhaps in a fit of conscience, sent them a warm letter of recommendation on my behalf. Bill Lipscomb, my Ph.D. supervisor in Minnesota, also had recommended me, so I received a telephone call suggesting that I might like to drive up to Cambridge for an interview. (In England, one travels "up" to Cambridge, and suspended students are "sent down." The only exception to this barographic hierarchy is London: *Everyone* goes "up to London.") I visited them, and we all went (up) to London to see David Phillips, who was collaborating with Kendrew on myoglobin data collection at the Royal Institution. There was a meeting of minds, and at the end of 1957, Lola and I moved all our worldly goods, which fitted into one very small British Ford "Popular," south to Cambridge. (The next larger model, which we couldn't afford, was called the "Prefect." Ford Prefect later became famous as the adopted name of the alien hero of the *Hitchhiker's Guide to the Galaxy*, but that's another story.) I had written to Herb Gutowsky at the University of Illinois to see if they would wait one more year for their assistant professor, and if they would be happy having a protein crystallographer rather than an inorganic crystallographer. They would. I also wrote to Lipscomb for advice, and he replied, "It's a wonderful opportunity, but be careful—don't become a professional postdoc!"

Max and John's offices and wet laboratories were in a one-story corrugated prefabricated building in a courtyard of the Cavendish Physics Laboratory complex. This construction, known as "the Hut," had been built for Metallurgy during the Second World War. It was a great experience to sit in an office with fellow postdoctorals Roger Hart and Alver from Norway, with John Kendrew in the adjacent office separated only by a plasterboard partition, Sidney Brenner and Francis Crick across the corridor, and Max Perutz across from John. Eleven o'clock coffee was heralded each morning by Francis

Reprint requests to: Richard E. Dickerson, Molecular Biology Institute, University of California, Los Angeles, Los Angeles, California 90024.

Crick's unmistakable laugh ringing down the corridor. When he left for an extended visit of his own to the U.S., he was sorely missed.

At one end of the long building was the densitometer room, where several young women spent their days feeding precession films through a Joyce-Loebl double-beam microdensitometer and measuring peak heights by hand with a millimeter scale. At the other end was the sole wet lab, occupied by people such as Seymour Benzer and Leslie Barnett. Figure 1 shows part of the structure group outside the Hut in 1958. The universal suit and tie among men did not reflect any "dressing up" for a photograph; these were standard laboratory wear at the time. The French popular science journal *La Recherche* once paid Max a left-handed compliment by describing him as, "a person who gives the appearance of wearing clothing chiefly to keep warm." In England in 1958 this was no small matter!

The Hut still is to be seen in the Cavendish courtyard, a melancholy relic that stands open to the weather and is used only as a bicycle shed. No one there apparently realizes today that the Hut is an historical treasure, and should be bronzed. Once while I was there in 1959, two Russian visitors came to visit Perutz. As he hosted them in his office in the Hut, they exclaimed in puzzlement, "But where is your Institute?" It is typical of Max that he

took pleasure in telling them with a smile, "This is my Institute." Several years later (1962) he got his Institute, on the Addinbrooks Hospital site on Hills Road. But it just wasn't the same thing.

Fortunately, there was room for the rotating anodes and other X-ray data collecting equipment in the basement of the adjacent New Cavendish Building. Cambridge, unlike Leeds, actually had a computer of its own: EDSAC II, a marvel with 2,000 words of fast access core storage, plus magnetic drum and tape. People there told me how tedious the old EDSAC had been, and how much of an improvement the current machine was. There were three categories of customer: "Users," "Partially Authorized Users," and "Fully Authorized Users." Users could perform calculations only during the day, or under the watchful eye of one of the higher grades. Fully Authorized Users were competent to turn on EDSAC II at the beginning of the day, and turn it off at night when the last job was finished. I ultimately rose to the rank of Partially Authorized User: I wasn't allowed to power up the computer, but could work as late as I liked at night, shutting down the machine at the end by throwing off a set of wall switches in carefully prescribed order.

I elected to work with John Kendrew on the high-resolution structure analysis of myoglobin. When I arrived, the plan had been to work at 2.5-Å resolution, but every-

Fig. 1. Part of the myoglobin/hemoglobin structure team outside the Hut in 1958, with the brick New Cavendish Laboratory and the chimneys of Old Cavendish behind. From left to right: Larry Steinrauf and Dick Dickerson (postdoctorals), Hilary Muirhead (graduate student), Michael Rossmann (postdoctoral), Philip ? (face obscured), Anne Cullis (research assistant), Bror Strandberg (postdoctoral), ? Wiebenga and unknown (technicians), and Max Perutz (white coat). In front: Leslie Barnett and Mary Pinkerton (research assistants).

184

one worried whether the cylinders of the 6-Å map would resolve themselves into α-helices at 2.5 Å. The largest circle of data that could be collected on a 5-inch-square precession film cassette with our 90-mm film distance was 2.0 Å, so we decided in laboratory discussions to go for the limit and to jump to 2.0 Å in one step. We were still worried, however. At 2.0 Å resolution, would an α-helix be a solid rod, a hollow "garden hose," or a hollow tube with spiral pattern around the outside? No one was really sure. Roger Hart was an electron microscopist, and had come from London to Cambridge after the tragic and premature death of Rosalind Franklin, with whom he had been a postdoc. Alver and I were small-molecule crystallographers and shared the odd distinction of being the only people in the laboratory who had ever solved an X-ray structure at the atomic level, Perutz and Kendrew included. But soon our ranks were swelled by the arrival of Michael Rossmann, to work with Perutz. Michael had moved from J. Monteith Robertson's crystal structure group in Glasgow, to a postdoctoral position with Bill Lipscomb in Minnesota, and we had overlapped for the last few months of my doctoral work there.

Michael shared an office with David Blow, and the two became our computing experts. I don't recall whether Michael ever achieved Fully Authorized User status on EDSAC II, but his least-squares and rotation-translation programs became so complex that the Computing Center developed the practice of using one of them for morning computer checks; they put the EDSAC to a more strenuous test than did their own diagnostic routines! EDSAC II was a thermodynamic monster, with a panel in the rear containing an array of 4-foot-long pull-out vacuum tube racks, each with a large handle. A twist of its handle and a pull would bring one rack sliding out, with vacuum tubes arranged in double rank like soldiers. The wiring diagrams for EDSAC II were in pencil on a large bundle of mechanical drawings. Whenever a change in wiring was made, the old diagram would be erased and the new circuit pencilled in. There may have been proper ink diagrams somewhere, but I never saw them.

Input/output on the EDSAC II left something to be desired: standard British telegraph tape readers and punches. The reading was mechanical rather than optical. Little fingers in the tape reader pushed against the tape, looking for holes. We developed the sloppy habit of editing minor tape glitches with patches of Scotch tape, but this gave the Computing Center fits. The patch eventually would be pushed off the computer tape, and the adhesive would gum up the tape reader. The Computing Center periodically fulminated against such a practice ("Any input tape found to be patched manually will be confiscated!") but could not stop it completely. The 2-Å myoglobin data set consisted of nearly 10,000 reflections from native myoglobin and each of three heavy atom derivatives. We wrestled with the problem of how to sort and merge 40,000 data points, in a computer with only 2,000 words of rapid-access memory. Finally we decided to cut the four original data tapes into strips of a common h and k, sort them by thumbtacking them onto a grid on a large piece of bulletin board, and then merge them manually by running the strips through a tape reader and punching out a new one-piece data tape. Figure 2 illustrates this process.

The three derivatives were parachlormercuribenzene sulfonate (PCMBS), mercury diammine, and gold chloride. The two mercurials were freshly prepared, but the gold chloride came from a cache of heavy atom diffusion trials that had been set up 2 years earlier by Gerhard Bodo and Howard Dintzis. Several hundred 2-cc vials of crystals were stored in a laboratory cupboard, along with Bodo's and Dintzis' log books. Bror and I tested endless numbers of these on the precession camera, looking for those that gave usable intensity changes. Under gold chloride, Bodo or Dintzis had written, "Two weeks: No changes," and "Twelve weeks: No changes." But after 2 years, I found beautiful intensity changes in one single vial with gold chloride-soaked myoglobin crystals. This became our third isomorphous derivative, but one with unexpectedly tragic consequences.

I found crystal mounting to be a tedious and fiendishly difficult process. The myoglobin crystals in general were too large to fit into the 1-mm-diameter glass capillaries, and had to be cut into quarters under the microscope with a sharp razor blade. A fragment then had to be maneuvered into the capillary, a plug of mother liquor added at each end, and the ends sealed with hot wax. I broke one capillary after another trying to learn how to do it. It looked so easy when Kendrew had shown me how. One of my office mates, Roger Hart, was surveying hemoglobin derivatives for Perutz, like Strandberg and I with myoglobin. Roger's hands were crippled by polio, and he became a challenge—if Roger could learn to mount crystals, then by God I could too. I finally mastered the technique. Later I discovered that Roger didn't mount his crystals at all—Max came in evenings to mount a batch for him!

Max and John were utterly different in personality. Kendrew came in two or three mornings a week to discuss the progress of the research, and to give help where help was needed. He was a great mentor for someone who wanted to learn how to be an independent investigator. At other times he was busy as a science advisor to the British government (on the Polaris missile system as I recall), as an administrator of Peterhouse (college), and on other affairs. In contrast, Max was never so happy as when in the laboratory at the bench, doing science. One learned by talking with John, but by watching Max.

The tragedy of gold chloride occurred one black Monday morning when I came into the laboratory ready to mount more crystals for data collection and found that I had failed to screw down the cap of the 2-cc vial tightly the previous Friday. The tube with the precious crystals

Fig. 2. Bror Strandberg (left) and Dick Dickerson (right) returning from the EDSAC II computing center carrying the paper tape sorting board for the myoglobin 2-Å data set. Individual strips of tape contained intensities from minimum to maximum *l*, and were sorted manually on *h* and *k* by thumbtacking them to the proper square on a grid drawn on the sorting board (high tech!). Tape strips for native myoglobin and the various heavy atom derivatives were tacked together and later were read through a tape reader again to produce one long master data tape.

was completely dry, and an efflorescence of dried salt covered the outside of the tube. I was stunned. I walked into Kendrew's office, explained briefly what had happened, and tendered my resignation.

Kendrew didn't shout at me and didn't laugh at me, which would have been even worse. Instead, he said calmly that of course I wasn't going to resign over a thing like that; we would just have to see where we were with the gold derivative. As feared, there were no more vials of crystals in gold chloride, and the project obviously could not wait 2 years for more gold to diffuse in. So we went with what we had already collected. Hence to be accurate, the 2-Å resolution structure analysis of sperm whale myoglobin was not carried out with three isomorphous derivatives, but with 2.75.

Space forbids saying much about how the data were collected, derivatives refined, and phases analyzed. The final calculation of the three-dimensional electron density map on EDSAC II was an all-night party. Dave Phillips and some of his group came up (sorry, down) from London for the event. (One heavy atom data set had been collected at the Royal Institution.) It took literally all night to calculate the map. John was sufficiently worried about machine errors that he then repeated the entire Fourier synthesis calculation on a defense computer to which he had access at a military base. We plotted the

map sections on Plexiglas sheets, stacked the sheets over a light box, and threw a cocktail party at dusk on the Peterhouse lawn to celebrate. I vividly remember Sir Lawrence Bragg, director of the Royal Institution and the man who had brought Perutz and Kendrew to Cambridge, taking the elbow of guests at the party and propelling them to the light box, pointing at an α-helix that ran obliquely through the map sections, and saying excitedly: "Look! See, it's hollow!" Hollow it was indeed, and striped with a barber pole pattern that was clearly resolvable into a $-NH-C_\alpha-CO-$ polypeptide backbone. As a genuinely unexpected dividend, at 2.0 Å you could even use the carbonyl groups to figure out which way the backbone chain ran. A small thing today, perhaps; but in 1958 who knew what to expect? It had never been done before.

The map was calculated in August 1959. I literally threw the maps on Kendrew's desk and ran for the boat train to Southampton. Bror Strandberg stayed 3 months longer and helped interpret the map in detail. Herman Watson joined the group and carried on the interpretation.

One interesting postscript: J.D. Bernal of Birkbeck College once remarked that you would never be able to interpret the three-dimensional Patterson map of a protein unless you built a model big enough to walk through.

That was nearly true of the Fourier map of myoglobin. Kendrew built a "wire forest" model by sinking 4-foot steel rods into a 6-foot-square plywood base on a 1-inch grid, and then color-coding electron density with small spring clips: white for the highest density — yellow — orange — red — green — blue — black. He then invented what today are called "Kendrew models" — accurate skeletal wire atoms and groups held together by locking couplings — and built the myoglobin molecule into the forest of wires. When the International Union of Crystallography held its Congress in Cambridge in the summer of 1960, Kendrew's huge myoglobin model was one of the showpieces. But the saying about a prophet being without honor in his own country was true. The room in which the model was constructed was owned by Electrical Engineering, and John was told in no uncertain terms that he would have to get his junk out of that room before the Fall term commenced! Part of the wire forest model is preserved today in the Science Museum in South Kensington in London.

Max Perutz and John Kendrew were awarded the Nobel Prize in Chemistry in 1962 for their hemoglobin/myoglobin analyses, and Francis Crick, Jim Watson, and Maurice Wilkins shared the Nobel Prize in Medicine that same year for their DNA structure. Did we consciously think we were working on Nobel Prize projects in 1958? Not really. We were aware that we were breaking new ground, and that the results would be considered important, but it was not clear how important. Max and John worked hard because they were passionately interested in the answers. Hemoglobin was, and still is, Max Perutz' life (along with skiing, the structure of ice, and a few other diversions). There is an old cartoon of a young boy at a children's party, in suit and tie with party hat and streamers, saying self-consciously to another child, "I hate this kind of thing, but I want a happy childhood to look back on." I regard myself as having had a very happy scientific childhood: enjoyable at the time, and even more pleasurable in retrospect as the field moves on. It was true in 1958, and still is true today that, to paraphrase an Arabic saying about Grenada quoted by Washington Irving: "Allah gives to those whom he loves, the means of living in Cambridge."

-8-

THE KNOWLEDGE EXPLOSION

The myoglobin/hemoglobin work created quite a stir and led to rapid recognition by Stockholm. But progress with other proteins in other groups, trying to use the same methods that had worked in Cambridge, was slow. It was difficult to isolate, purify and crystallize a new protein. The search for usable heavy atom derivatives, in which the metal atom bound in the *same* place on each protein molecule, without altering crystal packing or otherwise changing the crystals, was arduous. Film-based data collection took months, rather than days as at present. The available computers, even so-called "main frame" giants, could be large in size but were small in computing power and storage capacity. And the convenience of computer graphics display and analysis of the resulting electron density maps did not then exist.

By 1963 several other globins had been spun off from the Cambridge work. But only three non-globin proteins had been analyzed in three dimensions, and these only at low resolution (1). Joe Kraut at the University of California, San Diego, had solved the bovine enzyme precursor *chymotrypsinogen* at 4 Å resolution. Harry Carlisle at Birkbeck College, London, had obtained a 6 Å map of bovine pancreatic *ribonuclease*, and David Harker at Brooklyn Poltechnic Institute had a 4 Å map of the same protein. Fred Richards of Yale was working on a low resolution structure of a variant, *ribonuclease S*. David Phillips of the Royal Institution in London, Robert Corey at Caltech, and Dickerson, then at the University of Illinois, had each calculated 5 Å or 6 Å maps of hen egg white *lysozyme* in various space groups, and Phillips was driving ahead toward a high resolution map.

And that was it. Other protein structure projects were in process (see Table III of reference 1), but none had reached even the low resolution stage. No protein of any kind other than the globins had been pushed to atomic resolution. Perutz became concerned about this apparent lack of progress in protein structure analysis. To help matters along, he organized a protein workshop in Hirschegg, Austria in late March of 1966 and invited approximately forty colleagues from all over the world to participate. In view of Max's love of skiing, it was no surprise that the workshop was scheduled in the Austrian Alps during ski season. Plate IX shows a general view of the vicinity of Hirschegg in the Kleines Walsertal, just across the Austrian border south of Munich. Plate X shows the conference center, a facility owned by the Technical Institute of Darmstadt and known as the Waldemar-Petersen Haus. Several workshop delegates also can be seen in their typical morning regalia. Max patterned the workshop after the Gordon Conferences, which hold scientific sessions morning and evening but leave afternoons free for recreation. But he decided that the conditions for good skiing demanded that science be relegated to afternoons and evenings, with mornings open to take advantage of good snow. Ski instruction, fortunately, was available for people from flat places such as Cambridge and the American midwest who were unfamiliar with the sport. In Plate XI, Max is flanked by four colleagues during one of the sessions, and Plate XII show a general view of the audience.

The first Hirschegg workshop was a great success as a morale-building exercise in the field; so much so that two

FIGURE 8.1 Low-tech protein model building at Caltech around 1968, in the pre-computer-graphics era. Printed computer output with electron density numbers on a grid would be contoured by hand, taped together to make larger sheets, and their density contours copied off onto large Plexiglas sheets which would be stacked in three dimensions. From left to right: Tsunehiro Takano, Olga Kallai, Kathy Stephenson, R. E. Dickerson, David Eisenberg, Robert Corey, Robert Stroud, Joan Varnum, 2 others (unknown). The caption on the picture on the back wall of a medieval scholar holding a small inscribed polyhedron read, "Is there *really* a heme in trypsin?" The board with rows of colored thumbtacks at left rear compared the amino acid sequences of cytochrome *c* from various species from man to plants and microorganisms.

more were held there in 1968 and 1970, and a fourth meeting was moved to Alpbach, Austria in 1972. The number of attendees roughly doubled at each new workshop. At this point Max felt that his "pump-priming" activities had succeeded. The tradition was continued by the East Coast and West Coast Protein Workshops in the U.S., and then by Gordon Conferences and by biennial "Conversations" organized by Ramaswamy Sarma at SUNY Albany. By 1968, high-resolution structures were available for six enzymes: lysozyme, ribonuclease, chymotrypsin, carboxypeptidase, papain and carbonic anhydrase (2), and many more projects were in motion.

Figures 8.1 and 8.2 are a reminder of the trauma involved in interpreting a protein electron density map in the late 1960's. Electron density values were plotted out on large sheets of computer paper, at regularly spaced grid points in the unit cell. These then were contoured by hand at "contouring parties" like that of Figure 8.1. This photo has been included principally because it shows one of the founding fathers of protein structure, Robert Corey, seated quietly at center watching the activities of a group of young contourers.

The next stage in interpretation is depicted in Figure 8.2. The contoured maps were copied off onto Plexiglas sheets, which then were stacked at the correct spacing to produce a three-dimensional map of protein electron density. Unfortunately, it was not a map within which one could build an atomic model of the protein. Kendrew's forest of wires and colored clips mounted on a plywood base was never imitated. Instead, people tried to look *through* the map sections, and build a separate model to one side that reproduced what was seen.

Then Fred Richards came up with a brilliant technical innovation that enabled one to "build the model within the map." This was the *Richards Box*, seen in Figure 8.3. Legend has it that he thought of the idea while sailing his own boat from New England to Old England for a sabbatical at Oxford. The key element of the Richards Box is a half-silvered mirror, set at an angle of 45° from the vertical. The plastic map sections with contours are stacked vertically behind the mirror in front of a light box, while the wire model is constructed below the mirror. As one looks horizontally into the mirror, one sees a direct view of the density sections, and a reflected view of the model. If one stacks the sections in inverse order, the model that is built from them has the correct handedness. Sections through the molecule can be emphasized by choosing particular map levels, and by illuminating slices through the molecule with collimated light sources, the flat boxes at left and right in Figure 8.3.

FIGURE 8.2   Stacked Plexiglas sheets with electron density contours of cytochrome *c*. The model parts shown were not Kendrew models, but those developed independently by Pauling and Corey from their studies of bond lengths and angles. Cytochrome *c* was the only protein ever built using Pauling/ Corey model parts. Left to right: Eisenberg, Dickerson, Takano and Stroud.

There is an interesting sidelight connected with the Richards Box. In 1976 while in Moscow I visited Dr. Natalia Andreeva at the Institute of Crystallography. She proudly showed me the new Richards Box she had built. An American colleague had sent her a roll of half-aluminized plastic sheeting, and if a piece of this was wetted down and pressed tightly against a sheet of plate glass, the result was a very serviceable substitute for a true half-silvered mirror. She explained that it was absolutely impossible, even for a member of a government research institute, to obtain half-silvered mirrors in the Soviet Union. I told her that at home one needed only go down to the Pasadena Auto Glass Shop, describe what was wanted, and then pick it up a day or two later. I privately hypothesized that half-silvered mirrors were not available in the Soviet Union because the KGB used up all that were produced.

Progress in protein crystal structure analysis in the 1970's was truly explosive. The Protein Data Bank was set up at Brookhaven (now at Rutgers) as a repository for x-ray data and atomic coordinates from structure analyses. There was a lively controversy over

FIGURE 8.3   A Richards Box in use with the high-resolution cytochrome *c* molecule. The model is being constructed below, with the region being worked on illuminated from the sides by collimated light sources. (The flood lamp at left is only for photographic purposes.) The electron density map is seen at upper rear in front of a light box. All sections that do not correspond to the illuminated portion of the molecule are pulled out of the way to the right. The two images—map and model—are viewed superimposed in a half-silvered mirror set at a 45° angle. By this time the Pauling/Corey model parts had been abandoned in favor of the more easily measured Kendrew models.

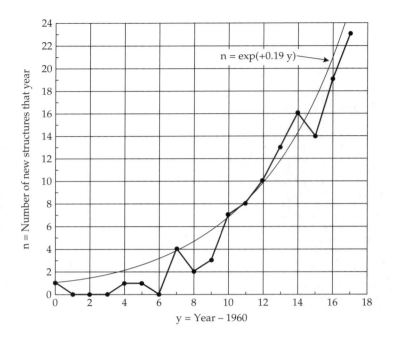

FIGURE 8.4   Plot of the rate of deposition of protein structure coordinates in the Protein Data Bank between 1960 and 1978. The data are fitted quite well by an exponential: n = exp(+0.19 y), where n is the number of coordinate sets deposited in a given year, and y is the year minus 1960. Note the sudden burst of activity in the years immediately following the 1966 Hirschegg meeting (y = 6).

requiring deposition of x-ray intensities as well as coordinates. A tradition in science is that one must make scientific data freely available at the time of publication, so others can check the validity of your results. Credit demands credibility. Some people referred to the atomic coordinates of a protein as the "data," but it was pointed out by others that these coordinates were not the experimental *data*; they were the *results*. The data were the measured x-ray diffraction intensities themselves; only with these could anyone possibly check the correctness of a published structure. One reason for the unwillingness of many to deposit their x-ray data was the understandable reaction that, if you had worked and struggled for five years or more to collect data and solve a protein structure, you were reluctant to release the data concurrently with its very first publication, and have someone who had never been involved with the project rush in, use your data, and see things that you had not yet noticed. (Shades of the DNA controversy!) But resistance died down as the labor in solving a protein structure diminished with improved technology, and today deposition of both x-ray data and coordinates is the norm.

In 1978, in preparation for a revision of Dickerson/Geis *Proteins* which never materialized, I plotted the number of protein coordinate sets deposited in the Protein Data Bank each year since 1960, and found that they followed a smooth curve, Figure 8.4, which could be fitted by the equation: n = exp(+0.19 y). Nothing more was done with this until twenty-four years later in 2002, when Arthur Arnone tested the formula against current depositions in the Protein Data Bank and to his surprise found that it still held up remarkably well: a total of 12,066 protein structures predicted vs. 12,123 structures actually in hand (3, 4). He named the equation "Dickerson's Law," and speculated about who might be responsible for the extra 57 structures. This is all very well, but it illustrates the perils of overextrapolation. Were this law to hold until the year 2033, there would have to be just over one million *new* protein structures that year, or three thousand per day! That would be megaproteomics with a vengeance!

References

1. R. E. Dickerson. 1964. In *The Proteins* (Hans Neurath, ed.), Academic Press, New York, 2nd edition, Vol. 2, pp. 603–778. "X-Ray Analysis and Protein Structure."
2. R. E. Dickerson and Irving Geis. 1969. *The Structure and Action of Proteins*, Benjamin/Cummings, Menlo Park, CA.
3. A. Arnone. 2002. *EnVision*, 18 (Jan/March), p. 10. Also available on-line at www.npaci.edu/online/v6.4/pdb.predict.html.
4. K. M. Reese. 2002. *Chemical and Engineering News*, May 13, p. 64.

–9–

EPILOGUE

The Age of Giants is behind us, and regrettably, few of the giants are left. Rosalind Franklin, of course, is the tragedy: she died in 1958, two months short of age 38, just as the DNA structure was coming into its own. She missed out on the Nobel Prize and the honors that were bestowed on Wilkins, Watson and Crick. William Astbury, the godfather of protein structure analysis, died three years later at age 63, after a long and illustrious career at Leeds University. During my first postdoctoral fellowship at Leeds in 1957, I spent a year in the same University with Astbury, but did not have the common sense or presence of mind to walk across campus to his Manmade Fibers Institute and meet him. At the time, as a structural inorganic chemist, I barely knew the difference between a protein and a proton. One was bigger than the other, but which was which? Max Perutz later told me that Bill Astbury had been a wonderful person and a joy to be around, as the limerick quoted from Lindo Patterson at the head of Astbury's 1938 review suggests.

When Sir Lawrence Bragg turned 80 in 1970, the Royal Institution (of which he was the former Director) gave a symposium in his honor, at which speakers described progress in the several areas of x-ray structure analysis in which Bragg had excelled. The historic lecture hall in which the meeting was held had hardly changed since it was depicted by Gillray in 1802 (see Figure 6.13). The U. S. National Academy of Sciences and the Soviet Academy each sent official words of congratulations and commemorative gifts. The U. S. gift was a handsome illuminated scroll extolling Bragg's achievements, signed by many American members of the Academy. But the Russians completely upstaged this by presenting Bragg with a geologist's hammer and chisel, constructed entirely from synthetic ruby grown by Russian scientists, housed in a velvet-lined walnut case. As I sat in the audience watching the Russian public-relations triumph, I smarted under the thought that with a little forethought we could have turned the tables. One of Bragg's most important early achievements was the x-ray crystal structure analysis of silicate minerals. Why didn't our Academy simply present Bragg with a few highly selected and beautifully housed specimens of Moon rocks?

All during the symposium Sir Lawrence sat in a wheelchair in the front row with a blanket over his knees, frequently appearing to be dozing. I wondered how much of the symposium he actually was following. But at the conclusion he came to the podium and gave a fine impromptu farewell in which he managed to say something intelligent and complimentary about each and every talk of the symposium. Contrary to appearances, he had heard every word. At the end he thanked us for his 80th birthday celebration and invited us all back for his 85th. Regrettably, a year later he was no more.

1971 also saw the passing of another of the giants, Robert Corey. I had the privilege of being a faculty colleague of Corey for eight years at Caltech until his death, sharing the macromolecular x-ray crystallography laboratory. During their earlier years of collaboration, Pauling was the firebrand and the fount of new ideas, Corey was the man who made things happen in the laboratory, and in effect, proved Pauling right (or wrong). Theirs was a complementary and productive partnership.

FIGURE 9.1 Francis Crick at a party in Del Mar, California in 2002, still smiling at the age of 86! (Photo courtesy of Odile Crick, La Jolla, CA.)

Pauling apparently lost interest in protein structure after his seminal work with Corey in the 1950's on the α-helix, β-sheets, and other types of polypeptide chain folding. He never actually solved a globular protein structure, and in 1964 left Caltech first to focus on issues of public policy and then to found an Institute of his own in Menlo Park. His later work on vitamin C cannot compare with his use of quantum mechanics to revolutionize theories of chemical bonding, and his application of these principles to the unraveling of the structures of polypeptide chains.

John Kendrew and Max Perutz offer an interesting contrast. John organized and managed (and did so brilliantly); Max worked in the laboratory. After their Nobel triumph in 1962, Max continued hemoglobin research and fostered the growth of the MRC Laboratory of Molecular Biology at Cambridge. In contrast, after the myoglobin work was completed, John left Cambridge to become the first Director of the EMBO Molecular Biology Laboratory at Heidelberg. Many years later he became the Master of St. Johns College, Oxford. His first love was administration; he quickly moved out of active scientific research.

A meeting of the International Union of Crystallography in Seattle in 1996 included a symposium featuring a group of Nobel Laureates, Kendrew among them. In a discussion period they were asked: "If you had to live your scientific life over again, would you do the same things, or would you do something differently?" Most of the Laureates responded predictably that they were quite happy with the way things went the first time. Kendrew took another tack. In Heidelberg he found that he enjoyed scientific administration much more than basic research *per se*. Had he realized at the outset how much he enjoyed administration, he probably would have gone directly into that, in place of the years working on myoglobin. In retrospect I am not sure that he really meant this; but in any event there is a flaw in his logic. Had he not earned a Nobel Prize for the myoglobin work, would he have been chosen to head the new EMBO laboratory?

John lived until he was 80, in 1997. Max did even better, making it to age 88 in 2002. In parallel with scientific research he showed his talents as a brilliant science writer, publishing reviews, essays and memoirs on science and science policies, many of which were collected in books that still make absorbing reading. In 2001 my colleagues David Eisenberg and Mary Kopka organized an October symposium at UCLA to commemorate my 70th birthday. Max was one of the speakers. But in the wake of the 9/11 terrorist attack he had second thoughts about traveling. Would it not be prudent, he suggested, to postpone the symposium for a few months until things calmed down? We considered this but concluded that it was too late to cancel the meeting with only four weeks' notice. To our regret, Max replied that he just did not feel it was wise to travel in October, and dropped out (as did two or three others). In retrospect we wondered whether it might have been better to delay the meeting for six months as he suggested. But it would have done no good; six months later Max was gone.

Peter Pauling I knew only casually, having talked with him at a few scientific meetings. We were the same age, born in 1931. He had a turbulent scientific and personal life, and was living in retirement in Wales when he died in 2003. Among all the Pauling children, Peter suffered most by choosing the same career as his father, x-ray crystallography, and living constantly under his shadow. It was after talking with Peter than I decided that for their own good, none of my five children should pursue exactly the same career in structural molecular biology as myself. One vivid image remains: At one meeting the woman with Peter told me about her first trip to the Pauling ranch at Big Sur, California. Everyone was friendly and made her feel at home. But she noticed that at breakfast so much vitamin C powder had been stirred into the orange juice that a spoon would stand straight up in the glass!

It was never my privilege to meet John Randall of King's College in London. But I had heard of him before I had any idea of who he was, or even of protein or DNA research. Several sources, including the BBC special of 1986, "The Race for the Double Helix," have implied that part of the conflict between Rosalind Franklin and Maurice Wilkins at King's arose because Randall, as director, was constitutionally averse to drawing lines of demarcation and making tough decisions. He wanted to remain "above the battle." In 1957, while a postdoctoral fellow in inorganic crystallography at Leeds University, I attended an Institute of Physics meeting in Cardiff. While there I was told a "Randall story." Supposedly

FIGURE 9.2   James Watson and the wire model of DNA made for him in 1953 by Tony Broad (see Figure 5.12). This photo was taken by Andreas Feininger in the 1950's for Time/Life, and used in 1999 by Apple Computers as part of an advertising campaign entitled "Think different". (Photo © Getty Images/Andreas Feininger and © Apple Computer Inc.)

Randall had met a young scientist at an earlier Institute of Physics meeting, and engaged him in conversation. At the end he remarked, "I have enjoyed talking with you, young man. Where are you studying?" When the student replied, "King's," Randall continued, "Indeed. I myself am at King's. With whom do you work?" To this the young man responded, "You, Sir!" True or not, the story indicates the kind of indecisiveness attributed to Randall by some of his contemporaries.

As these lines are being written, Maurice Wilkins, Francis Crick (Figure 9.1) and Jim Watson (Figure 9.2) are the three survivors of the Golden Age.* I never had the privilege of meeting Wilkins, and my acquaintance with Watson came later; he had left Cambridge before I arrived at the end of 1957. Francis Crick, however, was an inspiration to an aspiring young scientist. He and Sydney Brenner shared the office across from mine in the one-story, corrugated-roofed "Hut" where the protein structure group had its laboratories. The coffee hour at 11 am each day would be signaled by Crick's booming laugh in the corridor. (The same laugh that, according to Wat-

---

*In the spring of 2004 this was true, and I was fortunate in corresponding with Francis about the graduate course upon which this book is based. Regrettably, he succumbed to cancer on 28 July in La Jolla. This was an immense loss, both to those fortunate enough to have known him personally, and also to the larger audience who benefited, and will continue to benefit, from his contributions to science in many fields. One does not encounter his like often. But there is yet a new melancholy component to the story. Maurice Wilkins died in a London hospital on 5 October. He was the "quiet man" of the DNA Nobel trio, but was enormously respected both for his science and for his sense of social responsibility. After these two developments I can only say to the last survivor of the structural molecular biology revolution: "Jim, take care of yourself!"

son, drove Bragg crazy. But I welcomed it.) Francis did two significant things for me: He triggered a lifelong interest in the origin and evolution of life by loaning me the collected papers from an Origin of Life symposium in Moscow honoring A. I. Oparin, and he taught me how to see stereo pair drawings in three dimensions without the use of stereo glasses. (His method: All our life we have established a link between focus and convergence of our eyes. For close objects, focus close and converge the eyes; for distant objects, focus for distance and diverge the eyes. The trick to seeing in stereo is to decouple these two reflexes; to focus close while keeping the eyes diverged, so that each eye sees a different drawing. Difficult, yes, but it can be learned.)

There is a certain parallel between Kendrew vs. Perutz, and Watson vs. Crick. Both Perutz and Crick remained active research scientists all their lives, although Crick left molecular biology for neurobiology and the study of the mind. In contrast, both Kendrew and Watson moved from research into science administration, carving out distinguished careers: Kendrew at Heidelberg and Oxford, Watson at Harvard and then at Cold Spring Harbor, which he has headed for many years.

One last memory will illustrate the graciousness and magnanimity that was Francis Crick's hallmark. In 1980 my graduate student Horace Drew solved the first single-crystal x-ray structure analysis of B-DNA, the synthetic dodecamer: C-G-C-G-A-A-T-T-C-G-C-G. In those pre-computer-graphics days we plotted the electron density on Plexiglas sheets and stacked the sheets over a light box to create a three-dimensional image. Francis Crick had come to Caltech, not to visit the x-ray group, but to talk with the neurobiologist John Olds. However, since he was on the campus, he made a detour past our laboratory to see the new B-DNA structure. As it happened, when he arrived the only people in the laboratory were Horace, my long-time scientific colleague Mary Kopka, and myself. The four of us stood around the light box in a darkened room, staring at the image of the electron density contours and the atoms that had been marked on the map. Francis then turned to Horace and made a comment that Horace would never forget: "So THAT'S what it looks like!" Horace metaphorically floated to the ceiling and bobbed around the room like a hydrogen balloon. For weeks afterwards he would tell everyone who would listen what the great Francis Crick had said about his B-DNA structure! Crick had the gift of inspiring those around him.

-Appendix 1-

PIONEERS OF STRUCTURAL MOLECULAR BIOLOGY 1933–1963

Name	Situation in 1953						
	Position	Institution	Born	Died	Age in 1953	Field	
Sir Lawrence Bragg	Director	Cambridge	1890	1971	63	p/n	*
Dorothy Wrinch	Research Professor	Smith	1894	1978	59	p	
Robert B. Corey	Professor	Caltech	1897	1971	56	p	*
William T. Astbury	Professor	Leeds	1898	1961	55	p	
Linus Pauling	Professor	Caltech	1901	1994	52	p	*
Erwin Chargaff	Professor	Columbia	1905	2002	48	n	
John T. Randall	Director	Kings	1905	1984	48	n	
Dorothy C. Hodgkin	Professor	Oxford	1910	1994	43	p	*
Max F. Perutz	Faculty	Cambridge	1914	2002	39	p	*
Francis H. C. Crick	Graduate student	Cambridge	1916	2004	37	n	*
Maurice Wilkins	Faculty	Kings	1916	2004	37	n	
John C. Kendrew	Junior Faculty	Cambridge	1917	1997	36	p	*
Jerry Donohue	Postdoctoral	Cambridge	1920	1985	33	p	
Rosalind Franklin	Faculty	Kings	1920	1958	33	n	
James D. Watson	Postdoctoral	Cambridge	1928	——	25	n	*
Peter Pauling	Graduate student	Cambridge	1931	2003	22	p	*

Field abbreviations: p = worked mainly with proteins; n = worked mainly with nucleic acids; * = people whom I knew personally.

– Appendix 2–

HIGHLY RECOMMENDED READING

Many of the participants in this story, like Max Perutz and Jim Watson, are themselves good writers. Other writers such as Robert Olby and Horace Judson have stepped in to produce important chronicles of how it all came about, but with more focus on the DNA side of the story than on proteins. The following books are all absorbing reading and are highly recommended.

I. The history of what we today call molecular biology

1. **Robert Olby, *The Path to the Double Helix: The Discovery of DNA*.** Macmillan (London) and U. Washington Press 1974; Dover Reprints 1994.

   The best chronicle of the DNA story, including some afterthoughts. Lucid presentation of the science by a true scholar. Olby today is Francis Crick's official biographer.

2. **Horace F. Judson, *The Eighth Day of Creation: The Makers of the Revolution in Biology*.** Simon and Schuster 1979.

   Excellent journalistic account of how both the genetic *and* structural sides of molecular biology came about. Max Delbrück, for example, gets equal time with Max Perutz.

3. **Soraya de Chadarevian, *Design for Life: Molecular Biology After World War II*.** Cambridge University Press 2003.

   An authoritative and well-researched account timed to coincide with the "50 Years of DNA" celebrations at Cambridge in April 2003. Covers similar topics to Judson, but focuses more on structure studies at Cambridge than on genetics and molecular biology in general.

II. Personalities involved in the work

4. **James Watson, *The Double Helix: A Personal Account of the Discovery of the Structure of DNA*.** Text/Commentary/Reviews/Original Papers. Edited by Gunther S. Stent, UCB. Norton Critical Edition 1980.

   Watson's book first appeared in 1968, but this later edition is vastly more interesting because Gunther Stent has added commentaries, essays and reviews by Crick, Pauling, Aaron Klug, Robert Sinsheimer and others. Stent also includes the original 1953 Nature papers by all of the principals: Watson and Crick, Wilkins and coworkers, Franklin and Gosling.

5. **James Watson, *Genes, Girls and Gamow: After the Double Helix*.** Vintage (Random House) 2001.

   Picks up the story after "The Double Helix" and carries it forward to 1968. Highly entertaining.

6. **Victor K. McElheny, *Watson and DNA: Making a Scientific Revolution*.** John Wiley 2003.

   A biography of Watson through 2002. Somewhat worshipful but well-written.

7. **Anne Sayre, *Rosalind Franklin and DNA*.** Norton 1975.

   David Sayre at IBM was a well-known figure in x-ray diffraction; his "Sayre equations" constituted one of the earlier methods for solving crystal structures by direct methods. He and his wife Anne spent 1949–1951 in Dorothy Crowfoot Hodgkin's laboratory at Oxford.

While in England they became close friends with Franklin, who later visited them in the U.S. on more than one occasion. When Franklin fell ill with cancer in 1957, Anne Sayre returned to England to look after her until her death in 1958. This book was the first to tell Franklin's story, and was written less than 20 years after her death. It was specifically intended as a rebuttal to the negative picture of Franklin drawn by Watson in "The Double Helix."

8. **Brenda Maddox, *Rosalind Franklin: The Dark Lady of DNA*.** Harper Collins 2002.

   Maddox, who did not know Franklin personally, has written a more objective and more comprehensive biography 45 years after her death. Both Sayre and Maddox are well worth reading for the different slants they given on a fascinating and tragic character.

9. **Erwin Chargaff, *Voices in the Labyrinth: Nature, Man and Science*.** Seabury Press 1977.

10. **Erwin Chargaff, *Heraclitean Fire: Sketches from a Life before Nature*.** Rockefeller University Press 1978.

11. **Erwin Chargaff, *Serious Questions: An ABC of Skeptical Reflections*.** Springer Verlag 1986.

    Three collection of essays by an excellent scientist, brilliant writer and monstrous ego, who missed the boat on DNA and never quite got over it. Heraclitean Fire is more autobiographical than the others; Voices in the Labyrinth and Serious Questions are more polemical.

12. **Max Perutz, *Is Science Necessary? Essays on Science and Scientists*.** Dutton 1989.

13. **Max Perutz, *I Wish I'd Made You Angry Earlier: Essays on Science*.** Cold Spring Harbor Press 1998.

Perutz turned out to be a gifted writer and essayist, which is more than many scientists can say. These two books are collections of personal reminiscences, tales about the development of molecular biology, and essays on politics and society. Unlike Chargaff, Max is calm and reasoned [although still interesting], and has no personal axes to grind. He was wise as well as intelligent. Very highly recommended. A third book by him, "Science is Not a Quiet Life," is not a collection of essays in spite of its title; it is a volume containing reprints of all of his important papers. Of special interest are three papers on glaciology, one of which describes the proposed pykrete floating iceberg aircraft carrier that the British War Office hoped to build.

14. **Francis Crick, *Life Itself*.** Simon and Schuster, 1981.

15. **Francis Crick, *What Mad Pursuit: A Personal View of Scientific Discovery*.** Basic Books, 1988.

16. **Francis Crick, *The Astonishing Hypothesis: The Scientific Search for the Soul*.** Scribners, 1994.

    These three books are more far-ranging than those listed earlier, and reflect Crick's professional move from molecular biology into neurobiology, in particular the relationship between brain and mind. Provocative and highly recommended.

Molecular biologists (as you may have noticed) seem to have a weakness for bad verse. Robert Olby heads his Chapter 18, "Watson and Crick," with verses attributed to E. S. Anderson et al.:

> I'm Watson, I'm Crick; Let us show you our trick,
> We've found where the seed of life sprang from;
> We believe we're a stew of molecular goo,
> With a period of thirty-four Ångstroms.

- Appendix 3-

IRVING GEIS, THE MOLECULAR VESALIUS

KEY PAPERS

(A) 1997 R. E. Dickerson. *Structure* 5, 1247–1249. "Obituary: Irving Geis, 1908–1997."

(B) 1997 R. E. Dickerson. *Current Biology* 7, R740–R741. "Biology in Pictures: Molecular Artistry."

One cannot end an account of the rise of modern structural molecular biology without mention of the man who, more than any other, shaped the way we display and think about our results. This was Irving Geis, a New York artist who was brought into the fold by being asked to illustrate John Kendrew's 1961 *Scientific American* article on myoglobin (see frontispiece of this book). He went on from there to define a style of representing macromolecules that has had a large influence on modern computer graphics programs. His life and achievements are described in an obituary (A) from *Structure*. Reprint B displays a small sample of Irv's many talents as an artist of macromolecules. Plates III–VI and XIII–XVI of the color insert are also by Geis. The scientific books and textbooks listed in the second paragraph of A are a veritable gallery of Geis art, worth finding and preserving on this account even aside from their scientific content. The clarity and economy of Irv's molecular drawings have seldom been equaled. One of his memorable aphorisms was, "Color is a language; and as with any other language, one mustn't babble." Irv's drawings never babbled. The Geis Archives, containing a vast amount of his art, papers and correspondence, now are owned by the Howard Hughes Medical Institute and on display at their headquarters in Chevy Chase, Maryland.

A final assessment of Geis' talents appeared at the conclusion of (1):

> Irv was very taken with the importance of using art to put across scientific concepts. On more than one occasion, he likened himself to Andreas Vesalius, whose informative and artistic engravings taught the Renaissance public about the new field of human anatomy. Irv thought of his own role as that of a molecular Vesalius, using art to teach the modern public about the equally new field of molecular anatomy. But in view of his many-faceted talents, as artist, illustrator, interpreter, and scientific lecturer (an achievement of which he was very proud in his later years), it is more fitting to say that Irving Geis was the Leonardo da Vinci of protein structure.

References

1. R. E. Dickerson. 1997. Protein Science 6, 2483–2484. "Irving Geis, Molecular Artist."

Obituary: **Irving Geis, 1908–1997**

Richard E Dickerson

Address: Molecular Biology Institute, University of California, Los Angeles, 405 Hilgard Avenue, Los Angeles, CA 90094-1570, USA.

Structure 15 September 1997, 5:1247–1249
http://biomednet.com/elecref/0969212600501247

On 22 July 1997, macromolecular structure lost one of its pioneering artists: Irving Geis. After a cerebral hemorrhage the preceding Friday he was taken to a hospital in New York, where he lingered for four days in a coma and then succumbed, in the presence of his wife Miriam and daughter Sandy.

Geis was universally known as one of the earliest and most imaginative illustrators of macromolecular structures, providing detailed drawings for *Scientific American* articles on myoglobin (Kendrew, 1961), lysozyme (Phillips, 1966), cytochrome *c* (Dickerson, 1972), the serine proteases (Stroud, 1974), and DNA (Dickerson, 1983). His molecular paintings and drawings graced biochemistry textbooks by Zubay (1983), Mathews & Van Holde (1990), Voet & Voet (1990), Campbell (1991), and Garret & Grisham (1994). The author was fortunate to have had Irv as a friend and collaborator for 33 years, co-authoring *The Structure and Action of Proteins* (1969), *Chemistry, Matter and the Universe* (1976), and *Hemoglobin: Structure, Function, Evolution & Pathology* (1983). Irv also collaborated with Klainer in *Agents of Bacterial Disease* (1973), and with mathematician Darrell Huff in *How to Lie with Statistics* (1954) and *How to Take a Chance* (1959).

How did such a versatile artist prepare for a career in macromolecular illustration? The answer is that he didn't; it was all the fault of the great American Depression of the 1930's. Irv was born in New York in 1908 but lived in Anderson, South Carolina for a time, graduating from high school there as class valedictorian. He entered Georgia Tech to become an architect (1925–1927), and followed this with a Bachelor of Fine Arts from the University of Pennsylvania (1927–1929), majoring in Architecture and Fine Arts. But as Irv himself put it once, he finished just in time for the stock market crash, and found that there weren't many jobs around for inexperienced architects during the Depression. So he spent 1932–1933 at the University of South Carolina in the Department of Fine Arts, Design and Painting, and came to New York determined to make a living instead as a free-lance illustrator.

Irv worked regularly, with many contributions to *Fortune* Magazine, including working as illustrator for the *Fortune*

Survey. He and Miriam Artman were married in 1936, and the marriage survived 61 years until his death. When World War II came, Irv became Chief of the Graphics Section of the OSS (Office of Strategic Services), predecessor to the CIA, and Art Director of the domestic branch of the Office of War Information. After the war ended, he kept busy doing free-lance work for Standard Oil, Shell Oil, McCalls, Harpers, McGraw-Hill, Time, and others. But in 1948 Gerald Piel and Dennis Flanagan made a fateful decision that would alter the course of macromolecular illustration. They bought a century-old technical weekly called *Scientific American*, completely turned its design and typography on its head, and created a new magazine devoted to explaining science to the intelligent non-scientist, with Piel as publisher and Flanagan as Editor. Irv began illustrating regularly for them, specializing in astronomy, astrophysics, geophysics, chemistry and biochemistry. Among

Figure 1

Irving Geis (foreground) and the author going over page layouts for *Chemistry, Matter and the Universe* at the Benjamin Writing Center in Aspen, Colorado in the summer of 1973. As was typical of Irv's appreciation of printing, the book was consciously designed as a series of double-page spreads. Photo by Lee Hood, who was also part of the Writing Center. (Figure reproduced by courtesy of the Irving Geis Archives.)

other achievements, he made the first *Scientific American* drawings of orbiting Sputnik, continental drift, and the DNA double helix.

In 1961, Irv was asked by *Scientific American* to illustrate an article by John Kendrew on the first protein crystal structure, that sperm of whale myoglobin. The rest, as they say, is history. He became fascinated with the architecture of protein molecules (Georgia Tech training rising to the surface?), and five years later volunteered to illustrate an article by David Phillips on the first enzyme structure, that of egg white lysozyme.

Irv and I had met by chance following the 1964 Biochemistry Congress in New York, when I went to his apartment in upper Manhattan to purchase a print, and he asked me whether I had ever heard of a man named John Kendrew. I replied that John has been my postdoctoral mentor with

Figure 2

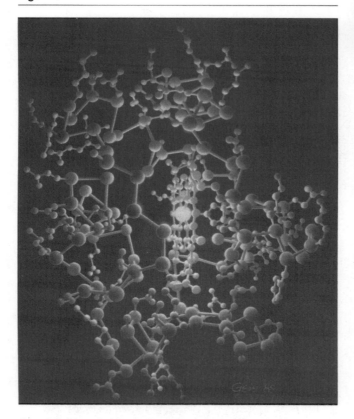

One of Geis' two 'molecular lantern' paintings of horse heart cytochrome *c*. Only the central heme group (orange), the alpha-carbon backbone (blue), and polar sidechains (green) are shown. Irv chose to make a glowing iron atom at the center of the heme the sole light source for the painting, to emphasize the key role of the heme in cytochrome function. The iron and the heme group around it are radiant, while the protein backbone and the polar surface sidechains are shrouded in darkness. (Figure reproduced by courtesy of the Irving Geis Archives.)

Figure 3

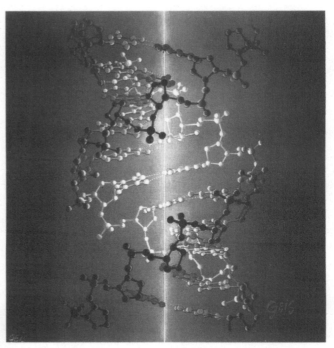

Geis' 'hot wire' painting of A-DNA. As with cytochrome *c*, a key aspect of structure is emphasized by making it the source of illumination. Irv was struck by the way in which the A-DNA double helix resembled a flexible ladder, wrapped obliquely around an imaginary central core. Hence the helix axis became a glowing wire, bathing the major groove edges of base pairs in light, but leaving the backbone and shallow minor groove in darkness. Irv struggled with B- and Z-DNA as artistic subjects, but found them less inspiring than A. Hence this portrait of A-DNA remains his nucleic acid artistic legacy. (Figure reproduced by courtesy of the Irving Geis Archives.)

myoglobin, Irv invited me in to see the original myoglobin painting, and a 33 year friendship began*. The year after the *Scientific American* lysozyme article appeared, Irv telephoned me to ask whether I was interested in collaborating with him on an *Atlas of Protein Structure*. He was quite up front about the situation: he had previously suggested the project to David Phillips, who was already involved in a similar atlas project elsewhere. So might I be interested instead? I agreed, and we went to work. As a warm up for the full *Atlas*, we decided to write a short monograph on protein structure at the freshman chemistry level. But the freshman chemistry monograph metamorphosed into a graduate-level *Structure and Action of Proteins* (1969), and our *Atlas* never was written. In retrospect, we should have written the *Atlas* when we had the chance; there were only eight protein structures known at that time. Today....!

---

*I regret to report that John Kendrew himself died in late August at the age of 80.

Figure 4

The humorous side of Irving Geis. This cartoon is from the chapter on electronic structure in the freshman chemistry textbook *Chemistry, Matter and the Universe*. The caption begins, 'An old-fashioned melodrama playing every night at the Palladium. Try to solve it first... then if you don't want to *Sulphur* any further, refer to electronic structures...' For translations of the cartoon itself the reader is referred to any periodic table. Geis enjoyed these mind games enormously. He used the same quirky but intellectually knowledgeable cartoon style to illustrate Darrell Huff's *How to Lie with Statistics*. He also once represented the hemoglobin molecule as four cherubs, ascending in formation to heaven carrying heart-shaped hemes. But the cherubs had the correct approximate 222 symmetry of the hemoglobin tetramer, and the hemes were oriented correctly. (Figure reproduced by courtesy of the Irving Geis Archives.)

Proteins was originally published with Harper & Row, but when our Editor moved to WA Benjamin, *Proteins* went with him. They suggested that Irv and I might try the freshman route once more, with an introductory chemistry textbook for non-majors. The inducement for this was three enjoyable summers at the Benjamin writing center in Aspen, Colorado. The center was an active one under editor James Hall, and included Lee Hood and Bill Wood (both then at Caltech) and David Eisenberg of UCLA. Lubert Streyer also showed up one summer, in the especially courageous position of writing a textbook in *Biochemistry* without a firm commitment from any publisher. He vacillated between Benjamin and Freeman, but ultimately chose the latter, and did rather well over the years. Figure 1 was actually taken by Lee Hood, and shows Irv and myself looking over page layouts for our freshman chemistry book, ultimately entitled *Chemistry, Matter and the Universe* (1976).

After Aspen, people kept telling us how much they had learned from *Proteins*, and asking when the second edition was coming out. A valiant effort was made in 1980 to get the revision on track, starting as before with basic principles of protein structure, and the pioneering protein structures of myoglobin and hemoglobin. But these early chapters grew beyond bounds, and developed into a book in their own right: *Hemoglobin* (1983). The second edition of *Proteins*, like the original *Atlas of Protein Structure*, would never see the light of day.

Irv continued to draw and paint, to illustrate textbooks, to write articles on macromolecular illustration, and to give invited lectures on illustrating macromolecular structures, especially of his first love, the globins. His home with

Miriam on Broadway near 200th street in upper Manhattan (where they had lived since 1945) was a treasure trove of his paintings and Miriam's sculptures. His studio was a renovated barber shop a block away; light and airy when occupied, closed with a massive folding-bellows protective fence and padlocks when unoccupied. Irv worried about break-ins from street gangs (which were a problem in a changing neighborhood). So one day he invited three or four of the young toughs into his studio to have a look around. They lounged around looking very bored, as their status demanded. Irv was pleased to hear the leader remark to the others, as they left, "There's nuttin in there; just a bunch a' molecules." The studio was never broken into. In later years he consolidated his studio in one room of his apartment, which made the apartment even more crowded and even more fascinating.

Towards the end of his life, Irv once remarked to me, "When we first met, I predicted that you would make me rich and famous. You met me half way." It is true that Irv and Miriam never got rich; Irv was too much the protein structure aficionado rather than the sharp businessman. But he did become famous; and his fame was both earned and deserved. He set standards for depiction of macromolecules that, consciously or unconsciously, have influenced the design of the computer programs that are ubiquitous today. There will never be another Irving Geis. He taught us all how to look, how to understand, and how to show others what we saw.

Biology in pictures

Molecular artistry

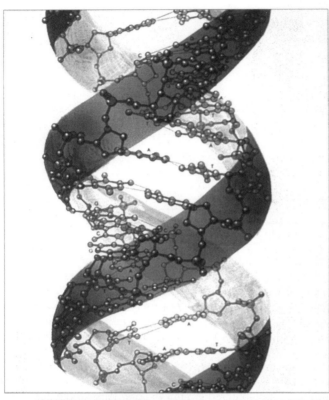

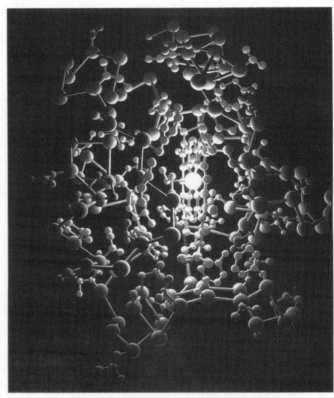

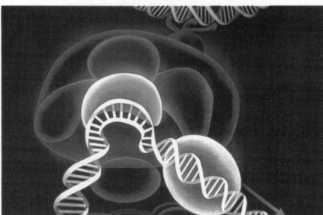

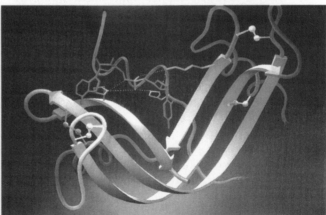

Irving Geis, who died in New York on 22 July 1997 at the age of 88, was a pioneer in the depiction of the structures of biological macromolecules. He was caught up in the subject 36 years ago, when *Scientific American* asked him to illustrate John Kendrew's December 1961 article on the first protein structure, that of sperm whale myoglobin. This was followed by illustrations for David Phillips' November 1966 *Scientific American* article on the first enzyme structure, that of hen egg lysozyme. In 1968 he coauthored *The Structure and Action of Proteins* with Richard Dickerson, creating a style that was instantly recognizable and that contributed much to the way in which later computer programs depicting protein folding were designed.

But Irving Geis was more than just a molecular draftsman. His

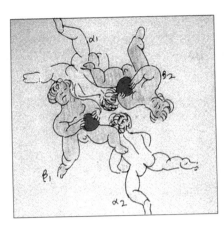

drawings inevitably contained aspects of artistic merit; not for art's sake, but for the sake of increased comprehension by the viewer. The four drawings on the previous page illustrate how Geis' macromolecular representations evolved from the literal to the abstract.

In the A-DNA helix at upper left, every atom is in place and correct, although Geis has added transparent bands of color that emphasize the double-helical backbone. The cytochrome c painting at upper right is more abstract, depicting only the heme group, α carbons connected by sticks that represent peptide bonds and polar sidechains extending out from the molecular surface. Again, what is drawn is precise, but Geis' artistic contribution was to emphasize the importance of the iron atom in the heme group by making it the sole source of illumination in the painting. If this suggested some of the paintings of the sixteenth century Flemish masters, so much the better. He called this his "molecular lantern" painting.

The ribonuclease-S painting at lower right is even more abstract, but again with a purpose. Extended chains participating in β sheets are represented by flattened arrows, in a style that has become conventional in computer graphics. The rest of the polypeptide chain is drawn as flexible wires, coiled here and there into α helices. Only the key amino-acid sidechains of the active site and disulfide bridges are drawn explicitly, and these are positioned accurately.

The result is an abstract and simplified molecule that nevertheless conveys precise scientific information.

A still higher level of abstraction is represented by the complex of DNA with the TATA-binding protein and other transcription factors at lower left. Here, the goals were twofold: to show the radical bending of DNA produced by transcription factor binding, and to illustrate the role of the large functional complex of many factors in bringing widely separated regions of DNA duplex into close proximity. For these purposes a detailed, atom-by-atom depiction of the factors was unnecessary, and Irv avoided obscuring the didactic point with unnecessary detail.

The ultimate in abstraction, the four cherubs bearing hearts, at first appears to have no connection at all with science. But their labels are a giveaway. The cherubs illustrate (quite accurately) the relative orientation of subunits in the hemoglobin molecule, with each heart representing a heme group (which also is red).

Irving Geis accumulated a lifetime of scientific paintings, drawings, preliminary sketches, models and correspondence with the scientists with whom he worked. It was his greatest wish that this mass of material be preserved in one place for reference and use by students, in the form of the Geis Archives. In 1987–88 he was awarded a Guggenheim Fellowship to catalogue the Archives. At present, efforts are being made to find a permanent home for the Geis Archives and some means of financial support for its acquisition and ongoing maintenance. If you have any ideas or suggestions, please contact Richard Dickerson (red@biop.ox.ac.uk).

Information kindly provided by Richard Dickerson. For more details, see *Structure* 1997, **5**:1247–1249. Images copyright Irving Geis, courtesy of the Geis Archives, 60 East 9th Street, New York 10003, USA. Photograph of Irving Geis by Sandy Geis.

Julius Rebek: bringing chemistry to life
Steven Dickman

Last year, Julius Rebek moved to La Jolla, California, to become the first director of the Skaggs Institute for Chemical Biology at the Scripps Research Institute. There was only one problem — what exactly is chemical biology?

For Rebek, a synthetic organic chemist whose own work had until the 1980s focused on straight chemistry, his new appointment — enabled by a $100 million grant from multimillionaire benefactor Sam Skaggs, the founder of a chain of drugstores — was an opportunity to begin to answer the question.

The new institute will stake its claim at the most exposed place on the frontier between chemistry and biology: using chemical components to make the real thing — life itself — in the test tube. "The question for me is not what happened when life began," he says, "because we can't know that, but how *could* it have happened?"

Rebek began on this path in 1990, when he was a professor at the Massachusetts Institute of Technology (MIT). He created a pair of quasi-biological molecules with a unique feature. Put them together in a bath of organic solvent and they would form a product that itself was a catalyst to bring the two molecules together. Two became four, four eight until there was practically an explosion in the Erlenmeyer flask. It was not life, not by any stretch, but the self-catalyzing reaction gave a clue as to how some early life-forming molecules might have reproduced. No less an evolution maven than Richard Dawkins wrote in a 1992 *Nature* commentary that Rebek's molecules brought to mind "whole

ANSWERS TO STUDY QUESTIONS

Chapter 2   Your Cells are not Micelles!

1. Fibers of wool, hair, silk and similar proteins, when pulled tight, give diffraction patterns indicating that they are built from units which repeat in a regular manner along the fiber direction. The repeat distance of a few Ångstroms is what would be expected if the repeating units are amino acids linked into a chain.

2. Many globular proteins, if first denatured, can then be drawn into fibers whose diffraction patterns are comparable to those obtained from fibrous proteins. Hence the denaturation process involves the unraveling of chains that have been folded in some manner, and not the breaking up of small spherical micelles.

3. If the dashed bonds in question had been strong covalent bonds, then the alpha fibers could not have been pulled out to give a beta pattern without breaking the chain.

Chapter 3   Workers of the World, Cast Off Your Chains!

1. Wrinch implied that Astbury claimed that the bonds holding short loops of chains together in hair keratin fibers were covalent bonds, rather than weaker interactions (what today we call hydrogen bonds) that could be broken upon stretching the fiber and formed again when tension was released. (We know today that these bonds actually were not between short loops of chains, but between turns of an alpha helix. But the point is the same.)

2. Wrinch proposed that her flat cyclol sheets could be bent or creased, and folded into hollow three-dimensional objects. This idea of globular proteins as being empty shells or cages turned out to be pure fantasy.

3. In Pauling's resonance model of a peptide bond, electron pairs that formerly were localized on the N atom and in the $C\!=\!O$ double bond were delocalized over all three N—C—O atoms. This increase in size of the "box" accessible to the electrons lowers their energy levels, making the structure more stable.

4. A thrown baseball, in principle, also has quantized energy levels dependent on the mass of the ball and the size of the space available to it. But baseballs are so much more massive than electrons, and ballparks so much larger than atoms, that the energy levels are effectively mashed down into a continuum. The concept of discrete energy levels of a baseball has no experimental meaning.

5. The closed cyclol loop does gain 49 kcal/mole of extra stability from its new N—C bond. But it loses 82 kcal/mole of stability when the peptide $C\!=\!O$ double bond becomes a single bond, and another 21 kcal/mole from loss of resonance energy.

6. According to Pauling, the central C—N bond of a peptide link is intermediate in character between a single and a double bond. A double bond would resist rotation by 45 kcal/mole, the difference in energy between double and single bonds. Pauling's partial double bond has only 21 kcal/mole of extra stability relative to a single bond, but the principle is the same.

7. The cyclol model of proteins would produce one free —OH group per cyclol ring, whereas an extended polypeptide chain has only the relatively few —OH groups that are contributed by serine, threonine and tyrosine side groups. Haurowitz' experimental finding (C) that intact proteins possessed few free —OH groups made the cyclol model unlikely.

8. A denatured polypeptide chain is roughly 1 kcal/mole of residues less stable than a folded protein, whereas it would be ca. 26 kcal/mole *more stable* than the cyclol structure. Cylols, if they existed in proteins, would denature spontaneously.

Chapter 4   Folding and Coiling of Polypeptide Chains

1. Globular protein crystals are typically ca. 50% solvent. If not sealed into capillaries they dry out. Loss of water between protein molecules disrupts their orderly packing and degrades their x-ray diffraction pattern.

2. Rapid or flash freezing immobilizes water molecules and keeps them from escaping from protein crystals. (But slow freezing which gives ice crystals time to nucleate and grow between molecules also disrupts the diffraction pattern.)

3. Bragg, Kendrew and Perutz, all experienced crystallographers, were so accustomed to the concept of crystals having helices with an integral number of residues per turn or unit cell length (twofold, threefold, fourfold, etc.) that they never even contemplated helices with non-integral numbers of residues (3.6-fold for example).

4. A severe cold kept Pauling in bed for a couple of days in Oxford, and gave him time to think about various ways of coiling a chain into a helix. Two key ideas not appreciated by his Cambridge rivals were the planarity of peptide links and the possibility of nonintegral helices.

5. Perutz tilted his fibers relative to the x-ray beam, in order to record the critical 1.5 Å reflection related to the repeat distance of amino acids along the helix axis direction. The Caltech paper on the alpha helix appeared in print on 15 April 1951 (C), and Perutz states that he only read it in the Cavendish library after publication (D). He rushed into the laboratory to test it immediately, and his experimental verification appeared in the 30 June issue of *Nature* (E). Hence the total elapsed time to publication was less than 11 weeks.

6. Pauling was concerned because his alpha helix model had a pitch of 5.4 Å per turn, whereas Astbury's photos of hair keratin showed a pitch of only 5.1 Å. Synthetic polypeptides did exhibit a 5.4 Å pitch, but the disagreement with natural fibers was unsettling. Pauling's decision ultimately to publish the alpha helix almost surely was triggered by the appearance in 1950 of the paper by his Cambridge rivals Bragg, Kendrew and Perutz, and their ultimately inability to reach any conclusions from their integral-turn helices.

7. The 4_{13} helix in Figure 12 of (B) can be converted into an alpha helix by twisting the helix about its axis until it has only 3.6 amino acids per turn rather than 4. This also straightens out the hydrogen bonds between turns of helix.

8. Francis Crick showed that hair keratin fibers have a helical repeat of only 5.1 Å instead of 5.4 Å because the fibers are built, not from isolated right-handed helices, but from coiled-coils in which two or three right-handed helices are wound around one another in a left-handed sense, foreshortening the repeat distance.

9. With L-amino acids, the common form in terrestrial life, a left-handed alpha helix would exhibit clash between each C=O group and the side chain attached to its neighboring alpha carbon atom. Hence our alpha helices are

right-handed. Life on other planets could use left-handed alpha helices, providing that their component amino acids were in the D- conformation. This is the case for the helix in Figure 2 of paper (C).

10. They are termed "pleated" because they are not flat. Instead, they have ridges and valleys like corrugated sheet metal. (See Figures 6 and 7 of paper [F].)

Chapter 5   The Race for the DNA Double Helix

1. Tilting the sieve around a horizontal axis effectively foreshortens the distance between wires in the vertical direction. This *decrease* in vertical spacings between scatterers produces an *increase* in vertical spacings between rows of reflections in the diffraction pattern. Hence the diffraction pattern is spoken of as being in "reciprocal space."

2. A 2 Å map of a complex protein will contain much more information than the same resolution map of a simple salt. Hence many more terms *should* be required to compute it.

3. Layer line spacing in the x-ray pattern is inversely proportional to the pitch, P, or distance per turn of helix. The strong axial reflection above and below the origin indicates the individual repeat distance, h, of the units that make up the helix. The number of repeats per turn is: $n = P/h$. Another way of looking at it is that the strong axial reflection in an n-fold helix is found on the nth layer line.

4. The front parts of successive turns of a helix can be regarded as a set of periodic fringes whose diffraction pattern is a row of spots in a direction perpendicular to the fringes, with spot spacing inversely proportional to the distance between them. This establishes one arm of the cross. Similar consideration of the back part of the helix explains the other arm of the cross. (In reality, just as the "front" and "back" sides of the helix are a planar approximation to a three-dimensional helix, so the "left" and "right" arms of the cross in the x-ray pattern are a planar approximation to a cylindrical arrangement of rings around the vertical axis. The two intersections of each of these rings with the plane of the photograph produce a pair of spots to left and right of the center line. See Figure 2 of Chapter 4.)

5. Only with the keto tautomers can one obtain base pairing in which the two paired bases are structurally equivalent, meaning that an A:T pair can be substituted for a G:C pair without disrupting the double-helical backbone.

6. Pauling and Watson: Model builders. Chargaff, Wilkins, and Franklin: Data analzyers.

7. The volume for one step along a cylindrical helix of correct dimensions was effectively three times the volume per nucleotide in other crystal structures. This suggested, but did not require, that each step contained three bases. This possibility was advocated at one time or another by Watson and Crick, by Bruce Frazer (C), and by Pauling and Corey (D). Rosalind Franklin at least considered the possibility although she never favored it (G).

8. As Jerry Donohue pointed out to Watson, phosphate groups would be fully ionized at pH 7, and hence would not have unionized —OH groups with which to hold a hydrogen-bonded central phosphate core together.

9. This is a tough one with no clear answer. Pauling and Corey somehow failed to realize by examining B-DNA diffraction patterns that the B helix had ten residues per turn. Either their diffraction patterns were abominable, or they simply were careless. Knowing the thickness of an organic ring, they knew that each base step advanced 3.4 Å along the helix axis. They thought the B

helix had eight steps per turn because they assumed that the pitch of the helix was 27.2 Å instead of the 34 Å that layer line spacings in a B-DNA photograph indicate. Where this erroneous 27.2 Å pitch came from is a mystery. The best guess is that they obtained it by misunderstanding a comment made by Rosalind Franklin about a different form, A-DNA.

10. The low-humidity A form has a lower water content. Its chains are packed together in a more regular manner, and the diffraction pattern consequently extends to higher resolution. Franklin therefore felt that the ultimate structure of the A form would be more detailed, and by implication more informative.

11. Franklin apparently never quite made the logical step from Chargaff's base ratios, to the concept of bases pairing off in purine/pyrimidine pairs: A with T, and G with C.

12. This is a very subjective question, which you must decide for yourself after reviewing the evidence. Klug estimates that it would have been only a matter of months.

13. Chargaff intimated that Perutz was wrong in passing on to Watson and Crick a Medical Research Council report which contained an excellent B-DNA photograph from the King's laboratory. Perutz and others pointed out that the report was public knowledge and was in no sense proprietary. If the King's people had not wanted the DNA photo seen they should not have included it in a public report. Having done so, they had no reason to blame Watson and Crick for seeing it and understanding more of its implications than they had.

14. Chargaff apparently deeply resented Watson and Crick for moving into what he regarded as his own field of expertise—DNA—and drawing radical new conclusions that Chargaff could not, or would not, see on his own.

15. The Cambridge workers were stung by the fact that they had lost the race for the polypeptide α-helix to Caltech, and were fearful that they would lose the DNA race as well.

Chapter 6 How to Solve a Protein Structure

1. Donohue's "paper chain paper chain" satirizes Francis Crick's coiled-coil for polypeptide chains in fibrous proteins such as hair keratin.

2. The sentence beginning, "It has not escaped our notice...." on page 6 of (A) is surely based on a sentence with a similar beginning in Watson and Crick's Nature paper of 25 April 1953. The fact that Donohue focuses on the complex equations that Crick derived in 1952/3 for fiber diffraction, but makes no mention at all of Perutz's revolutionary new MIR approach published in 1954 (hemoglobin paper IV), suggests that the parody was written some time in middle or late 1953.

3. The wave corresponding to a given x-ray reflection is directed along the line from the origin of the diffraction pattern to that reflection. Its wavelength is inversely proportional to the distance of the reflection from the origin. Its amplitude is proportional to the square root of the intensity or blackening of the reflection.

4. The diffraction pattern itself gives no information about the relative phase shift that must be given a particular wave before it is added into the synthesis.

5. For a given reflection, each derivative yields a twofold ambiguity in phase angle. Two different derivatives must be used, so the phase angle common to them both can be recognized as being the correct one. A third derivative is critically important in compensating for phase uncertainy arising from experimental errors.

6. The two hemoglobin maps were at low resolution, 6.5 or 6 Å, and were only projections of the full structure on two different planes. Structural overlap within these projections obscured all details regarding the hemoglobin molecules.

7. Because Kendrew calculated a full three-dimensional myoglobin map, he overcame the problem of overlap obscuring structural features in projection. He anticipated that myoglobin might contain alpha helices, and concluded from the dimensions of an alpha helix that it should be distinguishable as a solid rod at 6 Å resolution.

8. If myoglobin had contained no alpha helices, like an antibody molecule, it is unlikely that a 6 Å map would have shown any recognizable structural features other than the heme group.

9. Because the iron atom has so many electrons, the heme group was a dense ovoid of electron density, sitting in a V-shaped pocket between two straight rods which he interpreted, correctly, as being alpha helices.

Chapter 7   High Resolution Protein Structure Analyhsis

1. Even at low resolution, myoglobin and the alpha and beta chains of hemoglobin exhibited the same structure. This, and the similiarities in their functions as oxygen carriers, strongly suggested that they were evolutionary descendants from a common ancestor protein.

2. The "garden hose" model of an alpha helix, according to which it should change from a solid rod at 6 Å to a hollow cylinder at 2 Å, turned out to be correct. But in addition the polypeptide chain itself could be seen winding around the "garden hose," even with distinguishable carbonyl groups that defined the direction of the chain. (See figure B.1)

3. The 2 Å data set has $(6/2)^3 = 27$ times as many reflections as the 6 Å. The exponent "3" occurs because the diffraction pattern, like the protein, is three-dimensional.

4. A skeletal wire model of the protein chain could be built within the forest of rods, whereas this would be impossible with a pile of stacked Plexiglas sheets.

5. Because the two hemoglobin chains and myoglobin were folded in the same manner, Perutz could infer much about his low resolution hemoglobin structure by comparison with Kendrew's high resolution myoglobin structure.

6. Kendrew got the orientation of heme group wrong in his 6 Å structure. At low resolution the heme group is only a dense ovoid blob, and the planarity of the carbon and nitrogen rings around the iron is not visible. The unit cell of myoglobin contains two molecules. Kendrew used electron spin resonance at the time of the 6 Å map to establish the orientations of the two heme groups. But he had no way of knowing which orientation went with which molecule, and unfortunately made the wrong choice in his published paper. But the correct orientation was immediately obvious at high resolution. Compare Figure 5 of paper B (6 Å) with Figure 7.1 (2 Å). At high resolution, Kendrew had the flat heme group slipping into the pocket between alpha helices E and F. In the earlier low resolution stage he had it lying *across* the opening of the pocket.

7. The heme iron atom in deoxyhemoglobin has fourfold coordination to nitrogen atoms within the heme, and a fifth coordination to a nitrogen of histidine F8 (to use myoglobin numbering). It has no sixth coordinating group, and hence is pulled slightly out of the heme plane towards His F8. When an oxygen molecule binds and provides a sixth coordination, the iron atom is pulled back into the heme plane. His F8 and its supporting F helix are dragged along with it, and the change is propagated from one subunit to another by interactions of key

side chains in contact with one another. Helices F in subunits that have not yet bound oxygen are pushed toward the heme, driving their iron atoms into the heme plane and making them more susceptible to binding new oxygen atoms.

8. Because of these subunit interactions and changes in heme coordination, oxygen binding in hemoglobin tends to be an all-or-nothing proposition. In high-oxygen environments such as the lungs, a hemoglobin molecule will saturate rapidly and strongly. In low-oxygen environments at the tissues, hemoglobin will quickly give up *all* its oxygen atoms to myoglobin storage proteins. Myoglobin, with only one subunit, is only a passive receptacle for oxygen.

Appendix 3   Irving Geis, the Molecular Vesalius

A translation of Geis Figure 4 on page 289:

"Tonight: An Americium Tragedy"
"He was impossibly Boron, and besides he wouldn't pay no Antimony!"
"She may be an Indium, but she has a heart of Gold."
"If we can't Curium we'll have to Barium."
"Somebody call a Copper!"
"No thanks; Iodine later!"

CREDITS FOR THE KEY PAPERS

We are grateful to the following journals and publishers for their permission to reprint copyrighted material:

Annals of the New York Academy of Sciences

D. C. Hodgkin, 1979. *Ann. NY Acad. Sci.* 325, 121–148. "Crystallographic Measurements and the Structure of Protein Molecules as They Are."

Current Biology

R. E. Dickerson, 1997. *Current Biology* 7, R740–R741. "Biology in Pictures: Molecular Artistry."

Experientia

E. Chargaff, 1950. *Experientia* VI, 201–209. "Chemical Specificity of Nucleic Acids" [3 pages of 9].

Federation Proceedings

E. Chargaff, 1951. *Fed. Proc.* 10, 654–659. "Structure and Function of Nucleic Acid as Cell Constituents" [2 pages of 6].

Journal of the American Chemical Society

L. Pauling and C. Niemann, 1939. *J. Am. Chem. Soc.* 61, 1860–1867. "The Structure of Proteins;"

D. M. Wrinch, 1941. *J. Am. Chem. Soc.* 63, 330–333. "The Geometrical Attack on Protein Structure."

Journal of Chemical Education

M. M. Julian, 1984. *J. Chem. Edu.* 61, 890–892. "Dorothy Wrinch and a Search for the Structure of Proteins."

Journal of Structural Biology

R. D. B. Fraser, 2004. *J. Struct. Biol.* 145, 184–6. "The Sructure of Deoxyribose Nucleic Acid" (written in 1951 but not published then).

Proceedings of the Royal Society (A)

W. L. Bragg, J. C. Kendrew and M. F. Perutz, 1950. *Proc. Roy. Soc. London* A203, 321–357. "Polypeptide Chain Configurations in Crystalline Proteins" [24 pages of 37].

J. Boyes-Watson, E. Davidson and M. F. Perutz, 1947. A191, 83–132. "An X-ray Study of Horse Methaemoglobin, I."

M. F. Perutz, 1949. A195, 474–499. "An X-ray study of horse methaemoglobin, II."

M. F. Perutz, 1954. A225, 264–286. "The Structure of Haemoglobin, III. Direct Determination of the Molecular Transform."

D. W. Green, V. M. Ingram and M. F. Perutz, 1954. A225, 287–307. "The Structure of Haemoglobin, IV. Sign Determination by the Isomorphous Replacement Method."

E. R. Howells and M. F. Perutz, 1954. A225, 308–314. "The Structure of Haemoglobin, V. Imidazole-methaemoglobin: A Further Check of the Signs."

W. L. Bragg and M. F. Perutz, 1954. A225, 315–329. "The Structure of Haemoglobin, VI. Fourier Projections on the 010 Plane."

D. M. Blow, 1958. A247, 302–336. "The Structure of Haemoglobin, VII. Determination of Phase Angles in the Non-centrosymmetric [100] Zone."

A. F. Cullis, H. Muirhead, M. F. Perutz, F. R. S. and M. G. Rossmann, 1961. A265, 15–38. "The Structure of Haemoglobin, VIII. A Three-Dimensional Fourier Synthesis at 5.5 Å Resolution: Determination of the Phase Angles."

A. F. Cullis, H. Muirhead, M. F. Perutz, F. R. S. and M. G. Rossmann, 1962. A265, 161–187. "The Structure of Haemoglobin, IX. A Three-Dimensional Fourier Synthesis at 5.5 Å Resolution: Description of the Structure."

J. C. Kendrew, G. Bodo, H. M. Dintzis, R. G. Parrish, H. Wyckoff and D. C. Phillips, 1958. *Nature* 181, 662–666. "A Three-Dimensional Model of the Myoglobin Molecule Obtained by X-Ray Analysis."

G. Bodo, H. M. Dintzis, 1959. J. C. Kendrew and H. W. Wyckoff. *Proc. Roy. Soc. London* A253, 70–102. "The Crystal Structure of Myoglobin. V. A Low-Resolution Three-Dimensional Fourier Synthesis of Sperm Whale Myoglobin Crystals" [7 pages of 33].

Philosophical Transaction of the Royal Society

T. Astbury and H. J. Woods, 1934. *Phil. Trans. Roy. Soc. London* A232, 333–394. "X-Ray Studies of the Structure of Hair, Wool, and Related Fibres. II. The Molecular Structure and Elastic Properties of Hair Keratin" [8 pages of 62].

Proceedings of the National Academy of Sciences

L. Pauling, R. B. Corey and H. R. Branson, 1951. *Proc. Natl. Acad. Sci. USA* 37, 205–211. "The Structure of Proteins: Two Hydrogen-Bonded Helical Configurations of the Polypeptide Chain."

L. Pauling and R. B. Corey, 1951. *Proc. Nat. Acad. Sci. USA* 37, 729–740. "Configurations of Polypeptide Chains with Favored Orientations around Single Bonds: Two New Pleated Sheets" [5 pages of 12].

D. Eisenberg, 2003. *Proc. Nat. Acad. Sci. USA* 100, 11207–11211. "The Discovery of the α-helix and β-sheet, the Principal Structural Features of Proteins."

L. Pauling and R. B. Corey, 1953. *Proc. Nat. Acad. Sci. USA* 39, 84–97. "A Proposed Structure for the Nucleic Acids."

Protein Science

R. E. Dickerson, 1992. *Protein Science* 1, 182–186. "A Little Ancient History."

Science

E. Chargaff, 1968. *Science* 159, 1448–1449. "A Quick Climb up Mount Olympus."

M. F. Perutz, M. H. F. Wilkins, J. D. Watson, 1969. *Science* 164, 1537–1539. "DNA Helix."

Structure

R. E. Dickerson, 1997. *Structure* 5, 1247–1249. "Obituary: Irving Geis, 1908–1997."

Transactions of the Faraday Society

W. T. Astbury, 1938. *Trans. Faraday Soc.* 34, 378–388. "X-Ray Adventures Among the Proteins."

Nature

D. M. Wrinch, 1936. *Nature* 137, 411–412. "The Pattern of Proteins."

D. M. Wrinch, 1937. *Nature* 139, 972–973. "The Cyclol Theory and the 'Globular' Proteins."

M. F. Perutz, 1951. *Nature* 167, 1053–1054. "New X-ray Evidence on the Configuration of Polypeptide Chains: Polypeptide Chains in Poly-γ-benzyl-L-glutamate, Keratin, and Haemoglobin."

J. D. Watson and F. H. C. Crick, 1953. *Nature* pp. 737–738. "A Structure for Deoxyribose Nucleic Acid."

M. H. F. Wilkins, A. R. Stokes and H. R. Wilson, 1953. *Nature* pp. 738–740. "Molecular Structure of Deoxypentose Nucleic Acids."

R. E. Franklin and R. G. Gosling, 1953. *Nature* pp. 740–741. "Molecular Configuration in Sodium Thymonucleate."

A. Klug, 1974. *Nature* 248, 787–788. "Rosalind Franklin and the Double Helix."

M. F. Perutz, M. G. Rossmann, A. F. Cullis, H. Muirhead, G. Will and A. C. T. North, 1961. *Nature* 185, 416–422. "Structure of Haemoglobin. A Three-Dimensional Fourier Synthesis at 5.5 Å Resolution, Obtained by X-Ray Analysis."

J. C. Kendrew, R. E. Dickerson, B. E. Strandberg, R. G. Hart, D. R. Davies, D. C. Phillips and V. C. Shore, 1960. *Nature* 185, 422–427. "Structure of Myoglobin. A Three-Dimensional Fourier Synthesis at 2 Å Resolution."

Other

M. Perutz, 1998. In *I Wish I'd Made You Angry Earlier: Essays on Science.* Cold Spring Harbor Press, pp. 189–191.

E. Chargaff, 1978. In *Heraclitean Fire*, Rockefeller U. Press, pp. 100–103. "Gullible's Troubles."

Unpublished

J. Donohue and J. Briekopf, 1953. *Max F. Perutz reprint file.* "The X-Ray Structure Analysis of α-Globlglobin" [unpublished, 10 pages].

INDEX

Numbers in *italic* indicate information in a figure, figure caption, or table.

A

A-DNA, 151, 154, 290–291
adenine, 149, *152. see also* base pairing
α-helices
  "The discovery of the α-helix and β-sheet, the principle structural features of proteins" (Eisenberg), 135–138
  I Wish I'd Made You Angry Earlier (Perutz), 125–127
  myoglobin, 251, *251*
  Pauling's concept of, *60,* 60–61
  proposal, 58
  rise per turn, 63
amide bonds, 59, 130–134
amino acids, local environment, 59
Andreeva, Natalia, 277
Astbury, William T.
  on classes of fibrous proteins, 5
  on folded α-keratin chains, 20
  key articles by, 10–17, 23, 45–56
  x-ray diffraction studies, 7–8
  x-ray photographs, 151
Avery, Oswald, 148

B

B-DNA, 147, *147,* 151–154, *153,* 186–187
backbones, DNA, 149–150, *150*
backbones, ring, 59
base pairing
  Chargaff's ratios and, 149
  "Gullible's Troubles" (Chargaff), 195–198
  possibilities, *152*
  by Watson and Crick, 154–155
Baumeister, Wolfgang, 163–165
Bell, F. O., 151
Bernal, John Desmond, 5, 58, 200, *201*
beta bends, 20, *20*
β-sheets
  antiparallel, 64
  beta bends, 20, *20*
  "Configurations of Polypeptide Chains with Favored Orientations Around Single Bonds" (Pauling, Corey), 130–134
  "The discovery of the α-helix and β-sheet, the principle structural features of proteins" (Eisenberg), 135–138
  form of, 8
β states, 5, *7*
"Biology in Pictures: Molecular Artistry" (Dickerson), 290–291

Blow, David, 213, 233
Bodo, G., 236–240, 241–247
Bombyx mori, 7, 64
bond energies, *22,* 30–38
Boyes-Watson, Joy, 227
Bragg, William Lawrence
  background, 58, 279
  Cavendish Laboratories and, 200
  examination of helices, 59
  key articles by, 94–117, 232
  Perutz and, 202
  Royal Institution and, *212,* 213
  on the triple helix theory, 149
Branson, Herman R., 58, 61, 118–124, 135–138
Brenner, Sydney, 281
Briekopf, J., 217–226
Broad, Tony, 153
bromine, 207, *207*

C

carbonic anhydrase, 276
carboxypeptidase, 276
Carlisle, Harry, 275
Central Dogma, 255
Chargaff, Erwin
  attitude toward fiber diffraction, 210
  "DNA Helix" (Perutz, Wilkins, Watson), 192–193

on the double helix model, 153–154
key articles by, 158–160, 161–162, 190–191, 195–198
on nucleic acids, 148–149
on Perutz, 155
tragedy of, 23
on Watson and Crick, 155–156
Chargaff's ratios, 149, 158–162
Chase, Martha, 148
"Chemical Specificity of Nucleic Acids and Mechanism of their Enzymatic Degradation" (Chargaff), 158–160
chymotrypsin, 276
chymotrypsinogen, 275
cobalt atoms, 208
colloids, 3
"Configurations of Polypeptide Chains with Favored Orientations Around Single Bonds" (Pauling, Corey), 130–134
continuous phase, 3
Convolution Product, 143, 144, 144–145
Convolution Theorem, 145, 145–146
Corey, Robert B., 276
α-helix proposal, 58, 61
α-keratin form and, 8
atomic model for proteins, 250–251
β-sheet structures, 64
death of, 279
"The discovery of the α-helix and β-sheet, the principle structural features of proteins" (Eisenberg), 135–138
DNA model, 149–152, 150
in Donohue's parody, 201, 201
I Wish I'd Made You Angry Earlier (Perutz), 125–127
key articles by, 118–124, 130–134, 169–182
covalent bonds, 20. see also bond energies; hydrogen bonds
Crick, Francis, 280
on α-helices, 63–64
career of, 281–282
Central Dogma, 255
credit to, 153–156
"DNA Helix" (Perutz, Wilkins, Watson), 192–193
in Donohue's parody, 200–201

double helix and, 152–153
Franklin and, 23
on Fraser's helix model, 149
"Gullible's Troubles" (Chargaff), 195–198
polymer structure and, 8
"A Quick Climb Up Mount Olympus" (Chargaff), 190–191
"A Structure for Deoxyribose Nucleic Acid," 183–184
"The third man and the fourth paper" (Steven, Baumeister), 163–165
work on DNA structure, 149–152
cross-linking, 45–56
Crowfoot, Dorothy (Hodgkin). see Hodgkin, Dorothy Crowfoot
"The Crystal structure of myoglobin V. A low-resolution three-dimensional Fourier synthesis of sperm-whale myoglobin crystals" (Bodo, Dintzis, Kendrew, Wyckoff), 241–247
"Crystallographic Measurements and the Structure of Protein Molecules as They Are" (Hodgkin), 66–94
crystallography, description, 141–147
Cullis, Ann F., 234, 235, 257–263
cyclol hypothesis
controversy, 8–9
"The Cyclol Theory and the 'Globular' Proteins" (Wrinch), 28–29
"Dorothy Wrinch and a Search for the Structure of Proteins" (Julian), 42–44
Drum-Fold model, 254
"The Geometrical Attack on Protein Structure" (Wrinch), 38–41
"The Pattern of Proteins" (Wrinch), 26–27
"The Structure of Proteins" (Pauling and Niemann), 30–38
cyclol rings, structure, 20, 20
"The Cyclol Theory and the 'Globular' Proteins" (Wrinch), 28–29
cytochrome c, 276, 277
cytosine, 149, 152. see also base pairing

D
D-conformation, 61
Davidson, Edna, 227
Davies, D. R., 263–268

Davy, Humphrey, 213
denaturation
enthalpy and, 22
globular proteins, 21–22, 24
"X-Ray Adventures Among the Proteins" (Astbury), 45–56
deoxyhemoglobin, 252
deoxypentose nucleic acids, 148–149, 158–162, 184–186
Dickerson, Richard E., 275, 276
key articles by, 263–268, 269–273, 287–289, 290–291
Dickerson's Law, 278
diffraction
artifacts, 142
Fourier transforms and, 143, 203
patterns, 140–147, 141
set-up, 142
two-dimensional, 141, 142, 204
diketopiperazine ring, 20
Dintzis, H. M., 236–240, 241–247
direct reduplication, 255
discontinuous phase, description, 3
"The discovery of the α-helix and β-sheet, the principle structural features of proteins" (Eisenberg), 135–138
DNA. see also A-DNA; B-DNA
base composition, 148–149
"Biology in Pictures: Molecular Artistry" (Dickerson), 290–291
dilemma, 1–2
"DNA Helix" (Perutz, Wilkins, Watson), 192–194
double helix model, 152–153
forms, 151, 154
functional components, 151
interconvertibility of, 154
race for the double helix, 139–198
"Rosalind Franklin and the double helix" (Klug), 188–189
"DNA Helix" (Perutz, Wilkins, Watson), 192–194
Donohue, Jerry, 152, 200–202, 217–226
Dornberger-Schiff, Käte, 253, 255
"Dorothy Wrinch and a Search for the Structure of Proteins" (Julian), 42–44

double helices, 152–153, 188–189, 192–194. *see also* DNA

Drew, Horace, 282

Drum-Fold model, 253, *254*

Dunitz, Jack, 201

E

Edsall, John, 255

egg-white lysozyme, *203*, 275

Eisenberg, David, 64–65, 135–138, *276*, 280

elastic properties

"X-Ray Adventures Among the Proteins" (Astbury), 45–56

"X-Ray Studies of the Structure of Hair, Wool, and Related Fibres" (Astbury, Woods), 10–17

elastoidin, 5

electron density maps

display, 250

resolution, 205, *205*

enol tautomers, *152*, 152–153

enthalpy, 21, *22*

Ewald sphere, *62*, 63

F

"Faltentrommel," 253, *254*

Fankuchen, Isidore, 5

Faraday, Michael, 213

β-fibroin, *7*

fibrous proteins, description, 5

Florkin, Marcel, 4

folding, 59–61, 125–127

Fourier, M. Joseph, *201*

Fourier transform

"The Crystal structure of myoglobin V." (Bodo, Dintzis, Kendrew, Wyckoff), 241–247

data, 143

diffraction patterns and, 203

protein fibers, *62*

"The structure of haemoglobin: VIII." (Cullis, Muirhead, Perutz, Rossman, North), 234

"The structure of haemoglobin: IX." (Cullis, Muirhead, Perutz, Rossmann, North), 235

"Structure of Haemoglobin" (Perutz, Rossmann, Cullis, Muirhead, Will, North), 257–263

"Structure of Myoglobin" (Kendrew, Dickerson, Strandberg, Hart, Davies, Phillips, Shore), 263–268

"A Three-Dimensional Model of the Myoglobin Molecule Obtained by X-Ray Analysis" (Kendrew, Bodo, Dintzis, Parrish, Wyckoff), 236–240

Fourier waves, 205–206

"Frankendrew's Monster," 214–217, *215*

Franklin, Rosalind E.

B-DNA photograph, 147, *147*

credit to, 23, 153–156

death of, 155, 279

on DNA forms, 151, 154

"DNA Helix" (Perutz, Wilkins, Watson), 192–194

"Gullible's Troubles" (Chargaff), 195–198

key articles by, 186–187

"A Quick Climb Up Mount Olympus" (Chargaff), 190–191

"Rosalind Franklin and the double helix" (Klug), 188–189

"The third man and the fourth paper" (Steven, Baumeister), 163–165

work on DNA structure, 149

Fraser, Bruce, 149–150, 163–165, 166–168

G

γ-helix conformation, 61

Geis, Irving, 249, 286–289, *287*

gelatin, 5

"The Geometrical Attack on Protein Structure" (Wrinch), 38–41

Gillray, James, *212*

glass capillaries, 210, *210*

globglobin, 200–202, 217–226

globular proteins

beta bends, 20, *20*

crystallization, 58

denaturation, 21–22, 24

description, 5

"The Cyclol Theory and the 'Globular' Proteins" (Wrinch), 28–29

"The Pattern of Proteins" (Wrinch), 26–27

"X-Ray Adventures Among the Proteins" (Astbury), 45–56

Gortner, Ross, 4–5

Gosling, Raymond, 147, *147*, 186–187, 193

grease, aqueous environments, *4*

Green, D. W., 230

guanine, 149, *152. see also* base pairing

"Gullible's Troubles" (Chargaff), 195–198

Gutfreund, Herbert :"Freddy," *201*

H

"Habakkuk" project, 202

hair, 10–17, 45–56

Harker, David, 275

Hart, R. G., 263–268

heats of combustion, 30–38, 38–41

heavy metal atoms, 206–209

heavy metal binding, 206

Heisenberg, Werner, 20

helices. *see also* α-helices; α-helices

continuous, *146*

description of, 59

discontinuous, *146*

stabilization, 59

transform pattern, 146

hemoglobin

kinetic behavior, 252–253

"A little ancient history" (Dickerson), 269–273

MIR phasing, *213*

molecule arrangement, *210*

myoglobin and, 200, 209

"New X-Ray Evidence on the Configuration of Polypeptide Chains" (Perutz), 128–130

oxygen-binding curve, 252

"Polypeptide chain configurations in crystalline proteins" (Bragg, Kendrew, Pertuz), 94–117

saga of, 209–215

sources for, 200–201

"The structure of haemoglobin: IV." (Green, Ingram, Perutz), 230

"The structure of haemoglobin: VI."(Bragg, Perutz), 232

"The structure of haemoglobin: VII." (Blow), 233

"The structure of haemoglobin: VIII." (Cullis, Muirhead, Perutz, Rossman, North), 234

"The structure of haemoglobin: IX." (Cullis, Muirhead, Perutz, Rossmann, North), 235

"The structure of haemoglobin" (Perutz), 229

"Structure of Haemoglobin" (Perutz, Rossmann, Cullis, Muirhead, Will, North), 257–263

three-dimensional structure, 214, *214*

Hershey, A. D., 148

Hirschegg workshop, 275–276

Hodgkin, Dorothy Crowfoot, 5
key articles by, 66–94
on Linus Pauling, 60
move to Oxford, 58
vitamin B$_{12}$ structure, 207–208, *208*
x-ray crystallography by, 28

House Un-American Activities Committee, 65, 152

Howells, E. R., 231

hydrogen bonds, 118–124, 130–134, 152

I

I Wish I'd Made You Angry Earlier: Essays on Science (Perutz), 125–127

imidazole-methaemoglobin, 231

Ingram, V. M., 230

intermicellar liquid, 3

isomorphous replacement, 230

J

Journal of Irreproducible Results, 200

Journal of Molecular Biology, 213

JSTOR (journal storage) website, 7, 210

Julian, Maureen M., 23, 42–44

K

Kakllai, Olga, *276*

Kalaidjiev, Angel, 253

Kendrew, John
Bragg and, 58
career of, 282
cyclol hypothesis and, 20
in Donohue's parody, 200
examination of helices, 59
"Frankendrew's Monster," 214–217
hemoglobin research, 280
high-resolution analysis, 250
Journal of Molecular Biology and, 213
key articles by, 94–117, 236–240, 241–247, 263–268

myoglobin structure by, 209, 214

Nobel Prize, 253

protein crystal structure analysis, 200

search for protein sources, 200–201

α-keratin
acceptance of, 59
contracted form, 8
fiber diffraction data, 61, 63
fiber diffraction patterns, 6, *6*
helices, 59
stretched, 7

keratin-myosin ("k-m-e-f" group), 5

keratins. *see also* α-keratin
α/β transition, 7–8, 24
α-helices, 63–64
I Wish I'd Made You Angry Earlier (Perutz), 125–127
"New X-Ray Evidence on the Configuration of Polypeptide Chains" (Perutz), 128–130
"Polypeptide chain configurations in crystalline proteins" (Bragg, Kendrew, Pertuz), 94–117
"X-Ray Adventures Among the Proteins" (Astbury), 45–56
x-ray diffraction by fibers, 24
x-ray pattern, 5
"X-Ray Studies of the Structure of Hair, Wool, and Related Fibres" (Astbury, Woods), 10–17

keto tautomers, *152*, 152–153

Klug, Aaron, 151, 154, 188–189

Kopka, Mary, 280, 282

Kossel, Albrecht, 147

Kraut, Joe, 275

L

L-conformation, 61

light, diffraction, 140–147, *142*

Linderman glass capillaries, 58, *210*

liquid-in-liquid colloidal suspensions, 3

"A little ancient history" (Dickerson), 269–273

lysozyme, 206, 208, 275, 276

M

MacRae, Tom, 163–165

Maddox, Brenda, 163–165

Mayo, W. J., 4

mercury atoms, 206, 208

meromyocin, *6*

methemoglobin
5.5Å map of, 249
high-resolution analysis, 252
"The structure of haemoglobin: V." (Howells, Perutz), 231
"An X-ray study of horse methaemoglobin. I" (Boyes-Watson, Daficson, Perutz), 227
"An X-ray study of horse methaemoglobin. II" (Perutz), 228

metmyoglobin, 249, 252

micelles
description, 3
electrostatic charge, 4
molecular weight issue, 4
soap, 3–4, *4*

Miescher, Friedrich, 147

Mirsky, Alfred E., 23

model building, low-tech, *276*

"Molecular Configuration in Sodium Thymonucleate" (Franklin, Gosling), 186–187

molecular evolution, 250–253

"Molecular Structure of Deoxypentose Nucleic Acids" (Wilkins, Stokes, Wilson), 184–186

"Molecular Structure of Nucleic Acids," 183–187

Muirhead, Hilary, 234, 235, 257–263

Multiple Isomorphous Replacement (MIR) phase analysis, 200, *208*, 208–209, 214–217, 255

myocin, x-ray patterns, 5

myoglobin
α-helices, 251, *251*
"The Crystal structure of myoglobin V" (Bodo, Dintzis, Kendrew, Wyckoff), 241–247
heavy metal binding, 206
hemoglobin and, 200, 209
"A little ancient history" (Dickerson), 269–273
"Polypeptide chain configurations in crystalline proteins" (Bragg, Kendrew, Pertuz), 94–117
sources for, 200–201
"Structure of Myoglobin" (Kendrew, Dickerson, Strandberg, Hart, Davies, Phillips, Shore), 263–268

"A Three-Dimensional Model of the Myoglobin Molecule Obtained by X-Ray Analysis" (Kendrew, Bodo, Dintzis, Parrish, Wyckoff), 236–240

N

Nebbish trees, 145

"New X-Ray Evidence on the Configuration of Polypeptide Chains" (Perutz), 128–130

Niemann, Carl
    on cyclol theory, 20–21, 22–23
    key articles by, 30–38
    on protein structure, 200
    rebuttal by Wrinch, 38–41
    on x-ray methods, 24

nitrogen, bonds, 59

North, A.C.T., 234, 235, 257–263

Northrop, John H., 5

nucleic acids
    "Chemical Specificity of Nucleic Acids and Mechanism of their Enzymatic Degradation" (Chargaff), 158–160
    deoxypentose, 148–149
    interactions with, 150
    "Molecular Structure of Deoxypentose Nucleic Acids" (Wilkins, Stokes, Wilson), 184–186
    "Molecular Structure of Nucleic Acids," 183–187
    "A Proposed Structure for the Nucleic Acids" (Pauling, Corey), 169–182
    road to DNA helix, 147–149

O

"Obituary: Irving Geis, 1908–1997" (Dickerson), 287–289

oil, in aqueous environments, 4

Olby, Robert, 149

opals, 3

Oppenheimer, Robert, 20

oxyhemoglobin. *see* methemoglobin

P

papain, 276

Parrish, R. G., 236–240

The Path to the Double Helix (Olby), 149

"The Pattern of Proteins" (Wrinch), 26–27

Patterson, A. L., *201*, 211

Patterson vector maps, 59, 94–117, 211, *211*, 255

Pauling, Linus
    α-helix proposal, 58
    α-keratin form and, 8
    alpha helix concept, *60*, 60–61
    "analog computer," *60*
    atomic model for proteins, 250–251
    β-sheet structures, 64
    on cyclol theory, 20–21, 22–23
    "The discovery of the α-helix and β-sheet, the principle structural features of proteins" (Eisenberg), 135–138
    DNA model, 149–152, *150*
    House Un-American Activities Committee and, 65, 152
    I Wish I'd Made You Angry Earlier (Perutz), 125–127
    key articles by, 30–38, 118–124, 130–134, 169–182
    planar amide bond, 59
    on protein structure, 200
    rebuttal by Wrinch, 38–41
    resonance stabilization model, 21, *21*
    on x-ray methods, 24

Pauling, Peter, 152, 280–281

pearls, 3

peptide bonds, resonance stabilization model, *21*

Perutx, Max F.
    key articles by, 257–263

Perutz, Max F.
    α-helix proof, 63
    background of, 202
    Bernal and, 5
    career of, 282
    in Donohue's parody, 200, 202
    examination of helices, 59
    hemoglobin saga, 209–215, 280
    Hirschegg workshop and, 275
    I Wish I'd Made You Angry Earlier, 125–127
    key articles by, 94–117, 128–130, 192–194, 227–235
    lysozyme diffraction pattern, 208
    Nobel Prize, 253

on Pauling and Correy's helix data, 61, 63

protein crystal structure analysis, 200

search for protein sources, 200–201

wartime internment of, 202–203

Watson, Crick and, 155

x-ray crystallography by, 28

Phillips, David, 250, 255, 263–268, 275

phosphate group ionization, 152

pioneers, list of, *283*

pitch (P, rise), 59, 61

planar amide bonds, 59, 130–134

poly-γ-benzyl-L-glutamate, 61, 63, 128–130

poly-L-alanine chains, 6, *6, 7*

polymers, helices, 59

polynucleotide chains, 166–168

"Polypeptide chain configurations in crystalline proteins" (Bragg, Kendrew, Pertuz), 94–117

polypeptide chains
    "analog computer," *60*
    beta bends, 20, *20*
    "Configurations of Polypeptide Chains with Favored Orientations Around Single Bonds" (Pauling, Corey), 130–134
    "The Crystal structure of myoglobin V." (Bodo, Dintzis, Kendrew, Wyckoff), 241–247
    diffraction patterns, *62*
    "The discovery of the α-helix and β-sheet, the principle structural features of proteins" (Eisenberg), 135–138
    Drum-Fold model, *254*
    I Wish I'd Made You Angry Earlier (Perutz), 125–127
    "New X-Ray Evidence on the Configuration of Polypeptide Chains" (Perutz), 128–130
    "Polypeptide chain configurations in crystalline proteins" (Bragg, Kendrew, Pertuz), 94–117
    rise per turn, 63
    states of, 8
    "The Structure of Proteins" (Pauling, Corey, Branson), 118–124
    "The Cyclol Theory and the 'Globular' Proteins" (Wrinch), 28–29

"The Geometrical Attack on Protein Structure" (Wrinch), 38–41

"The Pattern of Proteins" (Wrinch), 26–27

"The Structure of Proteins" (Pauling and Niemann), 30–38

"X-Ray Adventures Among the Proteins" (Astbury), 45–56

"X-Ray Studies of the Structure of Hair, Wool, and Related Fibres" (Asbury, Woods), 10–17

Porter, George, 211

"A Proposed Structure for the Nucleic Acids" (Pauling, Corey), 169–182

Protein Data Bank, 277, *278*

protein dilemma, 1

protein fibers, *62*

protein packing, 211

Protein Workshops, East and West Coast, 276

proteins
  crystallization of, 5–6
  high-resolution analysis, 249–273
  solving structures, 199–247
  structural types, 5
  x-ray patterns, 5

pseudo-diketopiperazine rings, 8, 20

purines, 149, 166–168. *see also* base pairing

Pyke, Geoffrey, 202

pyrimidines, 149, 166–168. *see also* base pairing

Q

quantum mechanics, 20

"A Quick Climb Up Mount Olympus" (Chargaff), 190–191

R

Randall, John, 281

recommended reading, 284–285

resonance stabilization model, 21, *21*

ribonucleases, 275, 276, 290–291

Richards, Fred, 275, 276

Richards Box, 276, *277*

"Rosalind Franklin and the double helix" (Klug), 188–189

Rossmann, M. G., 234, 235, 257–263

Royal Institution, *212*

S

Sanger, F., 211

Schachman, Howard, *215*

Schrödinger, Erwin, 23

Segal, Jacob, 253, *254*, 255

selenium atoms, 208

Shomaker, Verner, *201*

Shore, V. C., 263–268

silk
  Bombyx mori, 64
  covalent bonds, 8
  fiber diffraction patterns, 5, *6*
  "X-Ray Adventures Among the Proteins" (Astbury), 45–56
  x-ray diffraction by fibers, 23

sine waves
  addition of, *140*
  molecule images and, *204*
  production, *144*

soap micelles, 4

soap molecules, 3–4

sodium stearate, structure, *4*

sodium thymonucleate, 151, 186–187

solid-in-gas colloidal suspensions, 3

solid-in-liquid colloidal suspensions, 3

solid-in-solid colloids, 3

Sommerfeld, Arnold, 20

sphere of reflection. *see* Ewald sphere

Standberg, B. E., 263–268

stearic acid, structure, *4*

Stephenson, Kathy, *276*

Steven, Alasdair C., 163–165

Stokes, A. R., 184–186

Strandberg, Bror, 250

Stroud, Robert, *276*

"Structure and Function of Nucleic Acid as Cell Constituents" (Chargaff), 161–162

"A Structure for Deoxyribonucleic Acid" (Watson, Crick), 183-184

"The Structure of Deoxyribose Nucleic Acid" (Fraser), 166–168

"Structure of Haemoglobin: A three-dimensional Fourier synthesis at 5·5-Å resolution, obtained by X-ray analysis" (Perutz, Rossmann, Cullis, Muirhead, Will, North), 257–263

"The structure of haemoglobin III: Direct determination of the molecular transform" (Perutz), 229

"The structure of haemoglobin: IV. Sign determination by the isomorphous replacement method" (Green, Ingram, Perutz), 230

"The structure of haemoglobin: V. Imidazole-methaemoglobin: a further check of the signs" (Howells, Perutz), 231

"The structure of haemoglobin: VI. Fourier projections on the 010 plane" (Bragg, Perutz), 232

"The structure of haemoglobin: VII. Determination of phase angles in the non-centrosymmetric [100] zone" (Blow), 233

"The structure of haemoglobin: VIII. A three-dimensional Fourier synthesis at 5·5Å resolution: determination of the phase angles" (Cullis, Muirhead, Perutz, Rossman, North), 234

"The structure of haemoglobin: IX. A three-dimensional Fourier synthesis at 5·5Å resolution: description of the structure" (Cullis, Muirhead, Perutz, Rossmann, North), 235

"Structure of Myoglobin: A three-dimensional Fourier synthesis at 2 Å resolution" (Kendrew, Dickerson, Strandberg, Hart, Davies, Phillips, Shore), 263–268

"The Structure of Proteins" (Pauling, Niemann), 30–38

"The Structure of Proteins: Two Hydrogen-Bonded Helical Configurations of the Polypeptide Chain" (Pauling, Corey, Branson), 118–124

Sumner, J. B., 5

T

Takano, Tsunehiro, *276*

tautomers, base pair possibilities, *152*

tetranucleotide hypothesis, 147–148

"The third man and the fourth paper" (Steven, Baumeister), 163–165

Thompson, Benjamin, 213

"A Three-Dimensional Model of the Myoglobin Molecule Obtained by X-Ray Analysis" (Kendrew, Bodo, Dintzis, Parrish, Wyckoff), 236–240

thymine, 149, *152. see also* base pairing

triple helix model, 149–152, *150*

true repeat distances, 8

turns, of helices, 59

Tussah silk fibroin, *7*

V

Varnum, Joan, *276*

vitamin B$_{12}$ structure, 207–208, *208*

W

Waldemar-Petersen Haus, 275

Watson, James

"A Structure for Deoxyribose Nucleic Acid," 183–184

career of, 282

Central Dogma, 255

credit to, 153–156

"DNA Helix" (Perutz, Wilkins, Watson), 192–194, *194*

in Donohue's parody, 200–201

double helix and, 152–153

Franklin and, 23

"Gullible's Troubles" (Chargaff), 195–198

on Pauling's manuscript, 152

"A Quick Climb Up Mount Olympus" (Chargaff), 190–191

"The third man and the fourth paper" (Steven, Baumeister), 163–165

work on DNA structure, 149–152

waves

diffraction patterns, 140–141, *141*

Fourier transform, 203

in phase diagram, *206*

phase shifts, *204*, *205*

sinusoidal, *204*

Weissenberg camera, 211

Wilkins, Maurice H. F., 147, 281

credit to, 153–156

"DNA Helix" (Perutz, Wilkins, Watson), 193–194

on Fraser's triple helix, 149

key articles by, 184–186

"The third man and the fourth paper" (Steven, Baumeister), 163–165

work on DNA structure, 149

Will, George, 257–263

Wilson, H. R., 184–186

Wolf, A., 253, *254*

Woods, H. J., *7*, 10–17

Wrinch, Dorothy M.

cyclol theory, 20, 22–23

"Dorothy Wrinch and a Search for the Structure of Proteins" (Julian), 42–44

key articles by, 26–27, 28–29, 38–41

Wyckoff, H., 236–240, 241–247

X

"X-Ray Adventures Among the Proteins" (Astbury), 23, 45–56

x-ray diffraction. *see also specific* authors, papers

"Crystallographic Measurements and the Structure of Protein Molecules as They Are" (Hodgkin), 66–94

experiments, 5–6

"Molecular Structure of Deoxypentose Nucleic Acids" (Wilkins, Stokes, Wilson), 184–186

"New X-Ray Evidence on the Configuration of Polypeptide Chains" (Perutz), 128–130

patterns, *6*, 140–147

protein fibers, *62*

"X-Ray Adventures Among the Proteins" (Astbury), 45–56

"X-Ray Studies of the Structure of Hair, Wool, and Related Fibres (Astbury, Woods), 10–17

"The X-Ray Structure Analysis of α-Globlglobin" (Donohue, Briekopf), 200–202, 217–226

"X-Ray Studies of the Structure of Hair, Wool, and Related Fibres; The Molecular Structure and Elastic properties of Hair Keratin" (Astbury, Woods), 10–17

"An X-ray study of horse methaemoglobin. I" (Boyes-Watson, Davidson, Perutz), 227

"An X-ray study of horse methaemoglobin. II" (Perutz), 228

X-rays, scattering, 140

Colophon

Editor: Andrew D. Sinauer

Production Editor: Chelsea D. Holabird

Production Manager: Christopher Small

Book Production and Design: Jefferson Johnson

Cover Design: Jefferson Johnson

Book and Cover Manufacturer: Courier Companies, Inc.